Qualitative Theory of Parabolic Equations. *Part 1*

QUALITATIVE THEORY OF PARABOLIC EQUATIONS

Part 1

*T.I. Zelenyak, M.M. Lavrentiev Jr.
and M.P. Vishnevskii*

Utrecht, The Netherlands, 1997

VSP BV
P.O. Box 346
3700 AH Zeist
The Netherlands

© VSP BV 1997

First published in 1997

ISBN 90-6764-236-3

Printed in The Netherlands by Ridderprint bv, Ridderkerk.

Contents

1

2

Introduction.

One of the basic problems of qualitative theory of differential equations is the investigation of the behavior of evolutionary problem solutions "as a whole" with respect to time.

The limit sets of these solutions may be stationary (not depending on time) solutions of the same problems, periodic solutions, or more complicated sets made up of such solutions. Very important is the problem of the behavior of the solution in the boundary of the domain of definition.

Unlike the dynamical systems in R_n, where each solution $x(t)$ may be extended up to the boundary of the domain of definition by the regular image of the right sides, the solutions of partial differential equations may be destroyed in one norm for $t \rightarrow T \leq \infty$ and be regular in the other weaker norm. Thus, in the qualitative theory of partial differential equations, arise specific problems connected with the fact that, unlike with R_n, infinite-dimensional linear spaces may be completed by non-equivalent norms.

A basic introduction to this theory was presented in the works of Hadamard, Cauchy, Poisson, Kovalevskaya, Bernstein (1960), Sobolev (1962), Petrovskii (1986, 1987) with the well-posedness and solvability of some classical problems of mathematical physics. For applications, the stability of solutions is of great importance.

In the qualitative theory of ordinary differential equations, the Liapunov methods play a fundamental role. To use their analogs for analysis of stability of solutions to parabolic, hyperbolic, and other nonclassical equations and systems, we have to devise time-invariant *a priori* estimates for solutions. In order to include in this investigation the important problems of mathematical physics, it is necessary to obtain such estimates that allow us to establish stability without imposing rigorous restrictions for the nonlinearities. Such possibilities for one-dimensional by spatial variables parabolic problems were noted by Zelenyak (1966, 1967, 1968, 1970, 1972) and were extended by Belonosov (1975) in the wide class of parabolic systems.

Very interesting is the paper of Eltysheva (1988), where the necessary and sufficient condition is obtained such that the smoothness of solution of linear hyperbolic system grows as t grows. This property, which is well-known in the theory of parabolic problems, allows us to establish stability with minimal restrictions on initial data. It allows us to describe for extended time the behavior of solutions in the optimal norms by the destruction of solution in the minimal norm (of the type $C(\Omega)$), its existence "as a whole", and the boundedness and stability in maximal possible norms (of the type $C^{2+\alpha,1+\frac{\alpha}{2}}$).

In this monograph we shall consider only parabolic problems. Here, mainly, lie the problems which nowadays have been investigated most thoroughly. This is the construction of Liapunov functionals which naturally generalize Liapunov functions for nonlinear parabolic equations of the second order with one spatial variable. We shall establish the theorems on stabilizing solutions, the necessary and sufficient conditions of general and asymptotic stability of stationary solutions including the so-called critical cases. We shall describe attraction domains for stable solutions of mixed problems for these equations. Estimates for the number of stationary solutions are obtained.

For the quasilinear second order equation with many spatial variables we shall construct the Liapunov functionals in the neighborhood of the stable stationary solution, and obtain the necessary and sufficient conditions of stability in the first approximation.

In the one-dimensional (with respect to spatial variables) case, since the regular global solutions are stabilized, investigation of stability of each solution may be reduced, as a rule, to investigation of stability of its limit stationary solution, if this limit exists. Parabolic problems of variational type with many spatial variables

$$\frac{\partial u}{\partial t} = -F_u + \sum \frac{\partial}{\partial x_i} F_{u_{x_i}},$$

$$u\big|_{\partial Q} = 0, \quad u\big|_{t=0} = \varphi(x)$$

do not have the property of stabilizing the solutions, which are bounded in the norms $C_{x,t}^{2+\alpha,1+\frac{\alpha}{2}}(Q)$, $Q = \Omega \times [0, \infty)$ uniformly with respect to t.

This fact for the above systems was established by Fokin (1981). Nevertheless, the limit set of a bounded solution to this problem consists of the connected set of stationary solutions. If this set has more than one element, then such solutions can not be asymptotically stable.

In chapters 3, 4, 5 we investigate limit sets of solutions to the problems which are multidimensional with respect to spatial variables.

We have complemented the book with the appendix where some statements that are used in the previous chapters are derived.

The review of results of the theory of parabolic equations may be found in Kruzhkov and Oleinik (1961), Solonnikov (1965), Ladyzhenskaya *et al.* (1967), Friedman (1968), Hale *et al.* (1984), Ladyzhenskaya and Ural'tseva (1986), Matano (1988), and Teman (1988).

Note, that the authors essentially used the classical results of S. N. Bernstein, I. G. Petrovskii, and S. L. Sobolev.

A complete survey of results essential, from our point of view, in the theory of parabolic equations and systems requires a separate monograph. Therefore, the authors give only the necessary references for the statements which were used, without examining the derivation of the problems.

Chapter 1

Local behavior of solutions of boundary-value problems for nonlinear parabolic systems in the neighborhood of a stationary or periodic solution.

This first chapter of the book will be rather technical. Here we consider in detail the behavior of solutions in the neighborhood of a stationary or periodic solution of the linear problem. Further, we shall use these results for investigating behavior of solution of one quasilinear parabolic equation in the neighborhood of a stationary or periodic solution.

In this chapter we shall also consider the general nonlinear parabolic system. We do this because to obtain these results for quasilinear parabolic equations requires the same effort.

One further point about the chapter may be noted. To obtain the basic results, we use the estimates for linear parabolic systems in the Hölder classes with the weight, which in the one-dimensional case were obtained by Belonosov V. S. (Belonosov and Zelenyak, 1975) and in the multi-dimensional case were obtained simultaneously by Belonosov

(1978, 1981) and by Solonnikov and Hachatryan (1980). The use of these estimates has allowed as in the theorems of chapter 1 to get rid of many, as it has turned out, unnecessary restrictions which usually are imposed on initial data and nonlinearities.

§1.1   The weight Hölder classes and some auxiliary lemmas.

1. Notations and definitions of the Hölder classes.   We shall use the following standard notations

R^n is the Euclidean space;

$x = (x_1, \ldots, x_n)$ are points of R^n;

(x, t) is a point of R^{n+1}, $x \in R^n$, $t \in R^1$;

Ω is a bounded domain from R^n;

$\partial\Omega$ is the boundary of Ω that coincides with the boundary of the set $\bar{\Omega}$.

Unless otherwise stated, we shall consider that $\partial\Omega \in C^{2m+\alpha}$, $(0 < \alpha < 1)$; where m is natural; $C^{2m+\alpha}$ is the Hölder space (Lavrentiev, 1980).

In addition to the domain Ω we shall consider the following finite and infinite cylinders

$$Q_T = \Omega \times (0, T),$$

$$Q = \Omega \times (0, +\infty),$$

$$Q^\tau = \Omega \times (\tau, +\infty),$$

$$Q^- = \Omega \times (-\infty, 0),$$

$$Q_0 = \Omega \times (-\infty, +\infty),$$

$$Q_T^\tau = \Omega \times (\tau, T), \quad \tau < T.$$

Lateral boundaries of these cylinders we shall denote by Γ_T, Γ, Γ^τ, Γ^-, Γ_0, Γ_T^τ respectively.

For finite differences of functions $f(x, t)$ determined in R^{n+1} we shall use the notations

$$\Delta_y^x f = f(x, t) - f(y, t);$$

$$\Delta_\tau^t f = f(x, t) - f(x, \tau).$$

For the derivative of the function f of the order $\beta, \mu + |\nu|$ (β and ν are the multi-indices $\nu = (\nu_1, \ldots, \nu_n)$ $\beta = (\beta_1, \ldots, \beta_n)$; $|\nu| = \sum_{i=1}^{n} \nu_i$) we shall use the notation

$$\frac{\partial^{\mu+|\nu|}}{\partial t^\mu \partial x_1^{\nu_1} \ldots \partial x_n^{\nu_n}} f = D_t^\mu D_x^\nu f, \quad D^\beta f = \frac{\partial^\beta}{\partial x_1^{\beta_1} \ldots \partial x_n^{\beta_n}}.$$

Following Belonosov (1978, 1979) we now introduce the weight Hölder classes.

Let s, r be real ($s \geq 0, r \leq s$), and m be natural. Assume $f(x)$ to be continuous in Q_T together with its derivatives $D_t^\mu D_x^\nu f$ of the order $2m\mu + |\nu| \leq s$.

Consider the following seminorms

$$[u]_{r,s,t}^{Q_T} = \sup \Theta^{\frac{s-r}{2m}} |\Delta_\tau^t D_t^\mu D_x^\nu u| |t - \tau|^{\frac{s-2m\mu-|\nu|}{2m}},$$

$$[u]_{r,s,x}^{Q_T} = \sup t^{\frac{s-r}{2m}} |\Delta_y^x D_t^\mu D_x^\nu u| |x - y|^{[s]-s}.$$

Here $\Theta = \min(t, \tau)$. The upper bound in the first equality is taken by all $(x, t) \neq (x, \tau)$ from Q_T and μ, ν such that $0 < s - 2m\mu - |\nu| < 2m$.

In the second equality we take the upper bound by all $(x, t) \in Q_T$ and μ, ν such that $2m\mu - |\nu| = [s]$. For the integer k we additionally set

$$\langle u \rangle_{r,k}^{Q_T} = \sup \left(t^{\frac{k-r}{2m}} \left| D_t^\mu D_x^\nu u \right| \right),$$

where the upper bound is taken by all $(x,t) \in Q_T$ and by all μ, ν such that $2m\mu + |\nu| = k$.

For the integer s, also determine the seminorms $[u]_{r,s}^{Q_T}$

$$[u]_{r,s}^{Q_T} = \langle u \rangle_{r,s}^{Q_T} + [u]_{r,s,t}^{Q_T}.$$

For noninteger s, set

$$[u]_{r,s}^{Q_T} = [u]_{r,s,x}^{Q_T} + [u]_{r,s,t}^{Q_T}.$$

Denote by $[u]_s^{Q_T}, [u]_k^{Q_T}$ the seminorms obtained from $[u]_{r,s}^{Q_T}$ and $\langle u \rangle_{r,k}^{Q_T}$ by neglecting the weight multipliers $t^{\frac{s-r}{2m}}$, $\Theta^{\frac{s-r}{2m}}$, $t^{\frac{k-r}{2m}}$.

Denote by $H^s(Q_T)$ the space of functions $u(x,t)$ which have in Q_T continuous partial derivatives $D_t^\mu D_x^\nu u$ up to the order $2m\mu + |\nu| \leq s$ and the finite norm

$$\| u \|_s = [u]_s^{Q_T} + \sum_{0 \leq k < s} |u|_k^{Q_T}.$$

It is easy to see that this is the usual Hölder space that arises when solving parabolic equations. In the literature it is denoted by $H_x^s{}^{\frac{s}{2m}}(Q_T)$ or $C_x^s{}^{\frac{s}{2m}}(Q_T)$ (Ladyzhenskaya *et al.* , 1968). We shall take the notation $H^s(Q_T)$ since m in our book, usually, will be the unit.

If a function u does not depend on t, then $|u|_s^{Q_T}$ becomes by the norm $|u|_s^\Omega$ in the standard Hölder space $C^s(\overline{\Omega})$.

Let $s \geq 0$ and $0 \leq r \leq s$. By the weight Hölder space $H_r^s(Q_T)$ is meant the set of functions $u(x,t)$ which have in Q_T continuous derivatives $D_t^\mu D_x^\nu u$ for $2m\mu + |\nu| \leq s$ and the finite norm

$$\| u \|_{r,s} = [u]_{r,s}^{Q_T} + \sum_{r < k < s} |u|_{r,k}^{Q_T} + \| u \|_r^{Q_T}.$$

If $r < 0$, then $H_r^s(Q_T)$ will be determined as the set of functions $u(x,t)$

with continuous derivatives $D_t^\mu D_x^\nu u$ of the order $2m\mu + |\nu| \le s$ and the finite norm

$$\| u \|_{r,s}^{Q_T} = [u]_{r,s}^{Q_T} + \sum_{r<k<s} \langle u \rangle_{r,k}^{Q_T}.$$

We shall say that $u(x,t)$ belongs to $H_r^s(Q)$, if $u(x,t)$ belongs to $H_r^s(Q_1)$ and $H^s(Q^1)$.

Determine the norm in this space as follows

$$\| u \|_{r,s}^{Q} = \| u \|_{r,s}^{Q_1} + \| u \|_s^{Q^1}.$$

Note, that such weight Hölder classes were introduced by Solonnicov and Hachatryan (1980), independently of Belonosov (1978, 1981).

2. Some auxiliary lemmas. While considering nonlinear problems we shall use the superpositions of the form $f(x, t, u_1(x,t), \ldots, u_\ell(x,t))$, where u_j belongs to $H_{r_j}^{s_j}(Q_T)$ $0 \le r_j \le s_j; j = 1, \ldots, \ell.$

Consider certain properties of these superpositions.

Lemma 1.1. Let $D_u^\beta f$, $|\beta| \le s$ as the functions of (x,t) belong to $H^{s-|\beta|}$ and $\| D_u^\beta f(x,t,u) \|_{s-|\beta|}^{Q_T}$ be uniformly bounded for $\max_j |u_j| < \eta$.

Assume, also, that $D_t^\mu D_x^\nu D_u^\beta f$ for $2m\mu + |\nu| + |\beta| \le s$ satisfy for $\max_j |u_j| < \eta$ the Lipschitz condition by u_j $(j = 1, \ldots, \ell)$ uniformly with respect to $(x,t) \in Q_T$. Then for each vector function $u(x,t) = (u_1(x,t), \ldots, u_\ell(x,t))$ all the components of which belong to $H_r^s(Q_T)$, $|u_j| < \eta$, we have the inclusion

$$f(x,t,u) \in H_r^s(Q_T).$$

Moreover,

$$\| f(x,t,u) \|_{r,s}^{Q_T} \le C \left(\| u \|_{r,s}^{Q_T} \right).$$

Proof   of this lemma is simple but rather tedious (the basic points could be found in Belonosov and Vishnevskii, 1977); therefore, we shall give only an idea of the proof. The vector function $u(x,t)$ is continuous up to $t = 0$; therefore, $f(x,t,\bar{u})$ will have same property. The smoothness conditions imposed on f yield that the Hölder constants with respect to variables x, t of $f(x,t,u)$ grow for $t \to 0$ not faster than the Hölder constants for the function $u(x,t)$.

If a function f satisfies the lemma conditions, then we may consider the map

$$F : u(x,t) \longrightarrow f(x,t,u(x,t)),$$

which acts from the vector function space (the functions belong to $H_r^s(Q_T)$) into the space $H_r^s(Q_T)$.

Lemma 1.2.   Let $f_{u_j}(x,t,u)$, $j = 1, \ldots, \ell$ satisfy the lemma 1.1 conditions. Then

$$\| F(u) - F(v) \|_{r,s}^{Q_T} \leq C \left(\| u \|_{r,s}^{Q_T}, \| v \|_{r,s}^{Q_T} \right) \| u - v \|_{r,s}^{Q_T},$$

where $C(\xi, \eta)$ - is bounded.

If, in addition, $f_{u_i u_j}$ satisfy the lemma 1.1 conditions for $i, j = 1, \ldots, \ell$ and

$$f_{u_j}(x,t,0) = 0 \qquad j = 1, \ldots, \ell,$$

then $C(\xi, \eta) \to 0$ for $\xi, \eta \to 0$.

Proof.   By the Taylor formula

$$F(u) - F(v) = \sum_{j=1}^{\ell} \int_0^1 f_{u_j}(x,t,v + \theta(u - v))(u_j - v_j)d\theta.$$

Applying lemma 1.1, we obtain the first part of lemma 1.2.

To prove the second part, we use again the Taylor formula

$$F(u + w) - F(u) = \sum_{j=1}^{\ell} f_{u_j}(x, t, u)w_j$$

$$+ \sum_{i,j=1}^{\ell} \int_0^1 (1 - \theta) f_{u_i u_j}(x, t, u + \theta w)w_i w_j d\theta.$$

Applying lemma 1.1, we obtain

$$\left| F(u + w) - F(u) - \sum_{j=1}^{\ell} f_{u_j}(x, t, u)w_j \right|_{r,s}^{Q_T} \le C \left(|w|_{r,s}^{Q_T} \right)^2.$$

Therefore, the differential $dF(u)$ of the map F in the point u is the linear map

$$w \longrightarrow \sum f_{u_j}(x, t, u)w_j,$$

which satisfies the estimate

$$\| dF(u) - dF(v) \| \le K \sum_{j=1}^{\ell} \| f_{u_j}(x, t, u) - f_{u_j}(x, t, v) \|_{r,s}^{Q_T}.$$

By the symbol $\| \ . \ \|$ we denote here the norm of the linear map from $H_r^s(Q_T)$ into $H_r^s(Q_T)$.

From the latter inequality we obtain that $dF(u)$ is continuous with respect to u, and the lemma statement follows from the equality $dF(0) = 0$. The lemma is proved.

Define by the space $H_r^s(\Gamma_T)$ of functions given in the lateral boundary of Q_T the set of traces in Γ_T of arbitrary functions from $H_r^s(Q_T)$. The norm in this space will be as follows

$$\| \varphi \|_{r,s}^{S_T} = \inf \| \Phi \|_{r,s}^{Q_T},$$

where the lower bound is taken by all $\Phi(x, t)$ from $H_r^s(Q_T)$ which coincide with $\varphi(x, t)$ in S_T. The statements of lemmas 1.1 and 1.2 hold also for the cases when we consider the functions superpositions given in Γ_T.

The weight Hölder spaces and their norms in the lateral boundaries of other considered cylinders are determined analogously.

3. The estimations of linear problem solutions in cylinders of finite height. Consider in the cylinder $Q_T = \Omega \times (0, T)$, $T \leq +\infty$ the general linear boundary-value problem

$$D_t u - A(x, t, D_x)u = D_t u - \sum_{|\alpha|+|\beta| \leq 2m} \tilde{A}_{\alpha\beta} D_x^\alpha (A_{\alpha\beta} D_x^\beta u)$$

$$= f(x, t), \quad (x, t) \in Q_T, \tag{1.1}$$

$$B_i(x, t, D_x)u = \sum_{|\beta| \leq m_i} B_{i\beta} D_x^\beta u = \varphi_i(x, t), \tag{1.2}$$

$$(x, t) \in \Gamma_T = \partial\Omega \times (0, T) \quad 1 \leq i \leq mN,$$

$$u(x, 0) = u_0(x). \tag{1.3}$$

Here $u(x, t)$, $f(x, t)$, $u_0(x)$ are vector functions with N components; $\varphi_i(x, t)$ are scalar functions; $A_{\alpha\beta}(x, t)$, $\tilde{A}_{\alpha\beta}(x, t)$ are square matrices $N \times N$; $B_{i\beta}(x, t)$ are matrices with the row length N; α, β are multi-indices. Assume that the following conditions hold.

(A.1) System (1.1) is uniformly parabolic in $\bar{Q}_T$ by Petrovskii and in each point $(x, t) \in \Gamma_T$ the complementary condition with respect to the boundary conditions (1.2) holds (see Belonosov, 1978, 1979).

(A.2) For a certain noninteger $s > 2m$ and a number $r \in [0, s]$

$$\partial\Omega \in C^s, \qquad \tilde{A}_{\alpha\beta} \in H^{s-2m}_{r_0}(Q_T),$$

$$A_{\alpha\beta} \in H^{s-2m+|\alpha|}_{r_\alpha}(Q_T), \qquad B_{i\beta} \in H^{s-m_i}_{r_i}(\Gamma_T),$$

where $r_0 = \max(r - 2m, 0)$, $r_\alpha = \max(r - 2m + |\alpha|, 0)$, $r_i = \max(r - m_i, 0)$.

(A.3) The right sides of equations (1.1) and (1.2) are such that

$$f(x, t) \in H^{s-2m}_{r-2m}(Q_T),$$

$$\varphi_i \in H^{s-m_i}_{r-m_i}(\Gamma_T) \quad i = 1, \ldots, N.$$

$(A.4)$ The initial data $u_0(x)$ are such that $u_0(x) \in C^r(\overline{\Omega})$ and, besides, the necessary compatibility conditions hold

$$D_t^k(P_i u)\Big|_{t=0, x \in \partial\Omega} = D_t^k \varphi_i(x, 0) \qquad 0 \le 2mk + m_i \le r. \qquad (1.4)$$

Remark. (Belonosov, 1978, 1981). Using (1.1) and (1.3), we may find for $t = 0$ the values of all derivatives $D_t^\mu D_x^\nu u$ of the order $2m\mu + |\nu| \le r$ by means of $f(x,t)$ and $u_0(x)$.

After substituting these values, we obtain the equalities (1.4), which will also be called the compatibility conditions of the order r.

As it is known (Solonnikov, 1965; Ladyzhenskaya *et al.*, 1967), if $r = s$ and $(A.1) - (A.4)$ hold, then problems (1.1)-(1.3) are uniquely solvable in the space $H^s(Q_T)$, and

$$\| u \|_s^{Q_T} \le C \left(\| u_0 \|_s^{\Omega} + \| f \|_{s-2m}^{Q_T} + \sum_{i=1}^{mN} \| \varphi_i \|_{s-m_i}^{\Gamma_T} \right) \qquad (1.5)$$

holds for solutions of this problem. Belonosov (1978, 1979 see, also, Solonnikov and Hachatryan, 1980) have obtained analogous result for the weight Hölder classes $H_r^s(Q_T)$.

Theorem 1.1 Let conditions $(A.1) - (A.4)$ hold. Then problem (1.1)-(1.3) is uniquely solvable in the space $H_r^s(Q_T)$, and the following estimate holds for the $u(x,t)$ solution of this problem

$$\| u \|_{r,s}^{Q_T} \le C \left(\| u_0 \|_r + \| f \|_{r-2m,s-2m}^{Q_T} + \sum_{i=1}^{mN} \| \varphi_i \|_{r-m_i, s-m_i}^{\Gamma_T} \right). \qquad (1.6)$$

In the case $r = 0$, estimate (1.6), generally speaking, fails. Nevertheless, in certain cases that are important for applications, Belonosov (1981) has obtained estimate (1.6) also for $r = 0$.

Let the operator $A(x, t, D_x)$ in (1.1) be as follows

$$A(x, t, D_x)u = \sum_{|\beta| + \max(|\alpha|, 1) \le 2m} \tilde{A}_{\alpha\beta} D_x^\alpha (A_{\alpha\beta} D_x^\beta u), \qquad (1.7)$$

condition $(A.1)$ holds, and the coefficients of equations are such that

$$(B.2) \quad \partial\Omega \in C^s, \quad \tilde{A}_{\alpha\beta} \in H_{r_0}^{s-2m}(Q_T),$$

$$A_{\alpha\beta} \in H_1^{s-2m+1}(Q_T) \quad \text{for} \quad |\alpha| + |\beta| = 2m.$$

$$(B.3) \quad f(x,t) = \sum_1^n D_{x_i} g + h, \quad \text{where} \quad g(x,t) \in H_{1-2m}^{s-2m+1}(Q_T),$$

$$h(x,t) \in H_{\varepsilon-2m}^{s-2m}(Q_T), \quad \text{where} \quad \varepsilon > 0,$$

$$\varphi_i \in H_{r-m_i}^{s-m_i}(\Gamma_T), \quad i = 1, \ldots, N.$$

Theorem 1.2 Let the conditions $(A.1), (B.2), (B.3), (A.4)$ hold for $r = 0$. Then problem (1.1) - (1.3) is solvable in $H_0^S(Q_T)$, where the solution $u(x,t)$ of this problem satisfies the following estimate

$$\| u \|_{0,s}^{Q_T} \leq C \left(\| u_0 \|_0^{\Omega} + \| g \|_{1-2m,s-2m+1}^{Q_T} \right.$$

$$\left. + \| h \|_{\varepsilon-2m,s-2m}^{Q_T} + \sum_{i=1}^{mN} \| \varphi_i \|_{-m_i,s-m_i}^{\Gamma_T} \right). \tag{1.8}$$

§1.2   Bounded solutions of linear parabolic systems.

In this paragraph we shall consider the boundary-value problem (1.1)-(1.3). We shall assume that $(A.1)-(A.4)$ hold, and, therefore, estimate (1.6) holds. The constant C in estimate (1.6) depends on T. In this case, $C(T)$ in general tends to infinity for $T \to +\infty$. The purpose of this section is to clarify under which conditions $C(T)$ will be uniformly bounded by T.

We shall investigate the dependence $C(T)$ in (1.8) for the system (1.1) with the operator of the form (1.7) under conditions $(A.1), (B.2), (B.3), (A.4)$.

1. The shift operator by the trajectory $V(t)$ and its properties. Let $\tilde{E}$ be a subspace in $H_r^{2m+\alpha}(\Omega)$, $0 < \alpha < 1$, which consists of functions satisfying the compatibility conditions (1.4) with the zero functions f and φ_i of the order r. Thus, if a vector function u_0 belongs to $\tilde{E}$, we may consider the homogeneous problem (1.1)-(1.3) with initial data $u_0(x) \in \tilde{E}$. Denote by $V(t)$ the operator which associates with the function $u_0(x)$ the solution $u(x,t)$ of the homogeneous problem in the moment t.

Assume, in addition, that

(A.5) all the coefficients of problem (1.1) and boundary conditions (1.2) ω - periodically depend on the time.

If, in addition to the conditions $(A.1) - (A.4)$, the condition $(A.5)$ holds, then $V(\omega)$ acts from $\tilde{E}$ into $\tilde{E}$.

Lemma 2.1. If $(A.1) - (A.5)$ hold true, then the operator $V(t)$ acting from $\tilde{E}$ into $\tilde{E}$ is completely continuous.

Proof. Let M be a bounded set in $\tilde{E}$. Then estimate (1.7) yields that $V(\omega)M$ is bounded in $H_r^{2m+\alpha}(Q_\omega)$. Therefore, the solution $u(x,t)$ is bounded in $H^{2m+\alpha}(Q_\omega^\varepsilon)$, where ε is sufficiently small. Using the inequality

$$|u(x,t)|_{2m+\beta}^\Omega \leq C(t)|u|_0^Q,$$

proved in Solonnikov (1965), we obtain that for $1 > \beta > \alpha$, $V(\omega)M$ is relatively compact in $C^{2m+\alpha}(\Omega)$, whence the lemma statement follows.

The complete continuity of $V(\omega)$ yields that the spectrum of $V(\omega)$ is no more than a countable set in the complex plane, which has the only limit point $\lambda = 0$ (see, for example, Agmon *et al.*, 1959, 1964).

One of the most important cases is that when the coefficients of (1.1) and (1.2) do not depend on time. Then, the spectrum of the operator $V(\omega)$ (as ω we may take any positive number) is connected with the spectrum of the elliptic problem which is obtained from (1.1)-(1.3) by the direct Laplace transformation with respect to t

$$\lambda v - A(x, D_x)v = F(x) \quad B_i(x, D_x)v = \Phi_i(x), \quad i = 1, \ldots mN. \quad (2.1)$$

Those parameter λ values when the problem has a unique solution in the class $C^{2m+\alpha}(\Omega)$ for each collection of the right sides $F(x)$ from $C^{\alpha}(\overline{\Omega})$, $0 < \alpha < 1$, $\Phi_i(x)$ from $C^{2m+\alpha-m_i}(\partial\Omega)$, $i = 1, \ldots, mN$ we shall call by the regular values, otherwise - by the spectrum points. This terminology is standard (Agmon *et al.*, 1959, 1964).

It was shown by Agmon *et al.* (1959, 1964), that if the spectrum of operator (2.1) has at least one regular point, then this spectrum is a countable set of eigenvalues without limit points. In this case, if λ is a regular point of problem (2.1), then

$$\| v \|_{2m+\alpha}^{\Omega} \in C(\lambda) \left(\| F \|_{\alpha}^{\Omega} + \sum_{i=1}^{mN} \| \Phi \|_{2m+\alpha-m_i}^{\Omega} \right). \qquad (2.2)$$

It is not clear whether the elliptic problem of the general form has regular points. But if elliptic system (2.1) is obtained from parabolic problem (1.1)-(1.3) by means of the Laplace transformation with respect to t, then problem (2.1) has regular points.

Lemma 2.2. There exists a real number λ_0, such that the whole half-plane $Re\lambda \leq \lambda_0$ consists of the regular points of problem (2.1).

Proof. Note, that the set $\{V(t)\}_{t \geq 0}$ forms a semigroup, i.e., $V(t_1) \cdot V(t_2) = V(t_1 + t_2)$ for each t_1 and t_2.

Denote by μ the exponential type (see, for example, Hille and Fillips, 1962) of the semigroup $V(t)$. Then for each positive ε for the norm of the operator $V(t)$ in $\tilde{E}$ the following estimate

$$\| V(t) \| \leq C_{\varepsilon} e^{(\mu+\varepsilon)t} \qquad (2.3)$$

holds. Inequalities (1.7) and (2.3) yield that the solution of autonomous problem (1.1)-(1.3) with the right sides f and Φ_i, which vanish for large t, satisfies the inequality

$$\| u(x,t) \|_{2m+\alpha}^{\Omega} \leq C_{\varepsilon}(u_0, f, \varphi_i) e^{(\mu+\varepsilon)t}. \qquad (2.4)$$

We must prove that problem (2.1) is uniquely solvable for each of the right sides $F(x)$ from $C^\alpha(\overline{\Omega})$, $\Phi_i(x)$ from $C^{2m+\alpha-m_i}(\overline{\Omega})$ if $\lambda = \lambda_0$, $Re\lambda_0 > \mu$.

Let $\zeta(t)$ be an infinitely differentiable function with the support in the interval $(0, +\infty)$ and its Laplace transform $\tilde{\zeta}(\lambda)$ does not vanish in the point λ_0.

Denote by $u(x,t)$ a solution to (1.1)-(1.3), where

$$f(x,t) = \frac{\zeta(t)F(x)}{\tilde{\zeta}(\lambda_0)} \quad , \quad \varphi_i(x,t) = \frac{\zeta(t)\Phi_i(x)}{\tilde{\zeta}(\lambda_0)},$$

$$u_0(x) \equiv 0.$$

From estimate (2.4) it follows that the Laplace transform $\tilde{u}(x,\lambda)$ of the function $u(x,t)$ is defined for $\lambda = \lambda_0$, belongs to $C^{2m+\alpha}(\overline{\Omega})$ and solves problem (2.1).

Assume that for a certain $\lambda, Re\lambda > \mu$, problem (2.1) has two different solutions $v(x)$ and $\omega(x)$. Then $u(x,t) = e^{\lambda t}(v(x) - \omega(x))$, will be the nontrivial solution to homogeneous system (1.1)-(1.2), which contradicts (2.4) since $Re\lambda > \mu$.

The lemma is proved.

Consider in the space $C^\alpha(\overline{\Omega})$, the linear operator

$$A \ : \ v \to A(x, D_x)v$$

the domain of definition of which consists of all functions v from $C^\alpha(\overline{\Omega})$ satisfying the conditions

$$B_i(x, D_x)v\Big|_{\partial\Omega} = 0 \ , \quad i = 1, \ldots, mN.$$

By lemma 2.2 and the considerations before it, the linear operator A is an extension of the infinitesimal operator $V(t)$ and has a discrete spectrum. Further, we shall find the connection between the eigenvalues of A and $V(t)$.

For this purpose, first, let us recall some notions of the general theory of linear operators.

Let L be a linear operator in a Banach space X. A vector $x \in X$ is called a eigenvector of the degree k corresponding to the eigenvalue λ, if

$$(L - \lambda I)^k x = 0,$$

where I is the identity operator.

Eigenvectors corresponding to different eigenvalues are linearly independent.

By the root subspace of degree k corresponding to the eigenvalue λ we shall call the set M_λ^k of all eigenvectors of the degree k which correspond to the eigenvalue λ.

It is clear that $M_\lambda^k \subseteq M_\lambda^{k+1}$. If L is a compact operator, then for $\lambda \neq 0$, all M_λ^k are finite-dimensional, and in the sequence $M_\lambda^1, M_\lambda^2, \ldots$ there are only a finite number of different subspaces. In this case the largest of M_λ^k will be called the root subspace.

Lemma 2.3. The set of nontrivial points of the operator $V(t)$ spectrum is the image of the operator A spectrum under the map $\lambda \to e^{\lambda t}$. Morever, for $t > 0$, the root space of degree k of the operator $V(t)$ corresponding to the eigenvalue $\mu \neq 0$ is the direct sum of the root subspaces of the degree k of the operator A corresponding to all eigenvalues λ such that $e^{\lambda t} = \mu$.

Proof. Denote by M_λ^k the root subspace of the degree k of the operator A corresponding to the eigenvalue λ. By substitution in (1.1) we can show, that $v \in M_\lambda^k$; then the function

$$u(x,t) = \sum_{n=0}^{k} \frac{1}{n!} t^n e^{\lambda t} (A - \lambda I)^n v$$

will solve homogeneous problem (1.1)-(1.3) with the initial data $v(x)$. In particular, for $k = 1$, we obtain $u(x,t) = e^{\lambda t} v(x)$.

Therefore, if $\omega_0 > 0$, then

$$V(\omega_0) v(s) = u(x, \omega_0) = e^{\lambda \omega_0} v(x),$$

i.e., $\mu = e^{\lambda \omega_0}$ is an eigenvalue of the operator $V(\omega_0)$. In this case all eigenfunctions of the operator A are eigenfunctions of the operator $V(\omega_0)$. If $k > 1$, then

$$V(\omega_0)v - \mu v = \sum_{n=1}^{k-1} \frac{1}{n!}\omega_0^n e^{\lambda \omega_0}(A - \lambda I)^n v \in M_\lambda^{k-1}.$$

By the induction method we obtain that $M_\lambda^k \subseteq J_\mu^k$, where J_μ^k is the root subspace of the degree k of the operator $V(\omega_0)$ corresponding to the eigenvalue μ.

As the subspaces J_μ^k are finite-dimensional, then the subspaces M_λ^k are also finite-dimensional. There exists a finite collection of eigenvalues $\lambda_1, \ldots, \lambda_r$ of the operator A, which satisfies the condition $e^{\lambda_j \omega_0} = \mu$.

It is clear that

$$M_{\lambda_1}^k \oplus \ldots \oplus M_{\lambda_r}^k \subseteq J_\mu^k.$$

Now we prove that $M_{\lambda_1}^k \oplus \ldots \oplus M_{\lambda_r}^k \supseteq J_\mu^k$. Let J_μ^k with $\mu \neq 0$ be a root subspace of the operator $V(\omega_0)$. If $t > 0$, then the operators $V(\omega_0)$ and $V(t)$ commute among themselves. Therefore, J_μ^k will be invariant relative to $V(t)$ for all $t > 0$.

Choose in J_μ^k an orthonormal (relative to the scalar product in $L_2(\bar{\Omega})$) basis $v_1 \ldots, v_s$. Let $v(x) \in J_\mu^k$, then $u(x,t) = V(t)v$. Since $u(x,t)$ for all t belongs to J_μ^k, then

$$u(x,t) = \alpha_1(t)v_1 + \ldots + \alpha_s(t)v_s.$$

In this case $u(x,t)$ is a continuously differentiable by t solution of homogeneous problem (1.1)-(1.3). Therefore, the functions

$$\alpha_i(t) = (u(x,t), v_i(x)), \quad i = 1, \ldots, s$$

will be continuously differentiable for $t \geq 0$.

Substitute the function $u(x,t)$ into (1.1) and go to the limit for $t \to 0$. As the result we obtain

$$Av = \sum_{i=1}^{s} \alpha_i'(0)v_i.$$

Therefore, $Av \in J_\mu^k$. This means that the space J_μ^k is invariant relative to the operator A.

Denote by A_1 the contraction of A onto the subspace J_μ^k. The subspace J_μ^k is finite-dimensional; therefore, it may be represented in the form

$$N_{\lambda_1}^{k_1} \oplus \ldots \oplus N_{\lambda_p}^{k_p} = J_\mu^k.$$

Here $N_{\lambda_i}^{k_i}$ is the root subspace of the degree k_i corresponding to the eigenvalue λ_i of the operator A_1. The subspace $N_{\lambda_i}^{k_i}$ is contained in the root subspace of the degree k_i which corresponds to the eigenvalue $e^{\lambda_i \omega_0}$ of the operator $V(\omega_0)$. Therefore, $\mu = e^{\lambda_i \omega_0}$ and $k_i \leq k$. Hence follows that $N_{\lambda_i}^{k_i}$ is contained in $M_{\lambda_i}^k$: i.e.,

$$M_{\lambda_1}^k \oplus \ldots \oplus M_{\lambda_p}^k \supseteq N_{\lambda_1}^{k_1} \oplus \ldots \oplus N_{\lambda_p}^{k_p} = J_\mu^k.$$

Thus, $M_{\lambda_1}^k \oplus \ldots \oplus M_{\lambda_p}^k = J_\mu^k$, and the lemma is proved.

2. The properties of $V(\omega)$ in the case when problem (1.1)-(1.2) ω - periodically depends on the time. Consider now the properties of the operator $V(\omega)$ in the case when the problem (1.1)-(1.3) ω-periodically depends on the time.

We have seen that if (1.1)-(1.3) is autonomous, then the eigenfunctions of $V(\omega)$ are connected with the eigenfunctions of A. It turns out that an analogous dependence holds for $V(\omega)$ in the periodic case.

Consider the inhomogeneous problem

$$D_t u - A(x,t,D_x)u - \lambda u = f(x,t) \quad (x,t) \in Q_\omega,$$

$$B_i(x,t,D_x)u = \varphi_i(x,t) \quad (x,t) \in \Gamma_\omega, \quad i = 1,\ldots,mN,$$

$$u(x,t+\omega) = u(x,t). \tag{2.5}$$

Problem (2.5) satisfies restrictions (A.1)-(A.3) of point 3 §1. Besides, the operators $A(x, t, D_x)$ and $B_i(x, t, D_x)$ together with the right sides $f(x, t), \varphi_i(x, t)$ are assumed to be periodically depending on time. Initial condition (1.1)-(1.3) is replaced by the condition of ω-periodicity of solution.

Following the well-known terminology, we shall denote as the regular points of the operator (1.1), (1.2) those parameter λ values for which problem (2.5) has a unique solution in $H^{2m+\alpha}(Q_\omega)$ for each collection of ω-periodic functions $f \in H^\alpha(\bar{Q}_\omega)$; $\varphi_i \in H^{2m+\alpha-m_i}(\Gamma_\omega)$. All the other points we shall denote as spectrum points. Let for a certain λ problem (2.5) have a nontrivial solution. We shall denote this λ as the eigenvalue of operator (2.5). The corresponding periodic solution we shall call the eigenfunction.

Set

$$L(x, t, D_x, D_t) = D_t - A(x, t, D_x).$$

We shall consider the operator L on the functions from $H^{2m+\alpha}(Q_\omega)$ which satisfy the boundary conditions

$$B_i(x, t, D_x)u = 0.$$

A periodic function $\varphi(x, t)$ will be called a root eigenvector of the degree k of operator (2.5) if

$$(L - \lambda I)^k \varphi = 0.$$

For problem (2.5) the statement holds, analogous to that proved in the first section.

Lemma 2.4. The set of the spectrum points of the operator $V(\omega)$ is the image of the problem (2.5) spectrum under the map $\mu \rightarrow e^{\lambda \omega}$, where λ is a spectrum point of problem (2.5). Morever, the root subspace of the degree k of the operator $V(\omega)$ corresponding to the eigenvalue $\mu \neq 0$ is the direct sum of the root subspaces of the degree k of operator (2.5) corresponding to all eigenvalues λ such that $e^{\lambda \omega} = \mu$.

Proof is just the same as the proof of lemma 2.3.

3. The spectral decomposition of the space $\tilde{E}$. We have shown above that the operator $V(\omega)$ spectrum is a discrete set of points from the complex plane with the unique limit point - zero.

Let us divide the spectrum σ of $V(\omega)$ into three sets: σ_- is the set of the spectrum points situated in the disk $|z| < 1$; σ_0 is the set of the spectrum points situated in the circle $|z| = 1$, and σ_+ is the set of the spectrum points lying in the domain $|z| > 1$.

Consider in the space $\tilde{E}$ (of functions satisfying the compatibility conditions (1.4) with the zero functions f and φ: of the order r) the spectral projectors which correspond to the sets $\sigma_-, \sigma_0, \sigma_+$

$$P_- = \frac{1}{2\pi i} \int_{\gamma_-} (\lambda I - V(\omega))^{-1} d\lambda,$$

$$P_+ = \frac{1}{2\pi i} \int_{\gamma_+} (\lambda I - V(\omega))^{-1} d\lambda,$$

$$P_0 = I - P_- - P_+. \tag{2.6}$$

The closed contour γ_- lies completely inside the unit disk and contains inside itself all the points of σ_-; the closed contour γ_+ lies outside the unit disk and contains inside itself all points of σ_+ and no other points of the spectrum. The ranges of values of the projectors P_+, P_0, P_- we denote correspondingly by $\tilde{E}^+, \tilde{E}^0, \tilde{E}^-$. It was shown in Riesz-Szökefalvi-Nagy (1972) that the spaces $\tilde{E}^+, \tilde{E}^0, \tilde{E}^-$ are invariant relative to the operator $V(\omega)$, $\tilde{E}^+ \oplus \tilde{E}^0 \oplus \tilde{E}^- = \tilde{E}$. The sets σ_0, σ_+ may be only finite, since (see sections 1, 2) the corresponding subspaces $\tilde{E}^+, \tilde{E}^0$ are finite-dimensional.

Denote by V_+, V_0, V_- the contractions of the operator $V(\omega)$ onto the subspaccs $\tilde{E}^+, \tilde{E}^0, \tilde{E}^-$ respectively. In this case the spectrum of operators V_+, V_0, and V_- coincide with σ_+, σ_0, and σ_- respectively.

Lemma 2.5. There exist such positive numbers λ and θ that for each positive number ε such constants $C_i(\varepsilon)$ $i = 1, 2, 3, 4$ exist that for each $r \geq 0$

$$\| V_-^n v_- \|_r^\Omega \le C_1(\varepsilon) e^{-(\lambda-\varepsilon)n\omega} \| v_- \|_r^\Omega, \tag{2.7}$$

where $v_- \in \tilde{E}^-$, $0 < \varepsilon < \lambda$, n a natural number,

$$\| V_+^{-n} v_+ \|_r^\Omega \le C_2(\varepsilon) e^{-(\theta-\varepsilon)n\omega} \| v_+ \|_r^\Omega, \tag{2.8}$$

where $v_+ \in \tilde{E}^+$, $0 < \varepsilon < \theta$, n a natural number,

$$\| V_+^n v_+ \|_r^\Omega \ge C_3(\varepsilon) e^{(\theta-\varepsilon)n\omega} \| v_+ \|_r^\Omega, \tag{2.9}$$

where $v_+ \in \tilde{E}^+$, $0 < \varepsilon < \theta$, n a natural number, and

$$\| V_0^n v_0 \|_r^\Omega \le C_5(\varepsilon) e^{\varepsilon|n|\omega} \| v_0 \|_r^\Omega, \tag{2.10}$$

where $v_0 \in \tilde{E}^0$, n an integer.

Proof. The spectral radii (Riesz-Szökefalvi-Nady, 1972) of the operators $V_-, V_+^{-1}, V_0, V_0^{-1}$ by the definition of these operators are less or equal to $e^{-(\lambda-\varepsilon)n\omega}, e^{-(\theta-\varepsilon)n\omega}, e^{\varepsilon\omega}, e^{\varepsilon\omega}$ respectively. Hence follows inequalities (2.7)-(2.10). The lemma is proved.

Note, that the set σ_- corresponds to that part of the spectrum of operator (1.1)-(1.2) which corresponds to the eigenvalues of the operator lying in the left half-plane of the complex plane. The set σ_0 corresponds to those eigenvalues of the operator which lie in the imaginary axis. Finally, the set σ_+ corresponds to that part of the spectrum which lies in the right half-plane of the complex plane.

4. Projectors $\tilde{P}, \hat{P}$. In order to investigate inhomogeneous problem (1.1)-(1.3) it is convenient to split the space $C^{2m+\alpha}(\bar{\Omega})$, $0 < \alpha < 1$ into the direct sum of two subspaces, one of which coincides with $\tilde{E}$. To obtain such subdivision, we need to prove that there exists a projector $\hat{P}$ with the kernel $\tilde{E}$. Denote by $\hat{E}$ the set of projector $\hat{P}$ values. As a the result we obtain

$$C^r(\bar{\Omega}) = \tilde{E} \oplus \hat{E}. \tag{2.11}$$

Here we shall construct the projector $\hat{P}$. We begin with an auxiliary statement. Recall, that by m_i we have denoted the order of the differential operator in the i-boundary condition (1.2). Assume that the set $\{i : m_i = 0\}$ is nonempty for certain i. We may consider, without loss of generality, that this set is nonempty for $i = 1, \dots, l$. Consider in the boundary $\partial\Omega$ the system of linear algebraic equations relative to unknown N-vector function $f(x)$

$$B_i(x, 0, D_x)f(x) = \sum_{j=1}^{N} P_{ij} f_j(x) = \psi_i(x) \quad i = 1, \dots, l, \qquad (2.12)$$

where $\psi_i(x)$ are arbitrary functions from $C^r(\partial\Omega)$, $r \geq 0$.

From the complementary condition it follows that the matrix rank of this system is equal to l for all $x \in \partial\Omega, l \leq N$.

Let $\xi \in \partial\Omega$ be a certain point in the boundary $\partial\Omega$. Let in this point the minor of the matrix P_{ij} formed by the columns $j_1, \dots, j_l$ be not zero. The same minor does not vanish in the closure of a certain neighborhood U_ξ of the point ξ. Define the operator W_ξ which maps each collection $\psi_1, \dots, \psi_l$ to the vector function $f(x)$ satisfying system (2.12) for $x \in \partial\Omega \cap U_\xi$. To this end equate to zero all the components of $f(x)$ whose numbers are not equal to $j_1, \dots, j_l$. The other components we shall find from the system

$$\sum_{k=1}^{l} P_{ij_k} f_{j_k} = \psi_j(x), \quad i = 1, \dots, l.$$

Such neighborhoods U_ξ and the operator W_ξ exist for each point $\xi \in \partial\Omega$. The system of neighborhoods $\{U_\xi\}_{\xi \in \partial\Omega}$ forms an open covering of the boundary $\partial\Omega$. Choose from it a finite subcovering $U_{\xi_1}, \dots, U_{\xi_s}$ and consider a decomposition of the unit $J_k(x)$, $k = 1, \dots, s$ corresponding to this subcovering.

Setting

$$W(\psi_1, \dots, \psi_l) = \sum_{k=1}^{s} J_k(x) W_{\xi_k}(\psi_1, \dots, \psi_l),$$

we obtain the linear operator W whose values for each $\psi_1, \dots, \psi_l$ satisfy system (2.12) everywhere in $\partial\Omega$.

Denote by P the bounded linear operator which with each function $f \in C^\alpha(\overline{\partial\Omega})$ associates the continuation of $F \in C^\alpha(\overline{\Omega})$ onto the whole domain $\overline{\Omega}$. Let $R = PW$. Taking into account the construction of the algorithm of the operator W, it is easy to verify that

$$\| R(\psi_1, \ldots, \psi_l) \|_\alpha^\Omega \le \text{const} \sum_{k=1}^{l} \| \psi_k \|_\alpha^{\partial\Omega} .$$

Let λ be any fixed regular value of the operator

$$\lambda v - A(x, 0, D_x)v = 0, \qquad P_i v|_{\partial\Omega} = 0.$$

If $w(x)$ is an arbitrary element from $C^{2m+\alpha}(\overline{\Omega})$, then as $\hat{P}w$ we shall take the solution of the problem

$$\lambda v - A(x, 0, D_x)v = f, \quad B_i(x, 0, D_x)v|_{\partial\Omega} = \varphi_i(x),$$

$$\varphi_i(x) = B_i(x, 0, D_x)w(x)|_{\partial\Omega}, \quad f = R(\psi_1, \ldots, \psi_l),$$

$$\psi_k(x) = B_k(x, 0)(\lambda w - A(x, 0, D_x)w)|_{\partial\Omega}, \quad k = 1, \ldots, l. \qquad (2.13)$$

Note, that $\hat{P}\hat{P}w = \hat{P}w = v$; therefore, $\hat{P}$ is a projector. Constructed projector $\hat{P}$ has two essential properties. First, the value $\hat{P}w$ may be determined by $B_i w$, $i = 1, \ldots, mN$, and $B_i(Aw)$, for those $i : m_i = 0$ in the boundary of Ω. Second, for each $w(x) \in C^{2m+\alpha}(\overline{\Omega})$ the following estimate

$$\| \hat{P}w \|_{2m+\alpha}^\Omega \le \text{const} \left(\sum_{i=1}^{mN} \| B_i w \|_{2m+\alpha-m_i}^{\partial\Omega} + \sum_{i:m_i=0} \| B_i(Aw) \|_\alpha^{\partial\Omega} \right)$$

$$(2.14)$$

holds.

Note, that the subdivision (2.11) and the corresponding projector are determined, generally speaking, not uniquely. Nevertheless, it is easy to show, that each projector with the kernel $\tilde{E}$ has the properties given above. Since we are interested in only these properties, the results obtained below will be valid for each projector choice.

Denote by $\tilde{P}$ the projector $I - \hat{P}$ with the range of values $\tilde{E}$. The following decomposition holds true

$$C^{2m+\alpha}(\bar{\Omega}) = \hat{E} \oplus E^- \oplus E^0 \oplus E^+.$$

In this case it will be convenient to extend the projectors P_-, P_0, P_+ onto the whole space $C^{2m+\alpha}(\bar{\Omega})$ setting $P_-w = P_-\tilde{w}, P_0w = P_0\tilde{w}$, $P_+w = P_+\tilde{w}$, where $\tilde{w} = \tilde{P}w$. Now, each element w from $C^{2m+\alpha}(\bar{\Omega})$ may be represented in the form

$$w = P_-w + P_0w + P_+w + \hat{P}w.$$

Consider problem (1.1)-(1.3) for $0 \leq r < 2m$, $\varphi_i \in H_{r-m_i}^{2m+\alpha m_i}(\Gamma_T)$. In this case the projector $\hat{P}$ may be obtained by closing the projector constructed before in $C^r(\hat{\Omega})$. Taking into account the definition of $H_{r-m_i}^{2m+\alpha+m_i}(\Gamma_T)$, we obtain that if $w(x) \in C^r(\bar{\Omega})$ and satisfies the compatibility conditions of the order r with the functions φ_i, then

$$\| \hat{P}w \|_r^\Omega \leq \text{const} \cdot \sum_{i:r-m_i \geq 0} \| \varphi_i \|_{i-m_i,2m+\alpha-m_i}^{\Gamma_T} . \qquad (2.15)$$

In this case any initial data $u_0(x) \in C^r(\bar{\Omega})$ of problem (1.1)-(1.3), satisfying the compatibility conditions of the order r, may be uniquely represented in the form

$$u_0(x) = \hat{v} + v_+ + v_- + v_0,$$

$$\hat{P}\hat{v} = \hat{v}, \ v_+ \in E^+, \ v_- \in E_r^-, \ v_0 \in E^0.$$

Projectors P_+, P_0 project in the finite-dimensional subspaces stretched on the C^3 functions.

Projector P_- projects in the infinite dimensional subspace. By (1.6), we may close P_- in the $C^r(\bar{\Omega})$ norm. As the result the range of values of new projector will be denoted by E_r^-.

5. Bounded solutions of inhomogeneous parabolic systems.
The main results of this paragraph may be summarized by the following
theorem.

Theorem 2.1.   Let the spectrum of $V(\omega)$ have no common points
with the unit circle in the complex plane. Then, for each collection of
right sides $f \in H^{\alpha}_{r-2m}(Q)$, $\varphi_i \in H^{\alpha}_{r-2m-m_i}(\Gamma)$, there exists a unique
function $v_+ \in E^+$ such that the solution of inhomogeneous problem
(1.1)-(1.3) with the initial data $u_0 = \hat{v} + v_+ + v_-$ exists for all $t \geq 0$,
belongs to the class $H^{2m+\alpha}_r(Q)$, and satisfies the inequality

$$\| u \|^Q_{r,2m+\alpha} \leq C \left(\| f \|^Q_{r-2m,\alpha} + \sum_{i=1}^{mN} \| \varphi_i \|^{\Gamma}_{r-m_i,2m+\alpha-m_i} + \| v_- \|_r \right),$$

$$\| \hat{v} \|^{\Omega}_r + \| v_+ \|^{\Omega}_{2m+\alpha} \leq C \left(\| f \|^Q_{r-2m,\alpha} + \sum_{i=1}^{mN} \| \varphi_i \|^{\Gamma}_{r-m_i,2m+\alpha-m_i} \right).$$

$$(2.16)$$

Here the constant C depends only on coefficients of problem (1.1)-(1.3)
and on the form of the projector $\hat{P}$.

Remark 2.1.   Theorem 2.1 for the case $r = 2m + \alpha$ was proved by
Belonosov and Vishnevskii (1977) when the spectra of $V(\omega)$ is situated
inside the unit sphere $(E^+ = \emptyset)$. However, using the scheme suggested
by Belonosov, it may be generalized for the case $E^+ \neq \emptyset$.

Proof.   Let $u(x,t)$ be a solution to (1.1)-(1.3) which belongs to
$H^{2m+\alpha}_r(Q_T)$ for each finite T. Let $t = n\omega$, where n is a natural number.
The function $u(x, n\omega)$ may be represented as follows

$$u(x, n\omega) = \hat{u}(x, n\omega) + \tilde{u}(x, n\omega).$$

Denote by M the value

$$M = \| f \|^Q_{r-2m,\alpha} + \sum_{i=1}^{mN} \| \varphi_i \|^{\Gamma}_{r-m_i,2m+\alpha-m_i}.$$

Then, (2.15) yields

$$\| \hat{u}(x, n\omega) \|^{\Omega}_{2m+\alpha} < CM, \quad n > 0, \quad \| \hat{u}(x, 0) \|^{\Omega}_{r} < CM,$$

where the constant C does not depend on n.

It remains to find v_+. Note, that it is sufficient to choose v_+ such that the solution of (1.1)-(1.3) with initial data $u_0 = v_+ + v_- + \hat{v}$ for any integer n satisfies the inequality

$$\| u(x, n\omega) \|^{\Omega}_{2m+\alpha} \leq \text{const}(M + \| v_- \|_r)$$

with a constant independent of n; with the help of (2.6), it is now not difficult to obtain (2.16).

Consider the function $\tilde{u}(x, n\omega)$. Let $w_k(x, n) = u(x, n\omega) - V(k\omega)\tilde{u}(x, (n - k)\omega)$. For $n \geq k > 0$ the function $w_k(x, n)$ is the value of the solution of inhomogeneous problem (1.1)-(1.3) for $t = n\omega$ with initial data $\hat{u}(x, (n - k)\omega)$ given for $t = (n - k)\omega$. It is easy to obtain that the function $w_1(x, n)$ satisfies the following inequality

$$\| w_1(x, n) \|^{\Omega}_{2m+\alpha} \leq \text{const} \cdot M,$$

where the constant does not depend on n.

Represent the function $u(x, n\omega)$ in the form

$$u(x, n\omega) = V(\omega)\tilde{u}(x, (n - 1)\omega) + w_1(x, n), \quad n > 0.$$

Let $z_n(x) = \hat{P}w_1(x, n)$, then

$$\tilde{u}(x, n\omega) = V(\omega)\tilde{u}(x, (n - 1)\omega) + z_n(x) = V^2(\omega)\tilde{u}(x, (n - 2)\omega)$$

$$+V(\omega)z_{n-1}(x) + z_n(x) = \sum_{k=1}^{n} V^{n-k}(\omega)z_k(x) + V^n(\omega)\tilde{u}(x, 0).$$

Note, that $z_k(x) \in \tilde{E}$; therefore,

$$V^{n-k}(\omega)z_k(x) = V^{n-k}_+ P_+ z_k(x) + V^{n-k}_- P_- z_k(x)$$

$$= V_+^n V_+^{-k} P_+ z_k(x) + V_-^{n-k} P_- z_k(x).$$

Thus,

$$\tilde{u}(x, n\omega) = V_+^n \left(v_+ + \sum_{k=1}^{n} V_+^{-k} P_+ z_k(x) \right) + \sum_{k=1}^{n} V_-^{n-k} P_- z_k(x) + V_-^n v_-.$$

Set $v_+ = - \sum_{k=1}^{\infty} V_+^{-k} P_+ z_k(x)$. From (2.9), the properties of $z_k(x)$ and P_+ it follows that the series in the right-hand side of v_+ is absolutely convergent in $C^{2m+\alpha}(\Omega)$ and

$$\| v_+ \|_{2m+\alpha}^{\Omega} \leq \frac{\text{const}}{1 - e^{-(\theta-\varepsilon)\omega}} \cdot M = CM.$$

Hence follows

$$\tilde{u}(x, n\omega) = \sum_{k=1}^{n} V_-^{n-k} P_- z_k(x) - \sum_{k=n+1}^{\infty} V_+^{n-k} P_+ z_k(x) + V_-^n v_-.$$

Using (2.10), (1.5), (1.6), we obtain $\| \tilde{u}(x, n\omega) \|_{2m+\alpha}^{\Omega} < CM$ for $n > 0$. This estimate and the obtained earlier estimate $\| \hat{u}(x, n\omega) \|_{2m+\alpha}^{\Omega} < CM$ yield the first inequality in (2.16). The second inequality in (2.16) follows from the inequalities $\| v_+ \|_{2m+\alpha}^{\Omega} < CM$ and $\| \hat{v} \|_{2m+\alpha}^{\Omega} < CM$. The theorem is proved.

6. Solutions of the linear inhomogeneous parabolic system, bounded in the whole axis .

Theorem 2.2. (Vishnevskii, 1982, 1986). Let all the conditions of the theorem 2.1 hold, the functions f, φ_i be determined for all $t \in (-\infty, +\infty)$, where $f \in H^{\alpha}(Q_0)$, $\varphi_i \in H^{2m+\alpha-m_i}(\Gamma_0)$, $i = 1, \ldots, mN$. Then there exists a unique solution of problem (1.1)-(1.2) $\eta(x, t)$ which for all $t \in (-\infty, +\infty)$ belongs to $H^{2m+\alpha}(Q_0)$ and satisfies the inequality

$$\| \eta(x, t) \|_{2m+\alpha}^{Q_0} \leq C \left(\| f \|_{\alpha}^{Q_0} + \sum_{i=1}^{mN} \| \varphi_i \|_{2m+\alpha-m_i}^{\Gamma_0} \right). \qquad (2.17)$$

Proof. Let $M = \parallel f \parallel_\alpha^{Q_0} + \sum_{i=1}^{mN} \parallel \varphi_i \parallel_{2m+\alpha-m_i}^{\Gamma_0}$. It is sufficient to show that there exists a unique function $\eta(x,t)$ that for all t satisfies equation (1.1), boundary conditions (1.2) and for each integer n satisfies the inequality $\parallel \eta(x,n\omega) \parallel_{2m+\alpha}^{\Omega} < CM$, where C does not depend on n. Using again the notations of theorem 2.1 we may remark that the functions $f, \varphi_i, \quad i = 1,\ldots,mN$ for all integer k allow us to find the function $z_k(x)$. Let $\eta^n(x,t)$ solve problem (1.1)-(1.2) with the initial data given for $t = n\omega$

$$\eta^n(x,n\omega) = \eta_+^n + \eta_-^n + \hat{\eta}^n, \quad \eta_+^n \in E^+, \quad \eta_-^n \in E^-,$$

$$\eta_+^n = - \sum_{k=n+1}^{\infty} V_+^{n-k} P_+ z_k(x), \quad \eta_-^n = \sum_{-\infty}^{n} V_-^{n-k} P_- z_k(x).$$

In this case, $\hat{\eta}^n(x)$ belongs to $\hat{E}$ and is uniquely determined by the collection of right sides and boundary conditions; n is an arbitrary integer. As before, we obtain for $z_k(x)$ from theorem 2.1

$$\parallel z_k(x) \parallel_{2m+\alpha}^{\Omega} \le \mathrm{const} \cdot M, \quad \parallel \hat{\eta}^n(x) \parallel_{2m+\alpha}^{\Omega_0} \le \mathrm{const} \cdot M$$

with the constant independent of n, k. Inequalities (2.9)-(2.10) yield

$$\parallel \eta_+^n \parallel_{2m+\alpha}^{\Omega_0} \le \frac{\mathrm{const}}{1 - e^{-(\theta-\varepsilon)\omega}} \cdot M = CM,$$

$$\parallel \eta_-^n \parallel_{2m+\alpha}^{\Omega_0} \le \frac{\mathrm{const}}{1 - e^{-(\theta-\varepsilon)\omega}} \cdot M = CM,$$

Note that

$$\eta^n(x,(n+1)\omega) = \hat{\eta}^{n+1}(x) + V_+\eta_+^n + V_-\eta_-^n + z_{n+1}$$

$$= \eta^{n+1}(x) + \sum_{-\infty}^{n+1} V_-^{n+1-k} P_- z_k(x) - \sum_{k=n+2}^{\infty} V_+^{n+1-k} P_+ z_k(x)$$

$$= \eta^{n+1}(x,(n+1)\omega).$$

This denotes that the solutions $\eta^{n+1}(x,t), \qquad \eta^n(x,t)$ coincide for

$t \geq (n+1)\omega$. Define the function $\eta(x,t)$ setting $\eta(x,t) = \eta^n(x,t)$ for $t \geq n\omega$. This function for all t will satisfy equation (1.1) and boundary conditions (1.2). In this case $\eta^n(x, n\omega) = \eta^n_+ + \eta^n_- + \hat{\eta}^n$ and, therefore, the constructed solution satisfies estimate (2.17).

Prove by contradiction the uniqueness of the constructed solution $\eta(x,t)$. Let $\eta_1(x,t)$ be another solution, where $\eta_1(x,t) \in H^{2m+\alpha}(Q_0)$. Then $w(x,t) = \eta(x,t) - \eta_1(x,t)$ is the solution of homogeneous problem (1.1)-(1.2),bounded in all the axis, $w(x,t) \in H^{2m+\alpha}(Q_0)$. From (2.7)-(2.10) it follows that $w(x,t) \equiv 0$. The theorem is proved.

Collolary. Let all conditions of theorem 2.1 hold, $f \in H^\alpha(Q_-)$, $\varphi_i \in H^{2m+\alpha-m_i}(\Gamma_-)$, $i = 1,\ldots,mN$. Then for each function $v_+ \in E^+$ there exist unique functions $\hat{v} \in \hat{E}$, $v_- \in E^-$ such that the solution to (1.1)-(1.3) is determined for all $t \leq 0$, belongs to $H^{2m+\alpha}(Q_-)$ and satisfies the inequalities

$$\| u \|_{2m+\alpha}^{Q_-} \leq C \left(\| f \|_\alpha^{Q_-} + \sum_{i=1}^{mN} \| \varphi_i \|_{2m+\alpha-m_i}^{\Gamma_-} + \| v_+ \|_{2m+\alpha}^\Omega \right),$$

$$\| v_- \|_{2m+\alpha}^\Omega + \| \hat{v} \|_{2m+\alpha}^\Omega \leq C \left(\| f \|_\alpha^{Q_-} + \sum_{i=1}^{mN} \| \varphi_i \|_{2m+\alpha-m_i}^{\Gamma_-} \right). \quad (2.18)$$

Proof. Extend the functions f, φ_i, $i = 1,\ldots,mN$ onto the whole cylinder Q_0 with the boundary Γ_0 preserving the corresponding norms. If we set $v_- = P_-\eta(x,0)$, $\hat{v} = \hat{P}\eta(x,0)$, then the statement follows from theorem 2.2; inequalities (2.8) follow from (2.7)-(2.10), (2.17).

§1.3 Bounded solutions of nonlinear parabolic systems.

1. Setting nonlinear problems. Consider in the cylinder Q the system which differs from system (1.1)-(1.3) by the nonlinear terms in the right sides and in the boundary conditions

$$Lu = D_t u - A(x,t,D_x)u = f(x,t,u,\ldots,D_x^{2m}u),$$

$$B_i u \Big|_{\Gamma} = \varphi_i(x, t, u, \ldots, D_x^{m_i} u), \quad m_i < 2m, \quad i = 1, \ldots, mN,$$

$$u(x, 0) = u_0(x). \tag{3.1}$$

Here A, B_i are the same operators as in the previous paragraph. By $D_x^k u$ we denote the collection of all partial derivatives with respect to $x_1, \cdots, x_n$ of vector function $u(x, t)$ of the order k.

Assume that the functions $f(x, t, u, \cdots, D_x^{2m} u)$ have the following form

$$\sum_{|\alpha| \le 2m-k} D_x^\alpha \left(\sum_{k < |\beta| \le 2m-\alpha} f_{\alpha\beta}(x, t, D_x^\nu) D_x^\beta u + f_\alpha(x, t, D_x^\nu u) \right),$$

where $0 \le k \le 2m$, $f_{\alpha\beta}, f_\alpha$ depend on x, t, u, and on all D_x^ν up to the order k, inclusively. Depending on k system (3.1) will be either quasilinear for $k < 2m$, or nonlinear of the general form ($k = 2m$). Assume that the functions $f_{\alpha\beta}, f_\alpha$ satisfy conditions of lemmas 1.1, 1.2. Assume that the first and second partial derivatives of $f_{\alpha\beta}, f_\alpha$ have similar properties by the arguments $D_x^\nu u_j$, where u_j are the components of the vector function u.

Assume that the boundary functions $\varphi_i(x, t, u, \ldots, D_x^{m_i} u)$ are as follows

$$\varphi_i = \sum_{|\alpha| \le m_i-k} D_x^\alpha \left(\sum_{k < |\beta| \le m_i-\alpha} \varphi_{\alpha\beta}(x, t, D_x^\nu) D_x^\beta u + \varphi_\alpha(x, t, D_x^\nu u) \right),$$

and the boundary data $\varphi_{\alpha\beta}, \varphi_\alpha$ satisfy the properties analogous to those imposed on the functions $f_{\alpha\beta}, f_\alpha$ with the corresponding change of formulations.

The most essential assumption is the following condition (smallness of nonlinearities). Let $r \in [k, s], s = 2m + \alpha, 0 < \alpha < 1, u(x, t) \in H_r^s(Q_{T+\tau})$, $v(x, t) \in H_r^s(Q_{T+\tau})$, r be an arbitrary, $T > 0$. Then

$$\| F(u) - F(v) \|_{r,s-2m}^{Q_{T+\tau}^\tau} + \sum_{i=1}^{mN} \| \Phi_i(u) - \Phi_i(v) \|_{r,s-m_i}^{\Gamma_{T+\tau}^\tau} \le l \| u - v \|_{r,s}^{Q_{T+\tau}^\tau}.$$

$$\tag{3.2}$$

Here by F we denote the map $u(x,t) \rightarrow f(x,t,u,\ldots,D_x^{2m}u)$ from the space $H_r^s(Q)$ into $H_r^{s-2m}(Q)$. By Φ_i we denote the map $u(x,t) \rightarrow \varphi_i(x,t,u,\ldots,D_x^{m_i}u)$, $(x,t) \in \Gamma$ from $H_r^s(\Gamma)$ into $H_r^{s-2m}(\Gamma)$. The number l in this paragraph will be assumed to be sufficiently small.

Initial data $u_0(x)$ in this paragraph are assumed to be from $C^r(\bar{\Omega})$ and satisfy the compatibility conditions of the order r.

2. Bounded in the whole axis solution of the nonlinear parabolic system. Noncritical case, when the spectrum of the operator $V(\omega)$ does not intersect with the unit circle.
Let $V(\omega)$ be the shift operator of linear problem (1.1)-(1.3) from the second paragraph, and let its spectrum not intersect the unit circle of the complex plane, i.e., $\sigma_0 = \emptyset$. The case $\sigma_0 \neq \emptyset$ will be considered further.

Theorem 3.1. (Vishnevskii, 1982, 1986). Let the above assumptions hold. Then, there exists a positive number l_0 such, that for all $0 < l < l_0$ the problem (3.1) without the initial condition $u(x,0) = u_0(x)$ has a unique solution $\eta(x,t)$ defined for all $t \in R$. This solution belongs to the class $H^s(Q_0)$ and satisfies the inequality

$$\| \eta(x,t) \|_s^Q \leq \frac{C}{1-Cl} \left(\| F(0) \|_{s-2m}^{Q_0} + \sum_{i=1}^{mN} \| \Phi_i(0) \|_{s-m_i}^{\Gamma_0} \right). \qquad (3.3)$$

Proof. Let H, a subset from $H^s(Q_0)$, consist of functions whose norm in $H^s(Q_0)$ does not exceed KM. By M we have denoted

$$M = \| F(0) \|_{s-2m}^{Q_0} + \sum_{i=1}^{mN} \| \Phi_i(0) \|_{s-m_i}^{\Gamma_0}.$$

The constant K will be chosen further.

Let W be the operator which a vector function $v(x,t) \in H$ maps into the vector function $u(x,t)$. The function $u(x,t)$ is a unique bounded in the whole axis solution to the problem

$$Lu = F(v), \quad P_i u = \Phi_i(v) \quad i = 1,\ldots,mN.$$

In theorem 2.2 we have established the uniqueness and existence of such solution. Estimates (2.17), (2.18) yield

$$\| u(x,t) \|_s^{Q_0} \leq C \left(\| F(v) - F(0) + F(0) \|_{s-2m}^{Q_0} \right.$$

$$\left. + \sum_{i=1}^{mN} \| \Phi_i(v) - \Phi_i(0) + \Phi_i(0) \|_{s-m_i}^{\Gamma_0} \right)$$

$$\leq Cl \| v \|_s^{Q_0} + CM \leq CM(Kl + 1),$$

$$\| u_1(x,t) - u_2(x,t) \|_s^{Q_0} \leq Cl \| v_1(x,t) - v_2(x,t) \|_s^{Q_0} .$$

If we choose K so that $C(Kl + 1) = K$ and take such l_0 that $Cl_0 < 1$, then W will map the set H into itself and will be a contraction operator. The unique fixed point of W will be the desired solution $\eta(x,t)$; $\eta(x,t) \in H$, $K = \frac{C}{1-Cl}$. Therefore, we shall have the inequality

$$\| \eta(x,t) \|_s^{Q_0} \leq \frac{CM}{1 - Cl}.$$

The theorem is proved.

3. Bounded for $t \geq 0$ and for $t \leq 0$ solutions of the nonlinear parabolic system in the noncritical case. In the noncritical case we assume that E_0 is empty, $\tilde{E} = E_r^- \oplus E^+$ (the definition of E_r^- see in §1.2 point 4). The following theorem holds.

Theorem 3.2. (Vishnevskii, 1982, 1986). Let the conditions from theorem 3.1 hold; and $l \leq l_0$. Then, there exists a map Ψ from E_r^- into $E^+ \oplus \hat{E}$ such that
1) each solution to problem (3.1) with the initial data

$$u_0(x) = \eta(x,0) + v_- + \Psi_-(v_-)$$

exists for all $t \geq 0$, belongs to $H_r^s(Q)$, and satisfies the inequality

$$\| u(x,t) - \eta(x,t) \|_{r,s}^{Q} \leq C_1 \| u_0(x) - \eta(x,0) \|_{s}^{\Omega} . \qquad (3.4)$$

Here v_- is an arbitrary function from E_r^-.

2) If v_-, v_-' are two functions from E_r^-, then

$$\| P_+ \Psi_-(v_-) - P_- \Psi_-(v_-') \|_s^{\Omega} + \| \hat{P} \Psi_-(v_-) - \hat{P} \Psi_-(v_-') \|_s^{\Omega}$$

$$\leq Cl \| v_- - v_-' \|_s^{\Omega} . \qquad (3.5)$$

3) $\Psi_- P_- \eta(x,0) = P_+ \eta(x,0) + \hat{P} \eta(x,0)$.

4) If $\tilde{u}(x,t)$ solves problem (3.1), is determined for $t \geq 0$, and belongs to the class $H_r^s(Q)$, then

$$\tilde{u}(x,0) = \eta(x,0) + v_- + \Psi_-(v_-),$$

where

$$v_-(x) = P_-(\tilde{u}(x,0) - \eta(x,0)).$$

Proof. Let $v(x,t) = u(x,t) - \eta(x,t)$. Using notations

$$F_1(v) = F(v + \eta) - F(v), \quad \Phi_i^1(v) - \Phi_i(v + \eta) - \Phi_i(v))$$

we note that from (3.2) it follows that

$$\| F_1(v_1) - F_1(v_2) \|_{r-2m,s-2m}^{Q_{T+\tau}^\tau} + \sum_{i=1}^{mN} \| \Phi_i^1(v_1) - \Phi_i^1(v_2) \|_{r-m_i,s-m_i}^{\Gamma_{T+\tau}^\tau}$$

$$\leq l \| v_1 - v_2 \|_{r,s}^{Q_{T+\tau}^\tau}, \quad +\infty \geq T > 0, \tau \in R.$$

$$F_1(0) = 0, \quad \Phi_i^1(0) = 0, \quad i = 1, \ldots, mN.$$

Denote by H the subset from $H_r^s(Q)$ consisting of the functions whose norm does not exceed M, and whose projection in E_r^- for $t = 0$ is equal to a fixed element $v_- \in E_r^-$.

Let $v(x,t) \in H$ and $u(x,t)$ solve the problem

$$Lu = F_1(v), \quad B_i(u)|_\Gamma = \Phi_i^1(v), \quad i = 1, \dots, mN,$$

$$u(x,0) = u_0(x) - \eta(x,0),$$

where $u(x,0)$ is such that $P_- u(x,0) = v_-$.

In theorem 2.1 it was proved that such solution exists and is unique. From estimate (2.16) it follows that

$$\| u \|_{r,s}^Q \leq C \left(\| F_1(v) \|_{r-2m,s-2m}^Q + \sum_{i=1}^{mN} \| \Phi_i^1(v) \|_{r-m_i,s-m_i}^\Gamma \right.$$

$$\left. + \| v_- \|_r^\Omega \right) \leq C l K + C \| v_- \|_r^\Omega .$$

Choose K so that $K = \frac{C \| v_- \|_r^\Omega}{1 - Cl}$. Then

$$\| u \|_{r,s}^Q \leq \frac{C \| v_- \|_r^\Omega}{1 - Cl}.$$

Let W map the function $v(x,t) \in H$ to the function $u(x,t)$ which solves the above problem. By the construction of $P_- u(x,0) = v_-$ and from the obtained estimate we have $u(x,t) \in H$. Since W is a contraction in H, it has in H a unique fixed point $u(x,t)$. In this case

$$\| u \|_{r,s}^Q \leq \frac{C \| v_- \|_r^\Omega}{1 - Cl}. \tag{3.6}$$

Set $\Psi_-(v_-) = u(x,0) - v_-(x)$. It follows from (2.16) that

$$\| P_+ \Psi_-(v_-) - P_+ \Psi_-(v_-') \|_s^\Omega + \| \hat{P} \Psi_-(v_-) + \hat{P} \Psi_-(v_-') \|_s^\Omega$$

$$\leq C l \| v_- - v_-' \|_s^\Omega$$

for each $v_-, v'_- \in E_r^-$. The obtained inequality yields (3.5). From (3.6), (3.5) for $v'_- = 0$ we obtain (3.4).

We must now prove this last point of the theorem, that if $\tilde{u}(x,t)$ is uniformly bounded for $t > 0$ solution to problem (3.1) from the class $H_r^s(Q)$, then

$$\tilde{u}(x,0) = \eta(x,0) + v_- + \Psi_-(v_-)$$

for a certain function $v_- \in E_r^-$. For this purpose consider an arbitrary solution $\tilde{u}(x,t) \in H_r^s(Q)$ such that $P_-\tilde{u}(x,0) = \tilde{v}_-$. Denote by $u(x,t;\tilde{v}_-)$ the solution to (3.1) which we have constructed when proving the theorem. We know that $u(x,t;\tilde{v}_-) \in H_r^s(Q)$, $u(x,t;\tilde{v}_-) = \eta(x,0) + \tilde{v}_- + \Psi_-(\tilde{v}_-)$. Denote by $w(x,t)$ the difference $\tilde{u}(x,t) - u(x,t;\tilde{v}_-)$. We need to establish that $w(x,t) \equiv 0$. We know that $w(x,t)$ solves the problem

$$Lw = F(\tilde{u}) - F(u),$$

$$B_i w = \Phi_i(\tilde{u}) - \Phi_i(u), \quad \big(i = 1,\ldots,mN\big), \quad P_-w(x,0) = 0.$$

From (2.16), (3.2) we obtain

$$\| w \|_{r,s}^{Q} \leq Cl \| \tilde{u} - u \|_{r,s}^{Q} = Cl \| w \|_{r,s}^{Q}, \quad Cl < 1;$$

therefore, the latter inequality holds only if $w(x,t) \equiv 0$. The theorem is proved.

Collolary. Let $0 < \lambda_1 < \lambda$, where λ is taken from the first inequality (2.7). Then, there exists such $l_0^1 > 0$ that for $0 < l < l_0^1$ each solution of (3.1) with the initial data $u_0(x) = \eta(x,0) + v_- + \Psi_-(v_-)$ will in addition satisfy the inequality

$$\| (u(x,t) - \eta(x,t))e^{\lambda_1 t} \|_{r,s}^{Q} \leq C_2 \| u(x,0) - \eta(x,0) \|_{r}^{\Omega}. \qquad (3.7)$$

Proof.   Change the function as follows

$$v(x,t) = (u(x,t) - \eta(x,t))e^{\lambda_1 t}.$$

As a result, we obtain the new problem

$$Lv + \lambda_1 v = \tilde{F}(v), \quad B_i v = \tilde{\Phi}_i(v), \quad v(x,0) = u(x,0) - \eta(x,0);$$

$$\tilde{F}(v) = \left(F(ve^{-\lambda_1 t} + \eta) - F(\eta)\right)e^{\lambda_1 t},$$

$$\tilde{\Phi}_i(v) = \left(\Phi_i(ve^{-\lambda_1 t} + \eta) - \Phi_i(\eta)\right)e^{\lambda_1 t}.$$

It is easy to verify that $\tilde{F}, \tilde{\Phi}_i$ satisfy inequality (3.2) with such constant l_1 that satisfies the relation $l_1 = C_2 l$ with a certain constant C_2. Applying to the new problem theorem 3.2, we obtain that the solution $u(x,t)$ with the initial data $u(x,0) = \eta(x,0) + v_- + \check{\Psi}_-(v_-)$ satisfies estimate (3.7). From the latter point of theorem 3.2 it follows that $\check{\Psi}_- = \Psi_-$. The corollary is proved.

Using the corollary from theorem 2.2 and the considerations from theorem 3.2 proof, we may easily obtain the analog of theorem 3.2 for the cylinder Q_-. Namely

Theorem 3.3.   (Vishnevskii, 1982, 1986).   Let the conditions of theorem 3.1 hold and $0 < l \leq l_0$. Then, there exists a map Ψ_+ from E^+ into $E_r^- \oplus \hat{E}$ such that
1) each solution to (3.1) with the initial data $u_0(x) = \eta(x,0) + v_+ + \Psi_+(v_+)$ exists for all $t \leq 0$ and belongs to the space $H^s(Q_-)$. Besides, this solution satisfies the inequality

$$\| u(x,t) - \eta(x,t) \|_s^{Q_-} \leq C_1 \| u_0(x) - \eta(x,0) \|_s^\Omega . \qquad (3.8)$$

Here v_+ is an arbitrary function from E^+.
2) If $v_+, v_+' \in E^+$, then

$$\| \Psi_+(v_+) - \Psi_+(v_+') \|_s^\Omega \leq Cl \| v_+ - v_+' \|_r^\Omega . \qquad (3.9)$$

3) $\Psi_+(P_+\eta(x,0)) = P_-\eta(x,0) + \hat{P}\eta(x,0)$.

4) If $\tilde{u}(x,t)$ solves problem (3.1), is determined for all $t \leq 0$, and belongs to $H^s(Q_-)$, then

$$\tilde{u}(x,0) = \eta(x,0) + v_+ + \Psi_+(v_+),$$

where $v_+ = P_+(\tilde{u}(x,0) - \eta(x,0))$.

5) If $0 < \theta_1 < \theta$, then estimate (3.8) for $0 < l \leq l'_0$ may be revised as follows

$$\| (u(x,t) - \eta(x,t))e^{-\theta_1 t} \|_s^{Q_-} \leq C_1 \| u_0(x) - \eta(x,0) \|_s^{\Omega} . \qquad (3.10)$$

The number θ is taken here from (2.8),(2.10).

4. Bounded solutions of the nonlinear parabolic system in the critical case.

Consider, now, the case when the operator $V(\omega)$ spectrum intersects the imaginary axis, i.e., $\sigma_0 \neq \emptyset$, $E^0 \neq \emptyset$. The analogs of theorems 3.1-3.3 in this case also hold true.

Theorem 3.4. (Vishnevskii, 1982, 1986). Let the conditions of theorem 3.1 hold. Then there exists a positive number $l_{10} > 0$ such that for $0 < l < l_{10}$ there exists a map Ψ_0 from E^0 into $E^- \oplus E_r^+ \oplus \hat{E}$ with the following properties.

1) Let $0 < \gamma < \min(\theta, \lambda)$, where θ and λ are from (2.10); $u(x,t)$ be a solution to (3.1) with the initial data $u(x,0) = u_0 + \Psi_0(v_0)$, $v_0 \in E^0$ be arbitrary. Then $u(x,t)$ is determined for all $t \in R$ and satisfies the inequality

$$\| u(x,t)e^{-\theta|t|} \|_s^Q \leq C(\gamma) \| v_0(x) \|_r^{\Omega} . \qquad (3.11)$$

2) If $v_0, v'_0 \in E^0$ are arbitrary, then

$$\| \Psi_0(v_0) - \Psi_0(v'_0) \|_s^Q \leq C(\gamma)l \| v_0 - v'_0 \|_r^{\Omega} . \qquad (3.12)$$

3) $\Psi_0(0) = 0$.

4) If $\tilde{u}(x,t)$ solves (3.1) and is determined for all t, where the function $\tilde{u}(x,t)e^{-\gamma t}$ belongs to $H^s(Q)$, then

$$\tilde{u}(x,0) = \tilde{v}_0 + \Psi_0(\tilde{v}_0), \quad \tilde{v}_0 = P_0\tilde{u}(x,0).$$

Remark.   In the case when $F(0) = 0,\ \ \Phi_i(0) = 0,\ \ i = 1,\ldots,mN$ theorem 3.1 is the corollary of theorem 3.4. Really, in this case the space E^0 has only the zero element, and the unique bounded in the whole axis solution of the boundary-value problem is the trivial solution.

The proof of theorem 3.4 we shall give further.

Theorem 3.5.   (Vishnevskii, 1982, 1986). Assume that conditions of theorem 3.4 hold. Then, for $0 < l < l_{10}$ there exists a map Ψ_0^- from $E^0 \oplus E_r^-$ into $E^+ \oplus \hat{E}$ with the following properties.

1) If $u(x,t)$ solves (3.1) with the initial data $u_0(x) = u(x,0) = v_- + v_0 + \Psi_0(v_- + v_0)$, then the solution $u(x,t)$ exists for all $t \geq 0,\ u(x,t)e^{-\gamma t} \in H_r^s(Q)$ and the inequality

$$\| u(x,t)e^{-\gamma t} \|_{r,s}^Q \leq C(\gamma)\left(\| v_- \|_r^\Omega + \| v_0 \|_s^\Omega\right) \tag{3.13}$$

holds.

2) If v_-, v_-', v_0, v_0' are arbitrary functions from E_r^-, E^0 correspondingly, then

$$\| \Psi_0^-(v_0 + v_-) - \Psi_0^-(v_0' + v_-') \|_s^Q \leq Cl\ \| v_0 + v_- \|_r^\Omega - \| v_0' - v_-' \|_r^\Omega. \tag{3.14}$$

3) $\Psi_0(0) = 0$.
4) If $\tilde{u}(x,t)$ solves (3.1), is determined for all $t \geq 0$, and $\tilde{u}(x,t)e^{-\gamma t} \in H_r^s(Q)$, then

$$\tilde{u}(x,0) = \tilde{v}_- + \tilde{v}_0 + \Psi_0(\tilde{v}_- + \tilde{v}_0), \quad \tilde{v}_- = P_-\tilde{u}(x,0),$$

$$\tilde{v}_0 = P_0\tilde{u}(x,0), \quad \tilde{v}_- = P_-\tilde{u}(x,0).$$

The following result is analogous to theorem 3.3 in the case in question.

Theorem 3.6. (Vishnevskii, 1982, 1986). Let the conditions of theorem 3.4 hold. Then, there exists a map Ψ_0^+ from $E^+ \oplus E^0$ into $E_r^- \oplus \hat{E}$ with the following properties.

1) Each solution to (3.1) with the initial data $u_0(x) = u(x,0) = v_0 + v_+ + \Psi_0^+(v_0 + v_+)$ exists for all $t \le 0$. The function $u(x,t)e^{-\gamma t}$ belongs to the class $H^s(Q_-)$ and satisfies the inequality

$$\| u(x,t)e^{\gamma t} \|_s^{Q_-} \le C(\gamma) \left(\| v_+ \|_r^\Omega + \| v_0 \|_r^\Omega \right). \qquad (3.15)$$

2) If v_0, v_0', v_+, v_+' are functions from E^0, E^+ correspondingly, then

$$\| \Psi_0(v_0 + v_+) - \Psi_0^+(v_0' + v_+') \|_s^Q \le C(\gamma)l \left(\| v_0 - v_0' \|_r^\Omega + \| v_+ - v_+' \|_r^\Omega \right). \qquad (3.16)$$

3) $\Psi_0^+(0) = 0$.

4) If $\tilde{u}(x,t)$ solves (3.1), is determined for all $t \le 0$, and $\tilde{u}(x,t)e^{\gamma t} \in H^s(Q_-)$, then

$$\tilde{u}(x,0) = \tilde{v}_0 + \tilde{v}_+ + \Psi_0^+(\tilde{v}_0 + \tilde{v}_+),$$

$$\tilde{v}_0 = P_0 u(x,0), \qquad \tilde{v}_+ = P_+ u(x,0). \qquad (3.17)$$

It is convenient to prove the theorems in the following order: theorem 3.5, 3.6, 3.4.

Proof of theorem 3.5. Let $u(x,t) = v(x,t)e^{\gamma t}$; then $v(x,t)$ solves the problem

$$(L + \gamma I)v = e^{-\gamma t}F(ve^{\gamma t}),$$

$$B_i v|_\Gamma = e^{-\gamma t}\Phi_i(ve^{\gamma t}), \qquad i = 1, \ldots, mN,$$

$$v(x,0) = u_0(x).$$

Consider the shift operator for solutions of the homogeneous problem

$$(L + \gamma I)w = 0,$$

$$B_i w|_\Gamma = 0, \quad i = 1, \ldots, mN, \quad w(x, 0) = w_0(x).$$

Denote this operator by $V(\omega, \gamma)$. The spectrum of $V(\omega, \gamma)$ has no common points with the unit circle from the complex plane and may be divided into two sets $\sigma_+(\gamma)$ and $\sigma_-(\gamma)$. The set $\sigma_-(\gamma)$, as earlier, lies inside the unit circle and $\sigma_+(\gamma)$ outside it.

Denote the corresponding spectral projectors by $P_\pm(\gamma)$ and their ranges of values by $E^\pm(\gamma)$. From the definition of projectors and the ω-periodicity in time it follows that

$$E^+(\gamma) = E^+, \quad E_r^-(\gamma) = E_r^- \oplus E^0.$$

Apply to the problem in question theorem 3.2, taking into account that $\eta(x, t) \equiv 0$. Note that $v_-(\gamma) = v_- + v_0$ and denote by $\Psi_0^-(v_- + v_0)$ the map $\Psi_-(v_-(\gamma), \gamma)$, which acts from $E_r^-(\gamma) = E_r^- \oplus E^+$ into E^+, the existence of which follows from theorem 3.2. It is easy to show that theorem 3.5 follows from theorem 3.2.

Theorem 3.6 may be proved analogously.

In order to prove theorem 3.4 show that for each $v_0 \in E^0$ there exist unique functions $v_- \in E_r^-$, $v_+ \in E^+$ such that

$$P_+\Psi_0^-(v_- + v_0) = v_+, \quad P_-\Psi_0^+(v_+ + v_0) = v_-. \tag{3.18}$$

Note, that the function $\hat{v} \in \hat{E}$ is uniquely determined by the set of the functions v_0, v_-, v_+ and by the form of nonlinear terms of (3.1). Therefore, if $u_0 = v_0 + v_- + v_+ + \hat{v}$ and (3.18) holds, then the solution to (3.1) with the initial data $u_0(x)$ simultaneously satisfies the conditions of theorems 3.5, 3.6. This solutions satisfies (3.14)-(3.17). Let v_+ be an arbitrary function from E^+. Consider the map

$$W(v_+) = P_+\Psi_0^-(P_-\Psi_0^+(v_+ + v_0) + v_0),$$

where v_0 is a fixed function from E^0. Let $\| v_+ \|_s^\Omega \leq M$, then

$$\| W(v_+) \|_s^\Omega = \| P_+ \Psi_0^- (P_- \Psi_0^+ (v_+ + v_0) + v_0) \|_s^\Omega$$

$$\leq Cl \| v_0 + P_- \Psi_0^+ (v_+ + v_0) \|_s^\Omega \leq Cl(1 + Cl) \| v_0 \|_s^\Omega + C^2 l^2 \| v_+ \|_s^\Omega .$$

Let

$$Cl(1 + Cl) \| v_0 \|_s^\Omega + C^2 l^2 \| v_+ \|_s^\Omega = M, \quad Cl < 1,$$

$$M = \frac{Cl(1 + Cl) \| v_0 \|_s^\Omega}{1 - (Cl)^2} = \frac{Cl \| v_0 \|_s^\Omega}{1 - Cl}.$$

Note, that $\| W(v_+) \|_s^\Omega \leq M$; therefore, W maps the ball of radius M in $H(\Omega)$ into itself. Show that W is a contraction, and hence, has in this ball a unique fixed point.

Really

$$\| W(v_+) - W(v'_+) \|_s^\Omega \leq Cl \| P_- \Psi_0^+ (v_+ + v_0)$$

$$- P_- \Psi_0^+ (v'_+ + v'_0) \|_s^\Omega \leq (Cl)^2 \| v_+ - v'_+ \|_s^\Omega, \quad Cl < 1;$$

therefore, W is a contraction. If $v_+ \in E^+$ is the fixed point of W, set $v_- = P_- \Psi_0^- (v_+ + v_0)$ and (3.18) will be hold. In this case $\| v_+ \|_s^\Omega \leq M$. An analogous estimate may be obtained for $\| v_- \|_s^\Omega$. Theorem 3.4 is proved.

Further, we shall need the analogs of theorems 3.1-3.3 in a form rather distinct from theorems 3.5-3.6. Let us formulate these statements.

Theorem 3.7. (Vishnevskii, 1982, 1986). Let all the conditions of theorem 3.4 hold, $0 < l < l_{10}$. Then, there exists a map Ψ_- from E_r^- into $E^0 \oplus E^+ \oplus \hat{E}$ with the following properties.

1) If $u(x, t)$ solves (3.1) with the initial data $u(x, 0) = u_0(x) = v_- + \Psi_-(v_-)$, then the solution $u(x, t)$ exists for all $t \geq 0$, belongs to $H_r^s(Q)$, and satisfies the inequality

$$\| u(x, t) e^{\lambda_1 t} \|_{r,s}^Q \leq C \| v_- \|_r^\Omega .$$

Here v is an arbitrary element from E_r^-, $0 < \lambda_1 < \lambda$. The number λ is taken from inequalities (2.7)-(2.10).

2) If v_-, v_-' are two arbitrary functions from E_r^-, then

$$\| \Psi_-(v_-) - \Psi_-(v_-') \|_r^\Omega \le Cl \| v_- - v_-' \|_r^\Omega .$$

3) $\Psi_-(0) = 0$.

4) If $\tilde{u}(x,t)$ solves (3.1) and there exists $\lambda_1 : 0 < \lambda_1 < \lambda$ exists that $\tilde{u}(x,t)e^{\lambda_1 t} \in H_r^s(Q)$, then

$$\tilde{u}(x,0) = \tilde{v}_- + \Psi_-(\tilde{v}_-), \quad \tilde{v}_- = P\tilde{u}(x,0).$$

The proof of theorem 3.7 has the same scheme as the proof of theorem 3.5, and we shall omit it.

The theorem analogous to theorem 3.6 may be formulated also.

Now we show how to get rid of the condition $F(0) = 0$, $\Phi_i(0) = 0$, $i = 1,\ldots,mN$.

Let

$$\| F(0) \|_{s-2m}^{Q_0} + \sum_{i=1}^{mN} \| \Phi_i(0) \|_{s-m_i}^{\Gamma_0} = K < +\infty.$$

We shall prove the analog of theorem 3.5. The remaining statements may be proved analogously. As before, set $u(x,t) = v(x,t)e^{\gamma t}$; then

$$(L + \gamma I)v = e^{-\gamma t}F(ve^{\gamma t}),$$

$$B_i v|_\Gamma = e^{-\gamma t}\Phi_i(e^{\gamma t}), \quad i = 1,\ldots,mN; \quad v(x,0) = v_0(x) = u_0(x).$$

In order to use theorems 3.1 and 3.4 we consider instead of maps $e^{-\gamma t}F(ve^{\gamma t})$ and $e^{-\gamma t}\Phi_i(e^{\gamma t})$ the maps $\hat{F}, \hat{\Phi}_i$ which coincide with $e^{-\gamma t}F(ve^{\gamma t})$ and $e^{-\gamma t}\Phi_i(e^{\gamma t})$ for $t \ge 0$ and are equal to $F(v)$ and $\Phi_i(v)$ for $t < 0$.

By theorem 3.1 the problem

$$(L + \gamma I)v = \hat{F}(v), \quad B_i v|_\Gamma = \hat{\Phi}_i(v)$$

is the unique bounded in the whole axis solution $\hat{\eta}(x,t)$. For the function $w(x,t) = v(x,t) - \eta(x,t)$ apply theorem 3.4. In this case, instead of (3.3) we have the inequality

$$\| u(x,t)e^{-\gamma t} \|_{r,s}^Q \leq C(\gamma) \left(\| v_- \|_r^\Omega + \| v_0 \|_r^\Omega + M \right). \qquad (3.19)$$

As in theorem 3.1, M is as follows

$$M = \| F(0) \|_{s-2m}^{Q_0} + \sum_{i=1}^{mN} \| \Phi_i(0) \|_{s-m_i}^{\Gamma_0}.$$

In a similar way we may prove the analog of theorem 3.6 and then the analog of theorem 3.4.

§1.4 Integral sets of the nonlinear parabolic systems. Stability of integral sets.

Consider again problem (3.1). The set $M(t)$ we shall denote as the integral set if from $(u_0(x,\tau),\tau) \in (M(\tau),\tau)$ it follows that $(u(x,t;u_0,\tau),t) \in (M(t),t)$ for all t from the domain of $u(x,t;u_0,\tau)$ definition. Here $u(x,t;u_0,\tau)$ solves (3.1) with the initial data $u_0(x,\tau)$ which are given not for $t = 0$ but for $t = \tau$. The variable t attains any values from the interval of existence of $u(x,t;u_0,\tau)$.

Note, that the maps $\Psi_\mp, \Psi_0, \Psi_0^\mp$ which were constructed for (3.1) for $t = 0$ may be constructed for (3.1) for $t = \tau$. Denote these maps by $\Psi_\mp(\tau), \Psi_0(\tau), \Psi_0^\mp(\tau)$.

Due to the ω-periodicity of (1.1)-(1.3) the spectral projections $P_-(\tau)$, $P_0(\tau), P_+(\tau), \hat{P}(\tau)$ and their ranges of values $E_r^-(\tau), E^0(\tau), E^+(\tau), \hat{E}(\tau)$ will ω-periodically depend on the time.

Consider in the spaces $C^r(\bar{\Omega}), C^s(\bar{\Omega})$ the following sets

$$M_0^-(\tau) = \left\{ v_-(x,\tau) + v_0(x,\tau) + \Psi_0^-(\tau, v_-(x,\tau) + v_0(x,\tau)); \right.$$

$$v_-(x,\tau) \in E_r^-(\tau); \qquad v_0(x,\tau) \in E^0(\tau)\Big\};$$

$$M_0^+(\tau) = \Big\{v_+(x,\tau) + v_0(x,\tau) + \Psi_0^+\left(\tau, v_+(x,\tau) + v_0(x,\tau)\right);$$

$$v_+(x,\tau) \in E^+(\tau); \qquad v_0(x,\tau) \in E^0(\tau)\Big\};$$

$$M^0(\tau) = \Big\{v_0(x,\tau) + \Psi_0\left(\tau, v_0(x,\tau)\right); v_0(x,\tau) \in E^0(\tau)\Big\};$$

$$M^-(\tau) = \Big\{v_-(x,\tau) + \Psi_-\left(\tau, v_-(x,\tau)\right); v_-(x,\tau) \in E_r^-(\tau)\Big\};$$

$$M^+(\tau) = \Big\{v_+(x,\tau) + \Psi_+\left(\tau, v_+(x,\tau)\right); v_+(x,\tau) \in E^+(\tau)\Big\};$$

Lemma 4.1. The sets $M_0^-(t), M_0^+(t), M^0(t), M^-(t), M^+(t)$ are the integral sets of problem (3.1).

Proof. Take, for example, the set $M_0^+(t)$. For the remaining sets, the proof is analogous.

Let $(u_0(x,\tau),\tau) \in (M_0^+(\tau),\tau)$. Assume that $u(x,t;u_0(x,\tau),\tau)$ is determined for $t = t_1$ and attains in this moment the value $u_1(x)$. We have

$$u(x,t;u_1(x),t_1) = u(x,t;u_0(x,\tau),\tau)$$

On the other hand, the solution $u(x,t;u_1(x),t_1)$ is determined for all $t \leq t_1$, where the function $u(x,t;u_1(x),t_1)e^{\gamma(t-t_1)}$ belongs to $H^s(Q_-^{t_1})$. Therefore, by theorem 3.5

$$u_1(x) = v_0(x,t_1) + v_+(x,t_1) + \Psi_0^+(\tau, v_0(x,t_1) + v_+(x,t_1)),$$

$$v_0(x,t_1) = P_0(t_1)u_1(x), \quad v_+(x,t_1) = P_+(t_1)u_1(x).$$

Consequently $u(x,t_1;u_0(x,\tau),\tau) \in (M_0^+(t_1),t_1)$. The lemma is proved.

Lemma 4.2. If maps F, Φ_i, $i = 1, \ldots, mN$ are ω-periodic with respect to time, then the integral sets $M_0^-(t), M_0^+(t), M^0(t), M^-(t)$, $M^+(t)$ ω-periodically depend on time.

If the maps F, Φ_i, $i = 1, \ldots, mN$ do not explicitly dependend on time, then the integral sets are not explicitly depend on time.

Proof of the lemma is evident.

One of the main results of the first chapter lies in the following theorem.

Theorem 4.1. (Vishnevskii, 1982, 1986). Assume that the conditions of theorem 3.7 hold. Assume, also, that $0 < l \le l_{10}$. If $u(x,t)$ is an arbitrary solution to (3.1), then there exists a unique solution $\xi(x,t)$ lying in $M_0^+(t)$ such that

$$\| (u(x,t) - \xi(x,t))e^{\lambda_1 t} \|_{r,s}^Q \le C(\lambda_1) \| u_0(x) - \xi(x,0) \|_r^\Omega, \qquad (4.1)$$

where $0 < \lambda_1 < \lambda$, λ is the same constant as in (2.7)-(2.10).

Proof. Denote by $\xi(x,t;v_0,v_+)$ a solution to (3.1) with the initial data

$$\xi(x,0) = \xi_0 = v_0 + v_+ + \Psi_0^+(v_0 + v_+).$$

This solution is determined for all $t \in R$ and lies in the integral set $M_0^+(t)$. Let $v(x,t) = u(x,t) - \xi(x,t;v_0,v_+)$. This function solves the following problem

$$Lv = F(v + \xi) - F(\xi), \quad (x,t) \in Q;$$

$$P_i v = \Phi_i(v + \xi) - \Phi_i(\xi) \quad (x,t) \in \Gamma; \quad i = 1, \ldots, mN;$$

$$v(x,0) = u_0(x) - \xi(x,0) = u_0 - v_0 - v_+ - \Psi^+(v_0 + v_+). \qquad (4.2)$$

We shall prove that if $u_0 \in C^r(\bar{\Omega})$ and satisfies the compatibility conditions of the order r, then there exist unique functions v_0, v_-, v_+ such that

$$P_- u_0 - v_- = P_- \Psi_0^+(v_0 + v_+),$$

$$P_0 \Psi_-(v_-) = P_0 u_0 - v_0,$$

$$P_+ \Psi_-(v_-) = P_+ u_0 - v_+. \qquad (4.3)$$

Express the function v_- from the first equality and substitute it into the second and the third equality. As a result we obtain

$$P_0 \Psi_-(P_- u_0 - P_-(\Psi_0^+(v_0 + v_+))) = P_0 u_0 - v_0,$$

$$P_+ \Psi_-(P_- u_0 - P_-(\Psi_0^+(v_0 + v_+))) = P_+ u_0 - v_+.$$

Denote by W the operator which each pair of functions (v_0, v_+) from E^0 and E^+ maps into the pair of functions

$$P_0 \Psi_-(P_- u_0 - P_-(\Psi_0^+(v_0 + v_+))) = P_0 u_0 - v_0,$$

$$P_+ \Psi_-(P_- u_0 - P_-(\Psi_0^+(v_0 + v_+))) = P_+ u_0 - v_+.$$

which also belong to E^0 and E^+.

Denote by $\hat{C}$ such constant that

$$\| P_0 w \|_s^\Omega + \| P_+ w \|_s^\Omega \leq \hat{C} \| w \|_s^\Omega,$$

where $w \in E^0 \oplus E^+$. If we set $B = \frac{\hat{C} \|u_0\|_s^\Omega}{1 - Cl}$, then W maps the set

$$\left\{ (v_0, v_+);\ v_0 \in E^0;\ v_+ \in E^+;\ \| v_0 + v_+ \|_s^\Omega \leq B \right\}$$

into itself and is a contraction there. The unique fixed point of W in this set determines the unique pair of functions (v_0, v_+). From equality (4.3) we find the function v_-.

If (4.3) holds, then

$$v(x,0) = u_0 - v_0 - v_+ - \Psi_0^+(v_0 + v_+) = v_- + \Psi_-(v_-).$$

Using theorem 3.7, we obtain the following estimate

$$\| v(x,t)e^{\lambda_1 t} \|_{r,s}^Q \leq C \| v_- \|_r^\Omega .$$

This inequality yields estimate (4.1).

Now establish the uniqueness of $\xi(x,t)$. Let $\tilde{\xi}(x,t) \in M_0^+(t)$, $u(x,t)$ solve (3.1), and (4.1) hold. By theorem (3.7) we obtain

$$u(x,0) - \tilde{\xi}(x,0) = u_0 - \tilde{v}_0 - \tilde{v}_+ - \Psi_0^+(\tilde{v}_0 - \tilde{v}_+) = \tilde{v}_- + \Psi_-(\tilde{v}_-).$$

Here $\tilde{v}_0 = P_0\tilde{\xi}(x,0)$, $\tilde{v}_+ = P_+\tilde{\xi}(x,0)$, the function $\tilde{v}_- \in E_r^-$ is uniquely determined by theorem 3.7. But then $(\tilde{v}_0, \tilde{v}_+)$ is a fixed point of W lying in the ball of radius B. By the uniqueness of the fixed point we have

$$\tilde{v}_0 = v_0, \tilde{v}_+ = v_+, \tilde{v}_- = v_-.$$

Hence, $\tilde{\xi}(x,t) = \xi(x,t)$ which was to be proved. (Note, that the solution $u(x,t)$ is fixed). Theorem 4.1 is proved.

Remark. Theorem 4.1 may be reformulated as follows: each solution to (3.1) exponentially converges to the unique solution which lies in the integral set $M_0^+(t)$.

Further, we shall often use the fact that all the solutions to (3.1) which lie in $M_0^+(t)$ may be regarded as solutions of the systems of ordinary differential equations. The number of differential equations is equal to the multiplicity of the unit eigenvalue of the shift operator of the corresponding linear problem. In particular, if the multiplicity is equal to unity (which often occurs if we consider parabolic equations or systems satisfying the maximum principle), then we obtain one non-linear differential equation of the first order.

§1.5   Local theorems of existence and continuous dependence on initial data in the Hölder classes of weight functions.

1. The local existence theorem for the quasilinear parabolic equation.   Consider in the cylinder $Q = \Omega \times (0, +\infty)$ the problem

$$u_t = \sum_{i,j=1}^{n} a_{ij}(x,t,u,Du)\frac{\partial^2 u}{\partial x_i \partial x_j} + a(x,t,u,Du), \qquad (5.1)$$

$$(x,t) \in Q,$$

$$Bu = \beta_1 \left(\sum_{i=1}^{n} b_i(x)\frac{\partial u}{\partial x_i} + b_0(x,t,u) \right) + \beta_0 u = 0, \qquad (5.2)$$

$$(x,t) \in \Gamma = \partial\Omega \times (0, +\infty), \quad \beta_1^2 + \beta_0^2 > 0, \quad \beta_1 \geq 0,$$

$$u(x,0) = u_0(x). \qquad (5.3)$$

Here $u(x,t)$ is a function depending on x and t; Du is its gradient with respect to the spatial variables x.

Assume that $\beta_1 \geq 0, \beta_0^2 + \beta_1^2 > 0$; the functions

$$a_{ij}(x,t,u,p), a(x,t,u,p), b_0(x,t,u), b_i(x) \qquad (5.4)$$

are three times continuously differentiable by all the variables and

$$\mu^{-1}|\xi|^2 \leq \sum_{i,j=1}^{n} a_{ij}(x,t,u,p)\xi_i\xi_j \leq \mu|\xi|^2, \quad \mu > 0 \qquad (5.5)$$

$$x \in \bar\Omega, \quad -\infty < u, \quad p_i < +\infty, \quad i = 1,\dots,n \ \ \xi \in R^n,$$

$$\sum_{i=1}^{n} b_i(x)\cos(n, x_i) \geq \mu_0 > 0, \quad x \in \partial\Omega. \qquad (5.6)$$

Here n is the exterior normal to $\partial\Omega$ in the point x.

Denote by E the subset of $C^1(\bar{\Omega})$ consisting of those functions $u_0(x)$ which satisfy the boundary condition (5.2) $Bu_0(x) = 0$ for $x \in \partial\Omega$.

Here we shall consider the classical solution to the problem (5.1)-(5.3). This means the following. The function $u(x,t)$ is twice continuously differentiable by x and one time by t in $\bar{\Omega} \times (0,\Gamma]$. The function $u(x,t)$ is continuous together with its partial derivatives by $x_i, i = 1,\ldots,n$ in Q_T. The function $u(x,t)$ satisfies in $\Omega \times (0,\Gamma]$ to equation (5.1) for $(x,t) \in \Omega \times (0,\Gamma]$, the boundary conditions (5.2), and for $t = 0$ the initial data (5.3).

The quasilinear equation (5.1), the functions $u(x,t)$ and $u_{x_i}(x,t)$ enter in the nonlinear way. These functions are continuous in $\bar{Q}_T$. The derivatives $u_{x_i x_j}(x,t), i, j = 1,\ldots,n;$ $u_t(x,t)$ may, generally speaking, grow for $t \to 0$. The following theorem holds.

Theorem 5.1. (Vishnevskii, 1987, 1994). Let for problem (5.1)-(5.3) the assumptions (5.4)-(5.6) hold and the initial data belong to E. Then such positive T exists that problem (5.1)-(5.3) has a unique classical solution $u(x,t)$ in the cylinder Q_T.

Proof. For each function $u_0(x)$ from $C^1(\bar{\Omega})$ there exists the function $\psi(x,t)$ from $H_1^{2+\beta}(\bar{Q}_T)$, where $0 < \alpha < \beta < 1$. This function coincides with $u_0(x)$ for $t = 0 : \psi(x,0) = u_0(x)$ and

$$\| \psi(x,t) \|_{1,2+\beta}^{Q_T} \le C \| u_0 \|_1^{\Omega} . \tag{5.7}$$

The constant C here does not depend on $u_0(x)$. Really, if $\psi(x,t)$ solves the homogeneous problem

$$\psi_t = \Delta\psi, \quad (x,t) \in \bar{Q}_T; \quad \frac{\partial\psi}{\partial n} = 0, \quad (x,t) \in \bar{\Gamma}_T,$$

$$\psi(x,0) = u_0,$$

then our statement follows from estimate (1.6).

Consider instead of $u(x,t)$ the new function

$$v(x,t) = u(x,t) - \psi(x,t).$$

This function will solve the following parabolic equation

$$v_t = L(x,t)v + F(v) + f(x,t), \quad (x,t) \in \bar{Q}_T,$$

$$B_0(x,t)v + \Phi(v) + \varphi(x,t) = 0, \quad (x,t) \in \bar{\Gamma}_T, \qquad (5.8)$$

$$v(x,0) = 0.$$

Here $L(x,t)$ is the following linear operator

$$\sum_{i,j=1}^{n} a_{ij}(x,t,\psi,D\psi) \frac{\partial^2}{\partial x_i \partial x_j} + \sum_{i,j=1,p=1}^{n} a_{iju_{x_p}}(x,t,\psi,D\psi)$$

$$\times \frac{\partial^2 \psi}{\partial x_i \partial x_j} \frac{\partial}{\partial x_p} + \sum_{i,j=1}^{n} a_{iju}(x,t,\psi,D\psi) \frac{\partial^2 \psi}{\partial x_i \partial x_j}$$

$$+ \sum_{p=1}^{n} a_{u_{x_p}}(x,t,\psi,D\psi) \frac{\partial}{\partial x_p} + a_u(x,t,\psi,D\psi);$$

$$f(x,t) = \sum_{i,j=1}^{n} a_{ij}(x,t,\psi,D\psi) \frac{\partial^2 \psi}{\partial x_i \partial x_j} + a(x,t,\psi,D\psi) - \psi_t(x,t);$$

$$F(v) = \sum_{i,j=1}^{n} a_{ij}(x,t,v+\psi,D(v+\psi)) \frac{\partial^2(v+\psi)}{\partial x_i \partial x_j}$$

$$+ a(x,t,v+\psi,D(v+\psi)) - f(x,t) - L(x,t)v,$$

$$F_{v_p}(0) = 0.$$

Evidently, $F(0) = 0$; therefore, from lemma 1.2 it follows that

$$\| F(v_1) - F(v_2) \|_{-1,\alpha}^{Q_T}$$

$$\leq C \left(\| v_1 \|_{1,2+\alpha}^{Q_T}, \| v_2 \|_{1,2+\alpha}^{Q_T} \right) \| v_1 - v_2 \|_{1,2+\alpha}^{Q_T}, \qquad (5.9)$$

where $C(\xi,\eta) \to 0$ for $\xi,\eta \to 0$. Let

$$B_0(x,t) = \beta_1 \left(\sum_{i=1}^{n} b_i(x) \frac{\partial}{\partial x_i} + b_{0u}(x,t,\psi) \right) + \beta_0,$$

$$\varphi(x,t) = B\psi(x,t),$$

$$\Phi(v) = B(v + \psi) - \varphi(x,t) - B_0(x,t)v.$$

As earlier, $\Phi(0) = 0$, $\Phi_{v_p}(0) = 0$; therefore,

$$\| \Phi(v_1) - \Phi(v_2) \|_{-\delta+1,2+\alpha-\delta}^{\Gamma_T}$$

$$\leq C_1 \left(\| v_1 \|_{1,2+\alpha}^{Q_T}, \| v_2 \|_{1,2+\alpha}^{Q_T} \right) \| v_1 - v_2 \|_{1,2+\alpha}^{Q_T}. \qquad (5.10)$$

Here $C_1(\xi, \eta) \to 0$ for $\xi, \eta \to 0$. We have $\delta = 1$ if $\beta_1 \neq 0$; $\delta = 2$ if $\beta_1 = 0$ and, therefore, the Dirichlet problem is considered.

Denote by ρ the following value

$$\rho = \max \left(\| v_1 \|_{1,2+\alpha}^{Q_T}, \| v_2 \|_{1,2+\alpha}^{Q_T} \right).$$

Combining inequalities (5.9), (5.10), we obtain

$$\| F(v_1) - F(v_2) \|_{-1,\alpha}^{Q_T} + \| \Phi(v_1) - \Phi(v_2) \|_{-\delta+1,2+\alpha-\delta}^{\Gamma_T}$$

$$\leq q(\rho) \| v_1 - v_2 \|_{1,2+\alpha}^{Q_T}, \qquad (5.11)$$

where $q(\rho) \to 0$ for $\rho \to 0$.

Let $\xi(x,t)$ be a solution to the following linear problem

$$\xi_t = L(x,t)\xi + f(x,t), \qquad (x,t) \in \bar{Q}_T,$$

$$B_0(x,t)\xi + \varphi(x,t) = 0, \qquad (x,t) \in \bar{\Gamma}_T, \quad \xi(x,0) = 0. \qquad (5.12)$$

From the results of the first paragraph it follows that (5.12) has a unique solution from the class $H_1^{2+\beta}(\bar{Q}_T)$ (since $\psi(x,t) \in H_1^{2+\beta}(\bar{Q}_T)$). This solution satisfies the following estimate

$$\| \xi(x,t) \|_{1,2+\beta}^{Q_T} \le C_2 \ \| \psi(x,t) \|_{1,2+\beta}^{Q_T} \le C_3 \ \| u_0 \|_1^\Omega \ . \tag{5.13}$$

The second inequality follows from estimate (5.7).

The function $\xi(x,t)$ is equal to zero for $t = 0 : \xi(x,0) = 0$; therefore, from the definition of the Hölder classes (see §1.1), we obtain

$$\| \xi(x,t) \|_{1,2+\alpha}^{Q_T} \le C_4 T^{\frac{\beta-\alpha}{2}} \ \| \xi(x,t) \|_{1,2+\beta}^{Q_T} \ .$$

Taking into account (5.13), we have

$$\| \xi(x,t) \|_{1,2+\alpha}^{Q_T} \le C_5 T^{\frac{\beta-\alpha}{2}} \ \| u_0 \|_1^\Omega \ . \tag{5.14}$$

Set

$$w(x,t) = v(x,t) - \xi(x,t).$$

For the function $w(x,t)$ we obtain

$$w_t = L(x,t)w + F(w+\xi), \qquad (x,t) \in \bar{Q}_T,$$

$$B_0(x,t)w + \Phi(w+\xi) = 0, \qquad (x,t) \in \bar{\Gamma}_T, \tag{5.15}$$

$$w(x,0) = 0.$$

We now prove that (5.15) has a unique solution in Q_T from the class $H_1^{2+\alpha}(\bar{Q}_T)$. We shall use the standard method of contracting mappings and the inequalities obtained previously.

To this end consider $H_1^{2+\alpha}(\bar{Q}_T)$ the set M consisting of the functions $\eta(x,t)$ such that

$$\eta(x,0) = 0, \qquad \| \eta(x,t) \|_{1,2+\alpha}^{Q_T} \le K \ \| u_0 \|_1^\Omega \ . \tag{5.16}$$

The value K we shall define further. Denote by σ the product $K \ \| u_0 \|_1^\Omega$. Define in the set M, the mapping W which the function

$\eta(x,t)$ from M maps into the function $\zeta(x,t)$, which solves the following problem

$$\zeta_t = L(x,t)\zeta + F(\xi + \eta), \qquad (x,t) \in \bar{Q}_T,$$

$$B_0(x,t)\zeta + \Phi(\xi + \eta) = 0, \qquad (x,t) \in \bar{\Gamma}_T, \qquad (5.17)$$

$$\zeta(x,0) = 0.$$

Now we shall choose the constants K and T so that the mapping $\zeta = W\eta$ will map M into itself and be a contracting mapping in this set. Then the unique fixed point of W will solve problem (5.15). The inverse statement is evident: if $w(x,t)$ solves (5.15), then $Ww = w$ is the fixed point of W.

Using inequality (5.11), we see that

$$\| \zeta(x,t) \|_{1,2+\alpha}^{Q_T} \leq C_6 q \left(\| \eta + \xi \|_{1,2+\alpha}^{Q_T} \right) \| \eta + \xi \|_{1,2+\alpha}^{Q_T} . \qquad (5.18)$$

Setting $\gamma = \frac{\beta - \alpha}{2} > 0$, from (5.16), (5.14), we obtain

$$\| \eta + \xi \|_{1,2+\alpha}^{Q_T} \leq \delta + C_5 T^\gamma \| u_0 \|_1^\Omega .$$

Therefore, (5.18) yields

$$\| \zeta(x,t) \|_{1,2+\alpha}^{Q_T} \leq C_6 q \left(\delta + C_5 T^\gamma \| u_0 \|_1^\Omega \right) \left(\delta + C_5 \| u_0 \|_1^\Omega \right). \qquad (5.19)$$

Using inequality (5.11), we also obtain

$$\| W(\eta_1) - W(\eta_2) \|_{1,2+\alpha}^{Q_T} \leq C_6 q(\rho) \| \eta_1 - \eta_2 \|_{1,2+\alpha}^{Q_T}, \qquad (5.20)$$

$$\rho = \max \left(\| \eta_1 + \xi \|_{1,2+\alpha}^{Q_T}, \| \eta_2 + \xi \|_{1,2+\alpha}^{Q_T} \right).$$

Choose the constant K such that $K = C_5 T^\gamma$. Hence, we have

$$\delta = K \| u_0 \|_1^\Omega = C_5 T^\gamma \| u_0 \|_1^\Omega$$

Now, choose $T > 0$ to satisfy the inequalities

$$2C_6 q(2\delta) < 1, \qquad q(2\delta) = q_0 < 1. \tag{5.21}$$

We may do this, since for $T \to 0$ we have

$$\delta \to 0, \quad q(2\delta) \to 0.$$

Estimates (5.19), (5.21) yield

$$\| \zeta \|_{1,2+\alpha}^{Q_T} \leq C_6 q(2\delta) 2\delta < \delta.$$

So, the mapping W maps the set M into itself. Inequalities (5.20), (5.21) yield for $\eta_1(x,t), \eta_2(x,t) \in M$ the following inequality

$$\| W(\eta_1) - W(\eta_2) \|_{1,2+\alpha}^{Q_T} \leq C_6 q(\delta \cdot 2) \| \eta_1 - \eta_2 \|_{1,2+\alpha}^{Q_T}$$

$$\leq q_0 \| \eta_1 - \eta_2 \|_{1,2+\alpha}^{Q_T}.$$

Therefore, the mapping W is a contractive mapping in M. Since $w(x,t) \in M$, we have

$$\| u(x,t) \|_{1,2+\alpha}^{Q_T} = \| \psi(x,t) \|_{1,2+\alpha}^{Q_T} + \| \xi(x,t) \|_{1,2+\alpha}^{Q_T}$$

$$+ \| w(x,t) \|_{1,2+\alpha}^{Q_T} \leq C \| u_0 \|_{1,2+\alpha}^{\Omega}. \tag{5.22}$$

This inequality proves theorem 5.1.

2. The theorem in continuous dependence of a solution on initial data. Here we shall consider problem (5.1)-(5.3), and all the conditions of the previous section will be assumed to hold (in particular, conditions (5.4)-(5.6)). Assume, also, initial data for (5.1)-(5.3) belong to the set E. We shall establish the following theorem.

Theorem 5.2. (Vishnevskii, 1986, 1994). Let for problem (5.1)-(5.3) assumptions (5.4)-(5.6) hold. Let $u(x, t, u_0)$ be a classical solution to (5.1)-(5.3), $u_0(x) \in E$, $u(x, t, u_0) \in H_1^{2+\alpha}(Q_T)$, $0 < T < +\infty$.

Then, there exists such a positive δ, that from $v_0 \in E$, $\| v_0 - u_0 \|_1^{\Omega} < \delta$ it follows that the solution $u(x, t, v_0)$ with the initial data v_0 exists in Q_T and satisfies the estimate

$$\| u(x, t, v_0) - u(x, t, u_0) \|_{1,2+\alpha}^{Q_T} \leq C \| u_0 - v_0 \|_1^{\Omega} . \tag{5.23}$$

Proof. Denote by $w(x, t)$ the difference $u(x, t, v_0) - u(x, t, u_0)$. Decompose the nonlinear terms of (5.1)-(5.3) by the degrees of $w(x, t)$ and its derivatives. Select the principal linear terms of these decompositions. As a result, we obtain the following boundary-value problem

$$w_t = L(x, t)w + F(w), \qquad (x, t) \in Q_T,$$

$$B_0(x, t)w + \Phi(w) = 0, \qquad (x, t) \in \Gamma_T, \tag{5.24}$$

$$w(x, 0) = v_0 - u_0 = w_0(x).$$

The operators $L(x, t), B_0(x, t)$ are determined almost similar to $L(x, t), B_0(x, t)$ in problem (5.8); only, instead of $\psi(x, t)$, we take $u(x, t, u_0)$.

The mappings $F(w), \Phi(w)$, which are obtained as a result, satisfy inequality (5.11).

Note, also, that

$$f(x, t) = \sum_{i,j=1}^{n} a_{ij}(x, t, u, Du) \frac{\partial^2 u}{\partial x_i \partial x_j} + a(x, t, u, Du) - u_t = 0.$$

Analogously, $\varphi(x, t) = 0$.

Denote by M_1 the subset from $H_1^{2+\alpha}(Q_T)$ which functions satisfy the following conditions

$$\| \eta(x, t) \|_{1,2+\alpha}^{Q_T} \leq \delta_1, \quad \eta(x, 0) = w_0(x).$$

The positive number δ_1 will be determined further. Define in the set M_1 the mapping W_1 which functions $\eta(x,t) \in M_1$ map into $\zeta(x,t)$, where $\zeta(x,t)$ solves the following problem

$$\zeta_t = L(x,t)\zeta + F(\eta), \qquad (x,t) \in Q_T,$$

$$B_0\zeta + \Phi(\eta) = 0, \qquad (x,t) \in \Gamma_T, \qquad (5.25)$$

$$\zeta(x,0) = w_0(x).$$

Show that $\zeta(x,t)$ belongs to M_1 and the mapping

$$\zeta(x,t) = W_1\eta(x,t)$$

is contractive in the set M_1. The unique fixed point of W_1 will solve problem (5.24). The inverse statement also holds: each solution of problem (5.24) from $H_1^{2+\alpha}(Q_T)$ is a fixed point of W_1.

So, from (5.11) we have

$$\| \zeta(x,t) \|_{1,2+\alpha}^{Q_T} \leq C_6 q \left(\| \eta \|_{1,2+\alpha}^{Q_T} \right) \| \eta \|_{1,2+\alpha}^{Q_T} + C_6 \| W_0 \|_1^{\Omega}. \qquad (5.26)$$

Choose such $\delta > 0$ that for this δ, W_1 maps M_1 into itself. Take $\delta_1 = 2C_6\delta$. Choose δ so that

$$q(2C_6\delta)C_6 \leq q_0 < 1.$$

Then, taking into account (5.26), we obtain

$$\| \zeta(x,t) \|_{1,2+\alpha}^{Q_T} \leq 2C_6\delta = \delta_1.$$

Inequality (5.11) yields

$$\| W_1(\eta_1) - W_1(\eta_2) \|_{1,2+\alpha}^{Q_T} \leq C_6 q(2C_6\delta) \| \eta_1 - \eta_2 \|_{1,2+\alpha}^{Q_T}$$

$$\leq q_0 \parallel \eta_1 - \eta_2 \parallel_{1,2+\alpha}^{Q_T} .$$

Therefore, W_1 is contractive in M_1. This proves theorem 5.2.

3. The global existence theorem. Here we shall establish the global existence theorem for problem (5.1)-(5.3).

Theorem 5.3. Let conditions of theorem 5.1 hold, and $u(x, t, u_0)$ be a classical solution to (5.1)-(5.3), $u_0 \in E$. If, in addition, the following *a priori* estimate

$$\parallel u(x, t, u_0) \parallel_1^{Q_T} \leq M_1 \tag{5.27}$$

holds, then the solution $u(x, t, u_0)$ belongs to $H_1^{2+\alpha}(Q_T)$, and is bounded in this class by a constant M_2. Moreover, if M_1 is bounded for $T \to +\infty$, then M_2 is also bounded for $T \to +\infty$.

Proof. By theorem 5.1, for each $u(x, t, u_0)$, there exists $T(\tau) \in [0, T]$, such that in the cylinder $Q_{\tau+T(\tau)}^\tau = \Omega \times (\tau, \tau + T(\tau))$ there exists a unique solution of problem (5.1)-(5.2) with initial data $u(x, \tau)$ given for $t = \tau$. This solution belongs to the class $H_1^{2+\alpha}(Q_{\tau+T(\tau)}^\tau)$. The solution may have the derivatives by x and by t which grow with respect to the time for $t \to \tau$. The theorem will be established, if we show that $T(\tau) \geq \tau_0 > 0$. Then the theorem will be the corollary of theorem 5.2.

Assume the contrary. Then there exists such a sequence τ_i, that $T(\tau_i) \to 0$ for $i \to \infty$. Select from this sequence a subsequence converging to τ^*. Without loss of generality we shall assume that $\tau_i \to \tau^*$. We have $\tau^* \in [0, T]$; therefore, $T(\tau^*) > 0$. By theorem 5.2, for sufficiently large i the intervals of existence $T^*(\tau_i)$ for the problem

$$u_t = \sum_{i,j=1}^{n} a_{ij}(x, \tau^*, u, Du) \frac{\partial^2 u}{\partial x_i \partial x_j} + a(x, \tau^*, u, Du),$$

$$(x, t) \in Q_{\tau_i + T^*(\tau_i)}^{\tau_i},$$

$$Bu = \beta_1 \left(\sum_{i=1}^{n} b_i \frac{\partial u}{\partial x_i} + b_0(x, \tau^*, u) \right) + \beta_0 u, \qquad (5.28)$$

$$(x, t) \in \Gamma_{\tau_i + T^*(\tau_i)}^{\tau_i},$$

$$u(x, \tau_i) = u(x, \tau_i, u_0)$$

satisfy the inequality

$$T^*(\tau_i) \geq T(\tau^*).$$

Substitute in (5.28) instead of τ^* the variable t and apply the contractive mappings methods as we have done in theorems 5.1, 5.2. As a result we obtain that $T(\tau_i) \geq T^*(\tau_i)$, which contradicts the above assumption.

In conclusion, consider the dependence of T from the $C^1(\bar{\Omega})$ norm of the initial data $u_0(x)$. We take the simplest case $n = 1$ and the autonomous equation (5.1)-(5.3).

Denote by $v(x, t)$ the function $u_x(x, t)$.

Assume that

$$\| u_x(x, 0) \|_0^{[0,1]} \leq C_1, \quad \| u_x(x, t) \|_0^{[0,1]} \leq 2C_1 \quad \text{for} \quad t \in [0, \tau],$$

and $\| u_x(x, \tau) \|_0^{[0,1]} = 2C_1$.

Evidently, under these assumptions, we have

$$\| u(x, t) \|_1^{Q_\tau} \leq C_2.$$

The function $v(x, t)$ solves the boundary-value problem

$$v_t = A(x, u, u_x) v_{xx} + B(x, u, u_x) v_x + C(x, u, u_x) v + D(x, u, u_x)$$

with the standard boundary conditions and

$$A(x, u, u_x) \geq a_0 > 0, \quad |C(x, u, u_x)| \leq L, \quad |D(x, u, u_x)| \leq K$$

for $x \in [0,1]$, $\| u(x,t) \|_1^{Q_\tau} \leq C_2$.

Applying the maximum principle (see, for example, Ladyzhenskaya *et al.*, 1967) we obtain

$$|v(x,\tau)|_0^{[0,1]} \leq |v(x,0)|_0^{[0,1]} e^{\lambda \tau} + \tau K,$$

or $2C_1 \leq C_1 e^{\lambda \tau} + \tau K$.

If $\tau \to 0$, then the right side tends to $C_1 < 2C_1$. Therefore, there exists such positive T that

$$2C_1 = C_1 e^{\lambda \tau} + \tau K \quad \text{and} \quad \tau \geq T.$$

So, the following statement holds.

Lemma 5.3. Consider problem (5.1)-(5.3) in the case $m = 1$, $n = 1$. Given $C_1 > 0$, there exists such $T > 0$ that if $\| u_0(x) \|_1^\Omega < C_1$, then the solution $u(x,t;u_0)$ of problem (5.1)-(5.3) exists in $[0,T]$. Moreover, for each $\varepsilon > 0$ there exists such $C(\varepsilon)$ that

$$\| u(x,t;u_0) \|_{3+\alpha}^{Q_T^\varepsilon} \leq C(\varepsilon) \| u_0 \|_1^{[0,1]},$$

where
$$Q_T^\varepsilon = [0,1] \times [\varepsilon, T], \qquad 0 < \alpha < 1.$$

Note, that the analogous result may be obtained for the autonomous equation (5.1)-(5.3) for $n > 1$.

Chapter 2

Construction of Liapunov's functionals in the case of one spatial variable.

§2.1 Liapunov's functionals of the first order.

In this paragraph we shall construct those functionals which are the analogs of Liapunov's functions from the theory of ordinary differential equations. The method suggested by Zelenyak (1968) and by Belonosov and Zelenyak (1975) based on constructing such functionals allowes us to create a complete nonlocal theory of behavior of bounded solutions of mixed problems for a nonlinear parabolic equation with one spatial variable. After the exact setting of the basis problem we shall return to analogs with the theory of ordinary differential equations and calculus of variations.

Consider the quasilinear problem

$$u_t = L(u) = a(x, u, u_x)u_{xx} + b(x, u, u_x),\tag{1.1}$$

$$\alpha_i u_x(i, t) + \psi_i(u(i, t)) = 0; \quad i = 0, 1; \quad t > 0,\tag{1.2}$$

$$u(x, 0) = u_0(x)\tag{1.3}$$

in the semiband $(x,t) \in Q = [0,1] \times [0,\infty)$ or in the rectangle $(x,t) \in Q_T = [0,1] \times [0,T]$. Further we shall assume that $a, b, \psi_i \in C^3$ by all arguments. This assumption may be weakened, but this is not a goal of our consideration. If we do not specify the contrary, then

$$a(x,\xi,\eta) \geq a_0 > 0, \quad (x,\xi,\eta) \in [0,1] \times R^2,$$

$$\alpha_i^2 + \psi_i'^2(\xi) \geq a_0 > 0, \quad \xi \in R, \quad i = 0,1, (\alpha_i = \text{const}), \tag{1.4}$$

$$u_0(x) \in E = \{u_0(x) \in C^1[0,1] : \alpha_i u_0'(i) + \psi_i(u_0(i)) = 0, i = 0,1\}.$$

Condition (1.2) for the above assumptions covers the boundary problems of three types.

Definition 1.1. We shall say that a pair of functions ρ, $\Phi(x,\xi,\eta)$ generate the Liapunov functional if for each solution of problem (1.1), (1.2) $u(x,t) \in C^{2,1}(Q_T)$ the identity

$$\frac{d}{dt} \int_0^1 \Phi(x,u,u_x)dx = -\int_0^1 \rho(x,u,u_x)L(u)u_t dx, \tag{1.5}$$

holds, where

$$\rho \geq 0. \tag{1.6}$$

Remark 1.1. Due to equation (1.1) the identity (1.5) may be rewritten in the form

$$\frac{d}{dt} \int_0^1 \Phi(x,u,u_x)dx = -\int_0^1 \rho(x,u,u_x)u_t^2 dx, \tag{1.7}$$

i.e., the functional

$$J(u) = \int_0^1 \Phi(x,u,u_x)dx, \tag{1.8}$$

in this case of positive $\rho > 0$, strictly decreases when t increases along the solution of dynamic problem (1.1)-(1.3) except for the equilibrium points - the stationary solutions.

This property of functional (1.8) serves as a basis for the analog with Liapunov's functions of the theory of ordinary differential equations and, thus, justifies the term, Liapunov's functional.

Remark 1.2. A stationary (not depending on time) solution $y(x)$ of problem (1.1)-(1.3) by the definition satisfies the equation

$$a(x, y, y')y'' + b(x, y, y') = 0. \tag{1.9}$$

As it will be seen further, this equation for $\rho > 0$ is equivalent to the Euler-Lagrange equation for the variational problem connected with functional (1.8). Therefore, the problem of constructing the generating Liapunov functional pairs of functions $\rho, \Phi(\rho > 0)$ is reduced to the following problem.

By equation (1.1) (equation (1.9)) to determine the variational factors $\rho(x, \xi, \eta)$ after multiplying by which the right side of equation (1.1) (equation (1.9)) becomes the Euler-Lagrange equation for the corresponding Lagrangian $\Phi(x, \xi, \eta)$.

This variational factor $\rho(x, \xi, \eta)$ together with the Lagrangian $\Phi(x, \xi, \eta)$ generates Liapunov's functional.

Definition 1.2. We shall say that for problem (1.1) - (1.3) the condition (B) holds if the solution $\varphi(x_0, x, y_0, y)$ of the Cauchy problem

$$y'' = -\frac{b}{a}(x, y, y'), \quad y(x_0) = y_0, \quad y'(x_0) = y_1 \tag{1.10}$$

is determined and regular for $x \in [0, 1]$ for all data $(x_0, y_0, y_1) \in [0, 1] \times R^2$.

Definition 1.3. We shall say that equation (1.1) satisfies condition (B_1), if there exists such point $x^1 \in [0, 1]$, that for each data set

$(x_0, y_0, y_1) \in [0,1] \times R^2$ the solution $\varphi(x_0, x, y_0, y_1)$ of the Cauchy problem (1.10) is determined and regular in the closed interval with the ends in x_0, x^1.

Note, that if (B) holds, then (B_1) holds for each $x^1 \in [0,1]$.

Theorem 1.1. Let (1.1) satisfy (B_1). Then there exists a pair of functions $\rho, \Phi(x, \xi, \eta) \in C^2$ which generates the Liapunov's functional.

Proof. Let $u(x,t) \in C_2(Q_T)$ solve problem (1.1) - (1.3) (the required smoothness, for convenience, is slightly overstated). Assuming that the functions ρ, Φ satisfying identity (1.5) exist, we shall obtain the equations for them. After differentiating in the integral and integrating by parts, the equation (1.5) will take the form

$$\int_0^1 \left(\frac{\partial \Phi}{\partial u} - \frac{d}{dx}\frac{\partial \Phi}{\partial u_x} \right) u_t dx + \frac{\partial \Phi}{\partial u_x} u_t \bigg|_{x=0}^1 = - \int_0^1 \rho u_t(a u_{xx} + b)dx. \quad (1.11)$$

For validity of (1.11) it is sufficient that

$$\frac{\partial^2 \Phi(x, \xi, \eta)}{\partial \eta^2} = \rho a(x, \xi, \eta), \quad (1.12)$$

$$\frac{\partial^2 \Phi}{\partial x \partial \eta} + \frac{\partial^2 \Phi}{\partial \xi \partial \eta}\eta - \frac{\partial \Phi}{\partial \xi} = \rho b(x, \xi, \eta), \quad (1.13)$$

$$\frac{\partial \Phi}{\partial u_x} u_t \bigg|_{x=0} = \frac{\partial \Phi}{\partial u_x} u_t \bigg|_{x=1} = 0 \quad (1.14)$$

hold. The first two are considered for all values $(x, \xi, \eta) \in [0,1] \times R^2$, and the latter, for the solution of problem (1.1) - (1.3).

Substituting the dependent of two arbitrary functions $z_i(x, \xi)$ solution of (1.12)

$$\Phi(x, \xi, \eta) = \int_0^\eta (\eta - \tau)\rho a(x, \xi, \tau)d\tau + \eta z_1(x, \xi) + z_2(x, \xi) \quad (1.15)$$

into (1.13), we obtain

$$\int\limits_0^\eta (\rho a)_x d\tau + \eta \int\limits_0^\eta (\rho a)_\xi d\tau - \int\limits_0^\eta (\eta - \tau)(\rho a)_\xi d\tau + z_{1x} + \eta z_{1\xi} - \eta z_{1\xi} - z_{2\xi} = \rho b.$$

After collecting similar terms the equation takes the form

$$\int\limits_0^\eta (\rho a)_x d\tau + \int\limits_0^\eta \tau(\rho a)_\xi d\tau + z_{1x} - z_{2\xi} = \rho b. \qquad (1.16)$$

To solve this equation relative to desired variational factor ρ, we use the following method. Let (1.16) be hold for the value $\eta = 0$

$$z_{1x}(x, \xi) - z_{2\xi}(x, \xi) = \rho b(x, \xi, 0). \qquad (1.17)$$

Differentiate equation (1.16) by η

$$(\rho a)_x + \eta(\rho a)_\xi = (\rho b)_\eta.$$

This relation may be written as a partial differential equation of the first order

$$a\rho_x + \eta a\rho_\xi - b\rho_\eta + (a_x + \eta a_\xi - b_\eta)\rho = 0. \qquad (1.18)$$

As a result we obtain system (1.17), (1.18) equivalent to initial equation (1.16).

Because the function $a(x, \xi, \eta)$ is assumed to be strictly positive (1.4), the system of characteristics for (1.18) may be resolved relative to derivatives by x

$$\frac{d\xi}{dx} = \eta, \quad \frac{d\eta}{dx} = -\frac{b}{a}(x, \xi, \eta), \quad \frac{d\rho}{dx} = -\rho\frac{a_x + \eta a_\xi - b_\eta}{a}. \qquad (1.19)$$

The first two equations of this system have no variable ρ, and, moreover, they represent the system equivalent to an ordinary differential equation of the second order (1.9). Therefore, by the condition (B_1) (definition 1.3), determined in the intervals with the ends

x_0, x^1 functions $\xi = \varphi(x_0, x, y_0, y_1), \eta = \varphi_x(x_0, x, y_0, y_1)$ (recall that by $\varphi(x_0, x, y_0, y_1)$ we denote the solution of the Cauchy problem (1.10)) solve the Cauchy problem $\xi(x_0) = y_0, \eta(x_0) = y_1$ for the system of two first equations of (1.19).

Setting

$$F(x_0, x, y_0, y_1) = -\left.\frac{a_x + \eta a_\xi - b_\eta}{a}\right|_{\substack{\xi = \varphi(x_0, x, y_0, y_1), \\ \eta = \varphi_x(x_0, x, y_0, y_1)}} \tag{1.20}$$

we may write the general solution of the latter equation of (1.19) as

$$\rho = K e^{-\int_{x_0}^{x} F(x_0, \lambda, y_0, y_1) d\lambda}$$

Let us return to equation (1.18). As the factor in the latter formula we may take an arbitrary first integral of (1.9), preserving its value along each solution of the system of first two equations of (1.19).

Therefore, using the notations

$$A(x, \xi, \eta) = \varphi(x, x^1, \xi, \eta),$$

$$B(x, \xi, \eta) = \left.\frac{\partial}{\partial \lambda}\varphi(x, \lambda, \xi, \eta)\right|_{\lambda = x^1}, \tag{1.21}$$

$$\rho_0(x, \xi, \eta) = e^{-\int_{x_0}^{x} F d\lambda}\bigg|_{\substack{x_0 = x^1, \\ y_0 = A(x, \xi, \eta), \\ y_1 = B(x, \xi, \eta).}}$$

we may represent a solution of (1.18) as follows

$$\rho(x, \xi, \eta) = \Psi(A, B)\rho_0(x, \xi, \eta), \tag{1.22}$$

where $\Psi(A, B)$ is an arbitrary smooth function of its arguments. As its arguments we may take any independent first integrals of (1.9), for example (1.21).

Functions ρ in (1.22) and, therefore, Φ in (1.15) are determined for all $(x, \xi, \eta) \in [0, 1] \times R^2$. Choosing in (1.22) $\Psi(A, B) \geq 0$, we satisfy (1.6).

Thus, if (1.17) holds, the functions ρ, Φ constructed by formulae (1.15), (1.20)-(1.22) satisfy (1.12), (1.13).

To complete the theorem proof, choose the function $z_1(x,u)$ in (1.15) so that (1.14) holds. Set

$$z_1(x,\xi) = -x \int_0^{h_1(\xi)} \rho a(1,\xi,\tau) d\tau + (x-1) \int_0^{h_0(\xi)} \rho a(0,\xi,\tau) d\tau, \qquad (1.23)$$

where

$$h_i(\xi) = \begin{cases} \dfrac{\psi_i(\xi)}{\alpha_i}, & \text{if } \alpha_i \neq 0, \\ \\ 0, & \text{if } \alpha_i = 0, \end{cases}$$

i.e., the functions $h_i(\xi)$ and, therefore, $z_1(x,\xi)$, are determined by the form of (1.2).

Formulae (1.15), (1.23) for each $i = 0,1$ yield

$$\left. \frac{\partial \Phi}{\partial u_x} u_t \right|_{x=i} = \left[\int_0^{u_x} \rho a d\tau - \int_0^{h_i(u)} \rho a(i,u,\tau) d\tau \right] u_t \Bigg|_{x=i}. \qquad (1.24)$$

If $\alpha_i = 0$, then, by means of (1.4), the corresponding condition (1.2) takes the form $u|_{x=i} = \text{const}$, i.e., $u_t(i,t) = 0$ and (1.14) holds. If $\alpha_i \neq 0$, then $u_x|_{x=i} = h_i(u)|_{x=i}$ and the right side of (1.24) becomes zero. Thus, (1.14) holds for all considered types of boundary values.

For given functions ρ and z_1 in (1.22), (1.23), we may solve (1.17) relative to $z_2(x,\xi)$

$$z_2(x,\xi) = \int_0^{\xi} [z_{1x}(x,\xi) - \rho b(x,\xi,0)] d\xi. \qquad (1.25)$$

Constructed by formulae (1.15), (1.20)-(1.23), (1.25) functions ρ, Φ satisfy system (1.12)-(1.14), and therefore equation (1.5). The smoothness of these functions depends on the coefficients of (1.1). The arbitrariness in determining the function $z_2(x,\xi)$ in the form of an additive function of the variable x, evidently, does not depend on the functional (1.8) properties. Thus, the theorem is proved.

Now we shall give some remarks on the generalization of theorem 1.1.

Consider the following problem for the nonautonomous equation depending on the parameter

$$u_t = a(\varepsilon, t, x, u, u_x)u_{xx} + b(\varepsilon, t, x, u, u_x), \qquad (1.26)$$

$$u\big|_{x=i} = \beta_i(t), \quad i = 0, 1,$$

$$u(x, 0) = u_0(x),$$

where $a, b, \beta_i \in C^2$ by all the arguments; all the derivatives of a, b are uniformly bounded for $\varepsilon > 0$ for bounded values of its arguments. Let, besides,

$$a \geq \delta(\varepsilon) > 0,$$

where $\delta(\varepsilon) \to 0$ for $\varepsilon \to 0$.

Formulate the analog of the condition (B_1) for problem (1.26)

Definition 1.4.   We shall say that for problem (1.26) the condition (B_2) holds if the solution $\varphi(\varepsilon, t, x_0, x, y_0, y_1)$ of the Cauchy problem

$$a(\varepsilon, t, x, y, y')y'' + b(\varepsilon, t, x, y, y') = 0,$$

$$y(x_0) = y_0, \quad y'(x_0) = y_1$$

is determined and regular in the interval with the ends x_0, x^1 for all admissible values of parameters $\varepsilon > 0, t \in (0, T]$ and the Cauchy data $(x_0, y_0, y_1) \in [0, 1] \times [-M_1, M_1] \times [M_2(\varepsilon), M_3(\varepsilon)]$.

Almost similarly we may show the in validity of the following theorem.

Theorem 1.2. Let a solution $u(x,t) \in H^{2+\alpha}(Q_T)$ to the problem (1.26) be twice continuously differentiable for $t > 0$, $u_t \in L_2(Q_T)$, the condition (B_2) and the estimates

$$|u(x,t)| < M_1, \quad M_2(\varepsilon) < u_x(x,t) < M_3(\varepsilon)$$

hold.

Then, there exist such functions $\rho, \Phi(\varepsilon, t, x, \xi, \eta) \in C^2$, that

$$\int_0^1 \Phi(\varepsilon, t, x, u, u_x) dx = \int_0^1 \Phi(\varepsilon, 0, x, u_0, u_0') dx$$

$$- \int_0^t \int_0^1 \rho u_t^2 \, dx \, dt + \int_0^t \int_0^1 \frac{\partial \Phi}{\partial t} dx \, dt + \int_0^t \left(\frac{\partial \Phi}{\partial u_x} u_t \right) \Big|_{x=0}^1 dt. \qquad (1.27)$$

These functions ρ, Φ are as follows

$$\Phi(\varepsilon, t, x, \xi, \eta) = \int_{M_2}^\eta (\eta - \tau) \rho a d\tau + \Phi_1(\varepsilon, t, x, \xi), \qquad (1.28)$$

$$\rho(\varepsilon, t, x, \xi, \eta) = \left(\Psi(\varepsilon, t, y_0, y_1) e^{- \int_{x_0}^x F dx} \right) \Bigg|_{\substack{x_0 = x^1, \\ y_0 = A(\varepsilon, t, x, \xi, \eta), \\ y_1 = B(\varepsilon, t, x, \xi, \eta).}} \qquad (1.29)$$

$$\Phi_1(\varepsilon, t, x, \xi) = - \int_0^\xi \rho b(\varepsilon, t, x, \xi, M_2) d\xi, \qquad (1.30)$$

$$F(\varepsilon, t, x_0, x, y_0, y_1) = \frac{a_x + \eta a_\xi - b_\eta}{a} \Bigg|_{\substack{\xi = \varphi(\varepsilon, t, x_0, x, y_0, y_1), \\ \eta = \varphi_x(\varepsilon, t, x_0, x, y_0, y_1)}} \qquad (1.31)$$

$$A(\varepsilon, t, x, \xi, \eta) = \varphi(\varepsilon, t, x, x^1 \xi, \eta), \qquad (1.32)$$

$$B(\varepsilon, t, x, \xi, \eta) = \frac{\partial}{\partial x^1} \varphi(\varepsilon, t, x, x^1 \xi, \eta), \qquad (1.33)$$

where $\Psi(\varepsilon, t, A, B)$ is an arbitrary smooth function of its arguments,

M_2 have bounded from below $u_x(x,t)$; $\varphi_x(\varepsilon, t, x_0, x, y_0, y_1)$ solves the Cauchy problem from definition 1.4.

This tedious variant of theorem 1.1 will be used for investigating the correctness of one degenerate problem. Note, that in identity (1.27) which is the analog of (1.7), the left side (the generalized Liapunov functional) is neither positively determined nor monotone along the solution of dynamic problem (1.26).

Remark 1.3. When solving equality (1.5) we do not use the fact that $u(x,t)$ satisfies (1.1). All calculations may be done for an arbitrary smooth function attaining the boundary values and approximating the generalized solution of (1.1). Theorem 1.1 remains valid for solutions $u(x,t) \in W_{x,t}^{2,1}(Q_T) \cap C(Q_T)$ of problem (1.1)-(1.3) if identity (1.7) is replaced by its integrated variant

$$\int_0^1 \Phi(x, u, u_x)dx \Big|_{t=t_1}^{t_2} = -\int_{t_1}^{t_2}\int_0^1 \rho u_t^2 dx dt,$$

where $0 \le t_1 < t_2 \le T$.

Remark 1.4. When we have proved theorem 1.1, we obtain the formulae which allow us to construct the set of Liapunov's functionals (the set of variational factors). This set is determined by an arbitrary smooth function of two variables (see (1.22)). This, as will be shown further, allows us to apply the Liapunov functionals to establish *a priori* estimates for derivatives of the solutions and, therefore, for the existence theorem's proof.

It should be noticed that if the more restrictive than (B_1) condition (B) holds, then the set of variational factors (the Liapunov functionals) has one more degree of freedom. Really, in this case, a solution $\varphi(x_0, x, y_0, y_1)$ to the Cauchy problem (1.10) is determined in the interval $x \in [0,1]$. Therefore, in formulae (1.21), as independent first integrals A, B of equation (1.9), we may take the values of φ and φ_x in an arbitrary point $x^1 \in [0,1]$. This additional degree of freedom will be used for investigating the equations which do not satisfy the condition (B).

Remark 1.5. If we know *a priori* that a solution $u(x,t)$ of (1.1)-(1.3) is bounded together with its derivative $|u(x,t)| + |u_x(x,t)| \leq K$, then (B) does not lose generality. Really, in this case, instead of (1.1) we shall consider the equation

$$u_t = a(x, u, u_x)u_{xx} + b_k(x, u, u_x),$$

where the finite by arguments u, u_x smooth function b_k coincides with the function b for $|u| + |u_x| \leq K$. Though not changing the dynamic problem along the considered solution, we essentially change the qualitative properties of (1.9). The realizability of (B) for the second order ordinary differential equation, resolved relative to its higher derivative is guaranteed, as is well-known, by the finiteness of the right side.

Remark 1.6. Though in definition 1.1 we have established the vanishing of the variational factor, ρ, this does not allow us generally with the functional to transfer one of the basis properties of Liapunov's functions - the strict monotonicity along the trajectory of a dynamic problem, except for the equilibrium points. As we have noted, the analog of this property also holds for the additional restriction $\rho > 0$. Nevertheless, the consideration of generalized Liapunov's functionals with the vanishing factors $\rho \geq 0$ is useful for investigating degenerate problems.

§2.2 The existence condition for Liapunov's functionals.

In this paragraph, we shall compare the condition (B) for the existence of sets of Liapunov functionals with the restrictions on the growth order of nonlinearity by the gradient variable which are well-known in the qualitative theory of differential equations.

First, note that we consider the bounded solutions of (1.1)-(1.3)

$$\sup_{x,t} |u(x,t)| \leq M; \tag{2.1}$$

therefore, without loss of generality, instead of (1.1) we shall consider the equation

$$u_t = a(x, u, u_x)u_{xx} + \omega_M(u)b(x, u, u_x),\qquad (2.2)$$

where nonnegative cutoff function $\omega_M(u) \in C^2$ is such that

$$\omega_M(u) = \begin{cases} 1, & |u| \leq M, \\ \\ 0, & |u| \geq M + 1. \end{cases}\qquad (2.3)$$

Such substitution changes the properties of the ordinary differential equation in the condition (B) (definition 1.2) and is more convenient for further formulations.

The properties of regular continuation of solutions of the second order ordinary differential equations were investigated by certain authors, begining with the works of Tonelli, 1923, Nagumo, 1929, Bernstein, 1960, among them in connection with the of calculus of variatios problems. So, it was established (Bernstein, 1960) that if the right side of the equation

$$y'' = f(x, y, y')$$

grows not faster than a quadratic function

$$|f(x, y, y')| \leq K(y)(1 + y'^2)$$

then each bounded solution may be regularly continued in each bounded interval of x. The condition of a solution's boundedness is essential, as the following simplest example shows:

$$y'' = 1 + y'^2.$$

The derivative of solutions here is $y'(x) = \tan x$.

To avoid such restrictions, we shall consider the equations

$$y'' = \omega_M(y)f(x,y,y') \qquad (2.4)$$

with the cutoff function $\omega_M(y)$ of the form (2.3). Note, that for

$$f(x,\xi,\eta) = -\frac{b}{a}(x,\xi,\eta) \qquad (2.5)$$

equation (2.4) will be present in the definition of the condition (B) for equation (2.2).

Similar restrictions on growth arise in the theory of nonlinear parabolic equations (Ladyzhenskaya *et al.*, 1967, Krylov, 1985). In the most general form they have been formulated by Kruzhkov (1972, 1979), where is assumed that

$$\left| \frac{b}{a}(x,u,u_x) \right| \le K(u)\Psi(u_x),$$

and a smooth positive function $\Psi(s)$ is such that

$$\int\limits_{0}^{\pm\infty} \frac{s\,ds}{\Psi(s)} = \infty. \qquad (2.6)$$

It is clear that integral (2.6) is convergent for the functions $\Psi(s)$ with the order $s^{2+\varepsilon}$, $\varepsilon > 0$, but remains divergent for superquadratic growth with the order $s^2 \ln s$ and similar to it. This restriction allows us to establish the derivative estimate $\sup |u_x(x,t)| \le M_1$ for the bounded $\sup |u(x,t)| \le M$ solution of problem (1.1)-(1.3). As in the case of ordinary differential equations for equivalence $\Psi(s) \sim s^{2+\varepsilon}$, $\varepsilon > 0$ at infinity, examples were given of bounded solutions with unbounded derivatives (Filippov, 1961; Ladyzhenskaya *et al.*, 1967).

We now show that these restrictions are sufficient for the condition (B) (definition (1.2)) to hold.

Lemma 2.1. If $\omega_M(\xi)|f(x,\xi,\eta)| \le \omega_M(\xi)\Psi(\eta)$, where $\Psi(\eta)$ satisfies condition (2.6), and f has the form (2.5), then for equation (2.2) the condition (B) holds.

Proof. Note, that to prove the lemma it suffices to obtain an *a priori* estimate for $|y| + |y'|$ by the Cauchy data y_0, y_1 for the solution of (2.4).

Consider an arbitrary x from the interval of existence of smooth solution $y(x)$ of the Cauchy problem $y(x_0) = y_0, y'(x_0) = y_1$ for equation (2.4). Choose the point x_2 from the same interval so that $y'(x)$ does not change the sign in the interval with the ends x, x_2 and $y'(x_2) = 0$ holds. If such $x_2 \neq x$ does not exist, then either $y'(x) = 0$ and the required estimate is trivial, or $y'(x)$ is of constant sign in the considered interval, and as x_2 we choose x_0. In the obtained interval the derivative y' may be considered as a function of y and by the lemma's condition (2.4) yields

$$-\omega_M(y)\Psi(y') \le y'\frac{dy'}{dy} \le \omega_M(y)\Psi(y').$$

From this double inequality by integrating we obtain the following estimate

$$\left|\int_0^{y'(x)} \frac{sds}{\Psi(s)}\right| \le \left|\int_{y(x_2)}^{y(x)} \omega_M(s)ds\right| + \left|\int_0^{y_1} \frac{sds}{\Psi(s)}\right|. \tag{2.7}$$

By property (2.6) from here follows the estimate $|y'| \le K(M, y_1)$ which, by means of integrating, gives the desired boundedness of $|y(x)| + |y'(x)|$. The lemma is proved.

Lemma 2.2. In a rectangle $|x - x_0| < \delta, |y - y_0| < \delta$ $(|y_0| < M)$ let the inequality

$$|f(x, \xi, \eta)| \ge \Psi(\eta)$$

hold, where at least one of the integrals (2.6) converges.

Then there exist such Cauchy data $y(x_0^*) = y_0^*, y'(x_0^*) = y_1$ that the solution of this problem for equation (2.4) can not be regularly continued in the interval $x \in [0, 1]$.

The proof is based on the inequality of type (2.7) and the possibility of performing the calculation of

$$\int_{|y_1|}^{\infty} \frac{s\,ds}{\Psi(s)}$$

arbitrarily small for sufficiently large $|y_1|$. The evaluation is purely computative, and is therefore not given here.

Lemmas 2.1, 2.2 yield

Corollary 2.3. Let $b/a(x, y, y') = g_1(x, y)g_2(y')$, where $g_1(x, y) \not\equiv 0$. Then the estimate $|g_2(s)| \leq \Psi(s)$, where $\Psi(s)$ satisfies (2.6), is equivalent to the condition (B) for equation (2.2).

Now we shall give some examples which show that the results of lemmas 2.1, 2.2 can not be generalized in terms of the restrictions on the growth. First, we establish the following lemma.

Lemma 2.4. Each bounded solution for the Cauchy problem for the Euler equation

$$\frac{d}{dx}\frac{\partial F(y, y')}{\partial y'} - \frac{\partial F(y, y')}{\partial y} = 0 \quad , \quad \left(\frac{d^2 F}{dy'^2} \geq \delta > 0\right),$$

$$y(x_0) = y_0, \quad y'(x_0) = y_1$$

may be regularly continued in each finite interval.

This may be easily proved if we multiply the equation by y' and integrate it from x_0 to x. Straightforward calculations show that the value $\frac{dF}{dy'}y' - F$ is the first integral of the equation. Its boundedness implies, as is easy to see, the boundedness of $|y'(x)|$, from which follows the lemma statement.

Example 2.1. Set $\frac{\partial^2 F(\xi, \eta)}{\partial \eta^2} = \pi + \tan^{-1}(\xi\eta) \geq \frac{\pi}{2} > 0$. As F we may take the function

$$F = \frac{\pi}{2}\eta^2 + \frac{\eta}{2\xi} + \frac{\xi^2\eta^2 - 1}{2\xi^2}\tan^{-1}(\xi\eta) - \frac{\eta}{2\xi}\ln(1 + \xi^2\eta^2).$$

It is easy to show that for $\xi = 0$, the function $F(\xi, \eta)$ may be determined as a smooth function. The Euler equation will take the form

$$[\pi + \tan^{-1}(yy')]y'' = \frac{\tan^{-1}(yy')}{y^3} - \frac{y'}{y^2}.$$

By means of the Taylor expansion of the right side of this equation we may see that it is determined as a smooth function in the point $y = 0$. Moreover, for $y \to 0$, the right side tends to the function $-\frac{1}{3}y'^3$, i.e., integrals (2.6) in the point $y = 0$ converge and the growth condition fails in the line $(x, 0)$ of the plane x, y. Nevertheless, by lemma 2.4, the condition (B) for the corresponding parabolic problem holds.

On the other hand, begining with Ball and Mizel (1984), in the calculs of variation theory, there some examples were obtained of regular problems with a strictly convex Lagrangian, the exact solutions of which are not smooth. Let us consider the first of them (Ball and Mizel, 1984).

Example 2.2. Consider the minimization problem on absolutely continuous functions with the values $y(0) = 0, y(1) = k$ for the following functional

$$\int\limits_0^1 (ry'^2 + (y^3 - x^2)^2 y'^{14})dx, \qquad (2.8)$$

where $r = (\frac{2}{3}k)^{12}(k^3 - 1)(7 - 13k^3)$. Lagrangian (2.8) is analytic for three variables, and for $k^3 \in (frac713, 1)$ is strictly convex for y'. The problem of its minimization satisfies the conditions of the classical existence theorem. It was proved (Ball and Mizel, 1984) that its exact minimum is attained at the function $y(x) = kx^{\frac{2}{3}}$, which is not regular in the closed interval $[0, 1]$.

Direct calculation shows that the Euler-Lagrange equation for the Lagrangian (2.8) may be represented in the form

$$y'' = -\frac{2(y^3 - x^2)y'^{13}[39y^2y' - 28x]}{2r + 13 \cdot 14(y^3 - x^2)^2 y'^{12}}. \qquad (2.9)$$

The right side of this equation grows no more than quadratically by y' for each fixed pair (x^*, y^*) with $y^* \neq (x^*)^{2/3}$ and vanishes for $x^{*2} = y^{*3}$. Thus, for all x^*, y^* the right side of (2.9) satisfies the estimate

$$|f(x^*, y^*, y')| \leq K(1 + y'^2),$$

but K can not be chosen uniformly, since it unboundedly increases for (x^*, y^*) approximating to the curve $y = x^{\frac{2}{3}}$. The growth condition fails in this sufficiently weak sense; the condition (B) fails because of the irregular solution $y = kx^{\frac{2}{3}}$.

We may show that the condition (B_1) holds and the solution to each Cauchy problem for equation (2.9) may be regularly continued to the point $x = 1$.

In the given case, the non-smooth solution of the variational problem may be considered as the generalized solution to the Euler equation, smooth in the interval $x \in (0,1)$ and attaining the boundary values. A further collection of examples, suggested by Sychyev (1992),

$$\int_a^b (\mu(k)y'^2 + (y^5 - x^3)^2 y'^{32})dx \to \min, \quad y(a) = ka^{\frac{3}{5}}, \ y(b) = kb^{\frac{3}{5}},$$

$(\mu(k) = (\frac{3k}{5})^{30}(1 - k^5)^3(31k^5 - 16))$, has the unique solution $y_0(x) = kx^{\frac{3}{5}}$, and shows that a solution of the variational problem may have a singularity inside the considered interval and not attain the ordinary interpretation of generalized solution of the Euler-Lagrange equation.

Note that in the case $a \cdot b < 0$, the corresponding problem does not satisfy the condition (B_1).

More detailed consideration of the problems of calculus of variations is beyond our consideration; therefore, we return to analysis of the condition (B).

Lemma 2.5. If a function $f(\eta) \in C^1$ has $+\infty(-\infty)$ as a point of zero concentration, then there exists a function $\Psi(\eta)$, satisfying the corresponding condition (2.6), majorizing $|f(\eta)|$ for $\eta > 0, (\eta < 0)$

$$|f(\eta)| \leq \Psi(\eta). \tag{2.10}$$

Proof. The proof is based on constructing the piecewise linear majorant in the neighborhoods of zeros and the direct estimate of the integrals. Consider the increasing sequence of zeros. Let $f(\eta_i) = 0$, where $\eta_i \to +\infty$ for $i \to \infty$. Because of the smoothness of $f(\eta)$, there exist such numbers $\varepsilon_i, \alpha_i > 0$ that

$$|f(\eta)| \leq \begin{cases} \alpha_i(\eta - \eta_i), & \eta \in [\eta_i, \eta_i + \varepsilon_i], \\ \\ -\alpha_i(\eta - \eta_i), & \eta \in [\eta_i - \varepsilon_i, \eta_i]. \end{cases} \tag{2.11}$$

For a fixed $\lambda \in (0, \varepsilon_i)$ we shall choose $\Psi(\eta)$ (2.10) as a linear function in the interval $\eta - \eta_i \in [\lambda, \varepsilon_i]$; and for the other values η we shall take it arbitrary, smooth, and satisfying (2.10).

We have

$$\int\limits_{\eta_i+\lambda}^{\eta_i+\varepsilon_i} \frac{\eta \, d\eta}{\Psi(\eta)} = \int\limits_{\eta_i+\lambda}^{\eta_i+\varepsilon_i} \frac{\eta \, d\eta}{\alpha_i(\eta - \eta_i)} = \frac{\varepsilon_i - \lambda}{\alpha_i} + \frac{\eta_i}{\alpha_i} \ln \frac{\varepsilon_i}{\lambda},$$

i.e., for decreasing λ, the integral (2.6) may be greater than unity in the interval $\eta \in [\eta_i + \lambda, \eta_i + \varepsilon_i]$. So, condition (2.6) will hold. Thus, the lemma is proved.

Remark 2.1. Lemma 2.5 shows that $\sup_{0 \leq |s| \leq \eta} \Psi(s)$ may grow arbitrarily fast for $\eta \to \infty$ while conserving the property of divergence of integrals (2.6). Considering separated from zero majorizing functions $\Psi(\eta)$ narrows the class of ordinary differential equations satisfying condition (B).

§2.3 *A priori* estimates of the first derivative.

Using the Liapunov functional sets constructed in §1, we shall derive the *a priori* estimate for the derivative of a smooth problem (1.1)-(1.3) solution, assuming its boundedness in C. We shall follow the scheme suggested by Belonosov and Zelenyak (1975).

Theorem 3.1. Let $u(x,t) \in C^{2,1}Q_T$ be a solution of (1.1)-(1.3), let condition (B) hold, and $\sup_{x,t} |u(x,t)| \leq M(T)$. Then

1) the following estimate

$$\sup_{x,t} |u_x(x,t)| \leq M_1(M, \sup_x |u_0'|, T), \qquad (3.1)$$

holds, where the function M_1 depends on the coefficients of the problem;

2) if M does not depend on T, then M_1 may be chosen independent of T; if problem (1.1), (1.2) has the trivial solution $u(x,t) \equiv 0$, then $M_1 \to 0$ for $M + \sup |u_0'(x)| \to 0$;

3) if $\overline{\lim}_{t \to T_0}(|u| + |u_x|) = \infty$, then $M(T) \to \infty$ for $T \to T_0$ and

$$\sup_x |u(x,t)| \xrightarrow[t \to T_0]{} \infty. \qquad (3.2)$$

Proof. First, we shall prove that if condition (B) holds (definition 1.2), then the integrals $A, B(x, \xi, \eta)$ in (1.21) of equation (1.9) for an arbitrary $x^1 \in [0,1]$ satisfy the following condition

$$|A(x, \xi, \eta)| + |B(x, \xi, \eta)| \to \infty \qquad (3.3)$$

for $|\eta| \to \infty$ uniformly for $x \in [0,1]$ in each interval $|\xi| \leq M$.

By the definition of the integrals A, B the solution $\varphi(x_0, x, y_0, y_1)$ of the Cauchy problem (1.10) satisfies the identity

$$\frac{d}{d\lambda} \varphi(x^1, \lambda, A(x, \xi, \eta), B(x, \xi, \eta)) \bigg|_{\lambda = x} \equiv \eta. \qquad (3.4)$$

If (3.3) does not hold, then such sequences $x_n \to \overline{x}, \xi_n \to \overline{\xi}, \eta_n \to \infty$ exist that $A_n = A(x_n, \xi_n, \eta_n) \to \overline{A}, B_n \to \overline{B}$, where $\overline{A}, \overline{B}$ are finite. The condition (B), the continuous dependence of solutions of ordinary differential equations on parameters and on the Cauchy data yield that the values $\varphi_0 = \varphi(x^1, \overline{x}, \overline{A}, \overline{B})$ and $\varphi_1 = \frac{d}{d\overline{x}}\varphi(x^1, \overline{x}, \overline{A}, \overline{B})$ are finite and

$$\left| \varphi(x^1, x_n, A_n, B_n) - \varphi_0 \right| \leq \varepsilon,$$

$$\left| \frac{\partial \varphi}{\partial x_n}(x^1, x_n, A_n, B_n) - \varphi_1 \right| \leq \varepsilon$$

for a fixed $\varepsilon > 0$ and sufficiently large n. These inequalities contradict identity (3.4), which proves property (3.3).

By (3.3) we may choose such smooth function $\Psi_0(A, B)$ so that (see (1.21))

$$|\Psi_0(A, B, (x, \xi, \eta))| \geq |\eta|,$$

$$\Psi_0(A, B(x, \xi, \eta)) \cdot a \cdot \rho_0(x, \xi, \eta) \geq 1. \tag{3.5}$$

Set in (1.22), $\Psi(A, B) = \Psi_0^{2n+1}(A, B)$ and consider the corresponding functions ρ_n, Φ_n, z_i^n obtained from (1.22), (1.15), (1.23), (1.25). There exists such constant C_1 that

$$\sup_{x, |\xi| \leq M} |z_i^n| \leq C_1^{2n+2},$$

where

$$C_1 \geq \sup_{|\xi| \leq M} (|h_0| + |h_1|)$$

and

$$C_1 \geq \max\{\Psi_0(A, B), \quad \Psi_0 a \rho_0(x, \xi, \eta)\}$$

for

$$x \in [0, 1], |\xi| \leq M, \quad |\eta| \leq \sup_{i, |\xi| \leq M} |h_i(\xi)|.$$

Therefore, (3.5), (1.15) yield the estimate

$$\Phi_n(x, \xi, \eta) \geq \frac{\eta^{2n+2}}{(2n + 1)(2n + 2)} - (1 + |\eta|)C_1^{2n+2} \tag{3.6}$$

for $x \in [0, 1], |\xi| \leq M$.

Further, it is easy to see that

$$|\Phi_n(x, u_0, u_0')| \leq C_2^{2n+2}, \tag{3.7}$$

where C_2 depends on the functions $\Psi_0, |F|, |A|, |B|$ maximum in the domain

$$0 \leq x \leq 1, \quad \xi^2 + \eta^2 \leq R_1^2 = \sup_x |u_0|^2 + \sup_x |u_0'|^2. \qquad (3.8)$$

Since the positiveness of the function ρ_n (3.5) provides the monotonicity of the functionals (1.8) (see (1.5)), we have

$$\int_0^1 \Phi_n(x, u, u_x)dx \leq \int_0^1 \Phi_n(x, u_0, u_0')dx$$

and, therefore, by (3.6), (3.7), the estimate

$$\int_0^1 u_x^{2n+2}dx \leq (2n+1)(2n+2)\left[C_2^{2n+2} + C_1^{2n+2}\int_0^1 (1 + |u_x|)dx\right] \quad (3.9)$$

holds.

Use the Cauchy inequality with a weight

$$2\int_0^1 |u_x| \, dx \leq \varepsilon \int_0^1 |u_x|^2 \, dx + \frac{1}{\varepsilon}$$

and inequality (3.9) for $n = 0$ to estimate $\| \, u_x \, \|_{L_1(0,1)}$. Finally, we obtain

$$\int_0^1 u_x^{2n+2}dx \leq (2n+1)(2n+2)C_3 \cdot (C_1 + C_2 + 1)^{2n+2}. \qquad (3.10)$$

It is known that for a continuous function $f(x)$

$$\| f(x) \|_{L_p(0,1)} \xrightarrow[p \to \infty]{} \sup_x |f(x)|.$$

Hence, extracting from both sides of (3.10) the root of $2n + 2$ degree and going to the limit for $n \to \infty$, we obtain the estimate

$$\sup_{x,t} |u_x(x,t)| \leq 1 + C_1 + C_2 = M_1. \qquad (3.11)$$

The statement 1) of the theorem is established.

If we suppose that M does not depend on T, then C_1 and, therefore, M_1 will not depend on T. Suppose, now, that problem (1.1)-(1.2) has trivial solution $u(x,t) \equiv 0$. Set in (1.22) $\Psi = \Psi_n = (A^2 + B^2)^n$. By the function A, B definition we have

$$A(x,0,0) \equiv 0, \quad B(x,0,0) \equiv 0. \tag{3.12}$$

After differentiating equation (1.10) we see that the derivatives of the integrals A, B may be obtained from the solution of the linearized equation (1.10) as follows

$$A_\xi(x,0,0) = Y_1(x,x^1), \quad B_\xi(x,0,0) = \left.\frac{\partial Y_1(x,\lambda)}{\partial \lambda}\right|_{\lambda=x^1},$$

$$A_\eta(x,0,0) = Y_2(x,x^1), \quad B_\eta(x,0,0) = \left.\frac{\partial Y_2(x,\lambda)}{\partial \lambda}\right|_{\lambda=x^1},$$

where the functions $Y_i(x,\lambda)$ are determined as solutions of the equation

$$\frac{d^2Y}{d\lambda^2} + \left.\frac{\partial}{\partial y'}\frac{b}{a}(\lambda,y,y')\right|_{\substack{y=0\\y'=0}}\frac{dY}{d\lambda} + \left.\frac{\partial}{\partial y}\frac{b}{a}(\lambda,y,y')\right|_{\substack{y=0\\y'=0}}Y = 0,$$

attaining the values

$$Y_1|_{\lambda=x} = 1, \quad \left.\frac{dY_1}{d\lambda}\right|_{\lambda=x} = 0; \quad Y_2|_{\lambda=x} = 0, \quad \left.\frac{dY_2}{d\lambda}\right|_{\lambda=x} = 1.$$

Since Y_1, Y_2 are independent, then

$$Y_1Y_2' - Y_2Y_1' = (A_\xi B_\eta - B_\xi A_\eta)|_{\substack{\xi=0\\\eta=0}} \neq 0. \tag{3.13}$$

Further, by (3.12) we have $d(A^2 + B^2)|_{\substack{\xi=0\\\eta=0}} = 0$ and

$$d^2(A^2 + B^2)\Big|_{\substack{\xi=0\\\eta=0}} = \left[2(A_\xi d\xi + A_\eta d\eta)^2 + 2(B_\xi d\xi + B_\eta d\eta)^2\right]\Big|_{\substack{\xi=0\\\eta=0}}. \tag{3.14}$$

The form $[(A_\xi\xi_1 + A_\eta\eta_1)^2 + (B_\xi\xi_1 + B_\eta\eta_1)^2]|_{\substack{\xi=0\\\eta=0}}$ of the variables ξ_1, η_1 is positive definite (see (3.13)) and, therefore, by (3.14) for $|\xi| \leq M, |\eta| \leq M_1$ there exist such numbers $\mu, \nu > 0$ that

$$\nu^n(\xi^2 + \eta^2)^n \leq \Psi_n = (A^2 + B^2)^n \leq \mu^n(\xi^2 + \eta^2)^n. \qquad (3.15)$$

Taking into account that we have $h_i(0) = 0$ (the function $u \equiv 0$ solves the problem), we may obtain (considering similarly as when deducing estimates (3.5)-(3.10)) the inequality

$$\int_0^1 (u^2 + u_x^2)^n dx \leq (2n+1)(2n+2)C_4^{2n} \left[\left(\frac{\mu}{\nu}\right)^n R_1^{2n} + (\sup_{x,t} |u|)^{2n} \right],$$

where R_1 is from (3.8). From here, acting as above (see (3.11)), we obtain

$$\sup_{x,t}(|u^2| + |u_x^2|) \leq \frac{\mu}{\nu} R_1^2 + \left(\sup_{x,t} |u| \right)^2,$$

i.e., we have established the point 2) of the theorem.

Let $\overline{\lim}_{t \to T_0} \sup_x(|u| + |u_x|) = \infty$ and for a certain function $\rho > 0$ in (1.22) we have

$$\int_0^t \int_0^1 \rho u_t^2 dx dt \xrightarrow[t \to T_0]{} \infty. \qquad (3.16)$$

Since $u_0(x) \in C^1$, integrating (1.5) yields

$$\int_0^1 \Phi(x, u, u_x) dx - \int_0^1 \Phi(x, u_0, u_0') dx = - \int_0^t \int_0^1 \rho u_t^2 dx dt,$$

This means, that from (3.16) it follows that

$$\lim_{t \to T_0} \int_0^1 \Phi(x, u, u_x) dx = -\infty.$$

The positiveness of $\rho a(x, \xi, \eta)$ and (1.15) yield $-\Phi(x, \xi, \eta) \leq -\eta z_1(x, \xi) - z_2(x, \xi)$ and, therefore,

$$\lim_{t \to T_0} \int_0^1 (-u_x z_1(x, u) - z_2(x, u)) dx = +\infty. \qquad (3.17)$$

Setting $z(x, u) = \int_0^u z_1(x, \xi) d\xi$ we have $\frac{d}{dx} z(x, u) = z_1(x, u)u_x + \frac{\partial z}{\partial x} u$, hence

$$-\int_0^1 (u_x z_1 + z_2) dx = -z(x, u)|_{x=0}^1 + \int_0^1 \left[\frac{\partial z}{\partial x}(x, u) - z_2(x, u) \right] dx,$$

and, as the result, we see that (3.17) may hold only when

$$\sup_x |u(x, t)| \to \infty \quad \text{for} \quad t \to T_0.$$

To complete the theorem proof, construct the function $\rho(x, \xi, \eta)$ in (1.22) such that (3.16) holds. Acting analogously, we may establish the convergence of (3.3) uniform by x for $|\xi| + |\eta| \to \infty$. Further, let a sequence $t_i \to T_0$ be such that $\sup_x (|u(x, t_i)| + |u_x(x, t_i)|) \to \infty$. Taking, if necessary, a subsequence, we may consider that

$$u_t^2(x, t_i) \geq \varepsilon_i, \quad u^2(x, t_i) + u_x^2(x, t_i) > i$$

in the rectangles $G_i = \{(x, t): |x - x_i| \leq \delta_i, |t - t_i| \leq \eta_i\}$ for certain $x_i \in [0, 1]$, $\varepsilon_i, \delta_i, \eta_i > 0$, where $G_i \subset [0, 1] \times [0, T_0)$ and $G_i \cap G_k = \emptyset$ for $i \neq k$. Consider a strictly positive function $\omega(v) \in C^1$, where

$$\omega(v) = \begin{cases} i/\varepsilon_i \delta_i \eta_i, & v \in [i, i + 1 - \delta], \quad i = 1, 2, \ldots \\ \\ 1, & v \in [0, 1 - \delta], \quad \delta \in \left(0, \frac{1}{2}\right) \end{cases}$$

and $\omega(v) = \omega(-v)$. Choosing, by (3.3), the function $\Psi(A, B)$ in (1.22) so that $\rho(x, \xi, \eta) \geq \omega(\xi^2 + \eta^2)$ holds, we shall obtain

$$\int_0^t \int_0^1 \rho u_t^2 dx dt \geq \int_{t_i - \eta_i}^{t_i} \int_{x_i - \delta_i}^{x_i + \delta_i} \rho u_t^2 dx dt \geq \frac{i}{\varepsilon_i \delta_i \eta_i} \cdot \varepsilon_i \cdot 2\eta_i \cdot \delta_i = 2i$$

for $t_i < t$ and, therefore,

$$\int\limits_0^t \int\limits_0^1 \rho u_t^2 \, dx \, dt \xrightarrow[t \to T_0]{} \infty.$$

The theorem is proved.

Remark 3.1. The prove of (3.3) was based on condition (B). In the case of the weaker restriction (B_1) (definition 1.3) the set of the functionals exists but there is no uniform convergence by x of (3.3). Theorem 3.1, as it follows from the example, is not valid.

Example 3.1. Here we shall briefly set forth the result of Fila and Lieberman (1994). Consider the problem

$$u_t = u_{xx} + e^{u_x} \ , \quad x \in [0, L], t \geq 0 \ ,$$

$$u(0, t) = u(L, t) = 0; \quad u(x, 0) = u_0(x)$$

with a smooth function $u_0(x)$ satisfying the compatibility conditions. By the maximum principle (Ladyzhenskaya *et al.*, 1967) the solution to the problem is finite on the period of existence.

Considering the Cauchy problem for the equation $y'' + e^{y'} = 0$ (in this case it is the problem (1.10)), it is easy to show that there exist solutions which can not be extended in the interval $x \in [0, L]$ (see also lemmas 2.1, 2.2 and corollary 2.3). Integrals (2.6), evidently, converge. Besides, a solution to each Cauchy problem may be regularly continued to the point $x = L$ (the derivative of the solution with the data $y'(x_0) = y_1$) and may be calculated by the formula $y'(x) = -\ln(x - x_0 + e^{-y_1})$.

Thus, the condition (B) fails whereas the condition (B_1) holds.

It is proved that for $L < e$ there exists a regular solution of the parabolic problem uniformly bounded together with the derivatives. For $L = e$, this solution has the property $\sup_x |u_x(x, t)| \xrightarrow[t \to \infty]{} \infty$. If $L > e$, then such T exists that $\sup_x |u_x(x, t)| \xrightarrow[t \to T-0]{} \infty$ (T depends on initial data and on L).

§2.4   Some generalization of the Liapunov functionals concept.

It has been noted that the analogy with the ordinary differential equations theory of the Liapunov function leads to consideration of a nonnegative functional $J(u)$ of the form (1.8), monotone along dynamic problem solutions

$$\frac{d}{dt}J(u) = -\ell(u) \ , \ \ \ell(u) \geq 0.$$

As follows from (1.15), $J(u)$ may change sign. Moreover, the formal construction of Liapunov function analogs of the other form and their application to nonclassical equation have led to the consideration of non-monotone functionals. To obtain *a priori* estimates for solutions, the study of these generalized Liapunov functionals turns out to be useful.

1.   Equations which are not quasilinear.   For $(x,t) \in Q_T$ consider the problem

$$u_t = C(x, u, u_x, u_{xx}), \tag{4.1}$$

$$\alpha_i u_x - \psi_i(u)|_{x=i} = 0 \ \ (i = 0, 1), \ \ \ u(x, 0) = u_0(x) \tag{4.2}$$

Let the smoothness conditions from §2.1 and restrictions (1.4) with the change of the first inequality by the parabolicity condition

$$\frac{\partial C}{\partial r}(x, \xi, \eta, r) \geq a_0 > 0, \tag{4.3}$$

$(x, \xi, \eta, r) \in [0, 1] \times R^3$ hold for this equation.

By condition (4.3), there exists such function $g(x, \xi, \eta) \in C^2$ that

$$C(x, \xi, \eta, g(x, \xi, \eta)) \equiv 0. \tag{4.4}$$

The ordinary differential equation which determines the stationary solutions of the problem has the form

$$C(x, y, y', y'') = 0$$

and, by (4.3), (4.4), it is equivalent to the equation

$$y'' = g(x, y, y').$$ (4.5)

Definition 4.1.   We shall say that for equation (1.4) the condition (B) holds (the condition (B_1)), if the solution to the Cauchy problem $y(x_0) = y_0, y'(x_0) = y_1$, for equation (4.5) may be regularly continued in the interval $x \in [0, 1]$ (in the interval with the ends x_0, x^1 for a certain $x^1 \in [0, 1]$) for any data $(x_0, y_0, y_1) \in [0, 1] \times R^2$.

Setting $a(x, \xi, \eta) \equiv 1$, $b(x, \xi, \eta) = g(x, \xi, \eta)$, after almost similar to §2 calculations we may show the validity of the following theorem.

Theorem 4.1.   Let equation (4.1) satisfy the condition (B_1). Then such pairs of functions $\rho, \Phi(x, \xi, \eta) \in C^2$ exist that for the classical solution $u(x, t)$ of (4.1), (4.2), the equalities

$$\frac{d}{dt} \int_0^1 \Phi(x, u, u_x)dx = -\int_0^1 \rho(x, u, u_x)(u_{xx} - g(x, u, u_x))u_t dx.$$ (4.6)

hold.

The right side of identities (4.6) does not contain the term u_t^2 under the integral sign, unlike the equalities (1.5), (1.7). Nevertheless, for solutions of (4.1)-(4.2) we have

$$(u_{xx} - g(x, u, u_x))u_t = (u_{xx} - g(x, u, u_x)) \cdot a(x, u, u_x, u_{xx})$$

$$= (u_{xx} - g) \cdot (a(x, u, u_x, u_{xx}) - a(x, u, u_x, g))$$

$$= \frac{\partial a}{\partial u_{xx}}(x, u, u_x, r) [u_{xx} - g(x, u, u_x)]^2.$$ (4.7)

Thus, the integral $\int_0^1 \Phi(x, u, u_x)dx$ conserves the main property of a

Liapunov functional - the strict decreasing with time (for the positive function $\rho(x, u, u_x)$), except for the case when $u_{xx} - g(x, u, u_x) \equiv 0$. This is the case when $u(x)$ is a stationary solution of the problem (see (4.5)).

Remark 4.1. By (4.6), (4.7), the argument of theorem 3.1 (on the derivative of a bounded solution and the properties of the constant in these estimates) is word-for-word applied to equation (4.1).

Remark 4.2. As was shown in §2, the sufficient restrictions for the condition (B) to hold are

$$|g(x, \xi, \eta)| \leq K(M) \cdot \Psi(\eta) \quad , \quad |\xi| < M,$$

where $\Psi(\eta)$ satisfies equations (2.6). In terms of equation (4.1) this restriction may be formulated as

$$\pm C(x, \xi, \eta, \mp K(M)\Psi(\eta)) \leq 0 \quad , \quad |\xi| < M. \tag{4.8}$$

Inequalities (4.8) in the case when $\Psi(\eta) = 1 + \eta^2$ were presented by Kruzhkov (1979).

2. Equations that are insoluble relative to derivative u_t. Recently, physical models have appeared which lead to mixed problems for the equations.

$$\beta(u_t) = a(x, u, u_x)u_{xx} + b(x, u, u_x) \quad , \quad a \geq a_0 > 0, \tag{4.9}$$

where $\beta(s)$ is a strictly monotone, piecewise continuous function, being of the sign of its argument. This function, for example, may be as follows

$$\beta(s) = s + \mathrm{sgn}\, s. \tag{4.10}$$

The following considerations hold for an arbitrary function $\beta(s) \in$

$C^2(R \setminus \{0\})$, having a discontinuity of the first kind at $s = 0$, strictly monotone and having the sign of its argument.

Consider a sequence of smooth strictly monotone functions $\{\beta_n(s)\}$, coinciding with $\beta(s)$, correspondingly, out of the intervals $\left(-\frac{1}{n}, \frac{1}{n}\right)$ and such that $\beta_n(0) = 0$. By $u^n(x, t)$ we shall denote solutions of (4.2) for equations

$$\beta_n(u_t^n) = a(x, u^n, u_x^n)u_{xx}^n + b(x, u^n, u_x^n). \tag{4.11}$$

Again, producing almost the same considerations as in §1 we may establish the following theorem.

Theorem 4.2. Let equations (4.11) satisfy the condition (B_1). Then there exist such pairs of functions $\rho, \Phi(x, \xi, \eta) \in C^2$ that for solution of problems (4.11), (4.2) $u^n(x, t) \in C^{2,1}(Q_T)$ the equality

$$\frac{d}{dt} \int_0^1 \Phi(x, u^n, u_x^n)dx = -\int_0^1 \rho(x, u^n, u_x^n) \cdot \beta_n(u_t^n)u_t^n dx. \tag{4.12}$$

holds.   By the choice of the functions $\beta_n(s)$, the value $\beta_n(s) \cdot s \geq 0$, i.e., the integral $\int_0^1 \Phi(x, u^n, u_x^n)dx$ is strictly decreasing along each solution of the dynamic problem in the case $\rho > 0$ except for the stationary solutions.

Remark 4.3. The fulfilment of (B) and (B_1) is determined by the ordinary differential operator being in the right side of the above equation. For (4.11) the fulfilment of each of these conditions depends only on the properties of $a, b(x, \xi, \eta)$ and, therefore, does not depend on the choice of sequence $\beta_n(s)$ or on the index n.

Moreover, when (B) holds, then property (3.3) for the first integral of the ordinary differential equation from the right side of equations (4.11) remains valid. Therefore, the following theorem holds.

Theorem 4.3. Let solutions $u^n(x, t) \in C^{2,1}(Q_T)$ of problems (4.11), (4.2) be bounded uniformly by n : $\sup_{x,t} |u^n(x, t)| \leq M$ and the condition (B) hold.

Then the derivative is bounded uniformly by n.

$$\sup_{x,t} |u_x^n(x,t)| \le M_1.$$

Remark 4.4.   Here we shall not consider the possibility of passing to the limit for $n \to \infty$ and existence theorems.

3. Asymptotically normal parabolic equations.   This term we shall use for the problem

$$u_t = a''(u_x)u_{xx} + 2\mu u u_x \ , \quad (x,t) \in Q_T$$

$$u(0,t) = u(1,t) = 0 \ , \quad u(x,0) = u_0(x), \tag{4.13}$$

where μ =const, $a(s) \in C^4$ and its second derivative is positive only for large values of the argument $a''(s) \ge a_0 > 0$ for $|s| > N$ and may change its sign for $|s| < N$. The simplest analytical example is the polynomial $a(s) = \alpha_0 s^4 - \alpha_1 s^2$, $\alpha_i > 0$.

The systematic investigation of such equations, called equations of mixed type, was pioneered in the 1970s years by N. N. Yanenko (Yanenko and Novikov, 1973; Zelenyak et al., 1974). The physical bakground, analytical investigations, the results of numerical research, the bibliography and further investigation may be found in Lar'kin et al. (1983). We shall give here only some results obtained by application of the generalized Liapunov functionals (Zelenyak et al., 1974, Lavrentiev, 1989, 1990, 1993).

Remark 4.5.   Note, that in spite of the rather strong a priori estimates of solutions (Lavrentiev, 1989, 1990, 1993) nowadays we only know the existence of so-called measure-valued solutions (Slemrood, 1991). More smooth solutions may be constructed (Lavrentiev, 1987), but the description of corresponding initial data is not satisfactory. Starting from the problem (4.13), we may formulate the questions even for linear parabolic problems, the answers to which are not known now.

The first a priori estimates may be obtained relatively simply by multiplying the equation by functions u and u_t with further integrating.

So, multiplying equation (4.13) by $u(x,t)$ and integrating it by x, we have

$$\frac{1}{2}\frac{d}{dt}\int_0^1 u^2 dx = \int_0^1 u u_t dx = \int_0^1 u \cdot \frac{d}{dx}(a'(u_x) + \mu u^2)dx$$

$$= -\int_0^1 a'(u_x)u_x dx. \qquad (4.14)$$

The restrictions imposed on the function $a(s)$ guarantee that

$$a'(s) \cdot s \geq \frac{a_0}{2}s^2 - K \quad, \quad a(s) \geq \frac{a_0}{4}s^2 - K \qquad (4.15)$$

with a certain constant K depending on N and the interval of negativeness of $a''(s)$. The equality (4.14) and (4.15) yield the estimate

$$\int_0^1 u^2 dx + \int_0^t \int_0^1 u_x^2 dx dt \leq K_1. \qquad (4.16)$$

Multiplying (4.13) by u_t we have

$$\int_0^1 u_t^2 dx = \int_0^1 \frac{d}{dx}a'(u_x)u_t dx + 2\mu \int_0^1 u u_x u_t dx$$

$$= -\int_0^1 a'(u_x)u_{xt}dx + 2\mu \int_0^1 u u_x u_t dx$$

$$\leq -\frac{d}{dt}\int_0^1 a(u_x)dx + \frac{1}{2}\int_0^1 u_t^2 dx + 2\mu^2 \int_0^1 u^2 u_x^2 dx.$$

Hence, (4.15) and the estimate $\int_0^1 u^2 dx \leq \int_0^1 u_x^2 dx$ yield the inequality

$$\int_0^t \int_0^1 u_t^2 dx dt + \int_0^1 u_x^2 dx \leq K_2 \left(1 + \int_0^t \left(\int_0^1 u_x^2 dx \right)^2 dt \right), \qquad (4.17)$$

where the constant K_2 depends on initial data and on a_0, K, μ.

Applying the Gronuall lemma and taking into account (4.16), we obtain that

$$\sup_x |u(x,t)| \le \int_0^1 u_x^2 dx \le K_3$$

and, finally

$$\int_0^t \int_0^1 (u_t^2 + a''^2 u_{xx}^2) dx dt + \sup_t \left(\int_0^1 u_x^2 dx + \sup_x |u| \right) \le K_4. \qquad (4.18)$$

Using generalized Liapunov functionals, we shall now establish *a priori* boundedness of the solutions derivative (Zelenyak *et al.*, 1974).

Theorem 4.4.   Let $u(x,t) \in C^{2,1}(Q_T)$ solve (4.13). Then

$$\sup_{x,t} |u_x(x,t)| \le M_1. \qquad (4.19)$$

Proof.   Note, that because the right side of (4.13) is divergent, the first integral of the ordinary differential equation $a''(y')y'' + 2\mu y y' = 0$ (the right side of (4.13)) is the function $a'(y') + \mu y^2$. Not giving all the heuristic arguments, choose as the not separated from zero variational factor the following function

$$\rho_n(u,v) = \begin{cases} \frac{1}{v}(a'(v) + \mu u^2)^{2n+1}, & \text{for} \quad |a'(v) + \mu u^2| \ge M + 1 \\ \\ 0, & \text{for} \quad |a'(v) + \mu u^2| \le M, \end{cases} \qquad (4.20)$$

where

$$M = \sup_{u^2 \le K_4^2, |v| < N} \left| a'(v) + \mu u^2 \right|,$$

the parameter N determines the interval of alternating of the derivative $a''(v)$, K_4 is the constant from (4.18). For the remaining arguments $|a'(v) + \mu u^2| \in (M, M+1)$ we determine the function $\rho_n(u,v)$ so that the products $\tilde{\rho}_n(a'(v) + \mu u^2) = v \cdot \rho_n(u, v)$ are monotone and infinitely differentiable functions.

According to (1.15) we set

$$\Phi_n(u, v) = \int_0^v (v - \eta) a''(\eta) \rho_n(u, \eta) d\eta \qquad (4.21)$$

and verify that the pairs $\rho_n, \Phi_n(u, v)$ from (4.20), (4.21) generate the generalized (i.e., not strictly monotone) Liapunov functional on smooth solutions of (4.13). Really

$$\frac{d}{dt} J_n(u) = \frac{d}{dt} \int_0^1 \Phi_n(u, u_x) dx = \int_0^1 \left(\frac{\partial \Phi_n}{\partial u} - \frac{d}{dx} \frac{\partial \Phi_n}{\partial u_x} \right) u_t dx$$

$$= \int_0^1 \left[\int_0^{u_x} (u_x - \eta) a''(\eta) \frac{\partial \rho_n}{\partial u} d\eta \right.$$

$$\left. - u_x \int_0^{u_x} a''(\eta) \frac{\partial \rho_n}{\partial u} d\eta - u_{xx} a''(u_x) \rho_n(u, u_x) \right] u_t dx$$

$$= \int_0^1 \left[- \int_0^{u_x} \eta a''(\eta) \frac{\partial \rho_n}{\partial u} d\eta - u_{xx} a''(u_x) \rho(u, u_x) \right] u_t dx.$$

Using the function determined above $\tilde{\rho}_n(a'(v) + \mu u^2) = v \cdot \rho_n(u, v)$, we obtain

$$\frac{dJ_n}{dt} = \int_0^1 u_t \left[- \int_0^{u_x} a''(\eta) \cdot 2\mu u \tilde{\rho}'_n(a'(\eta) + \mu u^2) d\eta \right.$$

$$\left. - a''(u_x) u_{xx} \cdot \rho_n(u, u_x) \right] dx$$

$$= \int_0^1 u_t \left[- \int_0^{u_x} 2\mu u \frac{d}{d\eta} \tilde{\rho}_n(a'(\eta) + \mu u^2) d\eta - a'' u_{xx} \cdot \rho_n(u, u_x) \right] dx$$

$$= \int_0^1 \left[-2\mu u u_x - a''(u_x) u_{xx} \right] \rho_n(u, u_x) u_t dx = - \int_0^1 \rho_n(u, u_x) u_t^2 dx.$$

$$(4.22)$$

Because of the nonnegativity of ρ_n (4.20) the analog (4.22) of equality (1.7) denotes that

$$J_n(u(x,t)) \leq J_n(u_0(x)). \tag{4.23}$$

Estimate the left side of (4.23) from below. For each $t > 0$ represent $[0,1] = \bigcup_{i=1}^{3} E_i(t)$, where

$$E_1(t) = \{x \in [0,1]: \quad u_x(x,t) < -N_1\},$$

$$E_2(t) = \{x \in [0,1]: \quad u_x(x,t) > N_1\},$$

$$E_3(t) = \{x \in [0,1]: \quad |u_x(x,t)| \leq N_1\};$$

the number N_1 is chosen so that $|a'(u_x) + \mu u^2| \geq M + 1$ (see (4.20)) for $|u_x| > N_1$. Then

$$J_n(u) = \int_{E_1(t)} \Phi_n dx + \int_{E_2(t)} \Phi_n dx + \int_{E_3(t)} \Phi_n dx. \tag{4.24}$$

The third term of the right side of (4.24), evidently, is bounded

$$\left| \int_{E_3(t)} \Phi_n(u, u_x) dx \right| \leq K_5^{2n+2}. \tag{4.25}$$

Consider the first term

$$\int_{E_1(t)} \Phi_n dx = \int_{E_1} \left[-\int_{u_x}^{-N_1} (u_x - \eta) a''(\eta) \frac{(a'(\eta) + \mu u^2)^{2n+1}}{\eta} d\eta \right] dx$$

$$\geq K_6^{2n+1} \int_{E_1(t)} \int_{u_x}^{-N_1} (u_x - \eta)\eta^{2n} d\eta \tag{4.26}$$

(here we have used that $a''(\eta) \geq a_0 > 0$ for $\eta < -N_1$).

The analogous estimate

$$\int_{E_1(t)} \Phi_n \, dx \geq K_7^{2n+1} \int_{E_2(t)} \int_{N_1}^{u_x} (u_x - \eta)\eta^{2n} d\eta \qquad (4.27)$$

holds also for the remaining term of the right side of (4.24).

The integrals in the right sides of (4.26), (4.27) may be estimated similarly. Now let us give the computations for $E_2(t)$

$$\int_{E_2(t)} \int_{N_1}^{u_x} (u_x - \eta)\eta^{2n} d\eta \, dx$$

$$= \int_{E_2(t)} \left[\frac{u_x^{2n+2}}{(2n+1)(2n+2)} + \frac{N_1^{2n+2}}{(2n+2)} - \frac{u_x N_1^{2n+1}}{(2n+1)} \right] dx$$

$$\geq \int_{E_2(t)} \left[\frac{u_x^{2n+2}}{2(2n+1)(2n+2)} - \frac{N_1^{2n+2}}{(2n+2)} \right] dx. \qquad (4.28)$$

The latter inequality we obtained using the Young inequality. Combining (4.24) - (4.28), we obtain that

$$J_n(u(x,t)) \geq \frac{1}{n^2} K_8^{2n+1} \int_0^1 u_x^{2n+2} dx - K_9^{2n+2}, \qquad (4.29)$$

where the numbers K_i do not depend on n. Estimating from above $J_n(u_0(x))$ and using (4.23), (4.29), we may show that

$$\int_0^1 u_x^{2n+2} dx \leq n^2 K_{10}^{2n+2},$$

whence, after extracting the $2n + 2$ degree root and going to the limit for $n \to \infty$, we obtain the statement of the theorem.

Remark 4.6. The results of theorem 4.4 is generalized on the approximation of equation (4.13), obtained by the change of differential operators with respect to x by finite differences for $\mu = 0$ (Lavrentiev, 1982). Namely, consider the system

$$\frac{d}{dt} u_i(t) = \frac{a'(\frac{u_{i+1}-u_i}{h}) - a'(\frac{u_i-u_{i-1}}{h})}{h} \; ,$$

$$u_0(t) = u_{M+1}(t) = 0 \; , \quad u_i(0) = v_i \; ,$$

where $i = 1, \ldots, M$, $u_i(t) = u(ih, t)$, $h = \frac{1}{M+1}$, $v_i = u_0(ih)$.

It is possible to show its solvability and the validity of the uniform by $h > 0$ solution estimate

$$\int_0^t \sum_{i=1}^M (u_i')^2 dt + \sup_{t,i} \left(|u_i| + \left| \frac{u_{i+1} - u_i}{h} \right| \right) \le K_{11}.$$

§2.5   Liapunov functionals of the second order.

The order of a Liapunov functional is defined as the order of the highest derivative on which $J(u) = \int_0^1 \Phi dx$ depends. Up to now, we have talked about first order functionals, apposite for investigating quasilinear equations in general. In the theory of multidimensional problems considered in later chapters, the criterion of spectrum nonnegativity of a linearized problem has been formulated in terms of the existence of a zero-order functional. By second order functionals we mean expressions such as

$$J(u) = \int_0^1 \Phi(x, u, u_x, u_{xx}) dx. \tag{5.1}$$

First they were constructed for investigating the problem of stabilizing (existence of the limit for $t \to \infty$) bounded solutions (4.1) (Kvasova, 1973). Nevertheless, the use of generalized form (4.6) allows us to solve this problem using only the first order functionals discussed previously. In the next paragraph we shall apply functionals (1.5) to establish *a priori* estimates for second derivatives of solutions.

Definition 5.1. We shall say that for equation (4.1) the condition (B_2) holds if the solution $\varphi(x_0, x, y_0, y_1, y_2)$ of the Cauchy problem

$$\frac{d}{dt}C(x, y, y', y'') = 0 \quad , \quad y(x_0) = y_0 \quad , \quad y'(x_0) = y_1 \quad , \quad y''(x_0) = y_2$$
(5.2)

is determined and regular in the interval $x \in [0, 1]$ for all $(x_0, y_0, y_1, y_2) \in [0, 1] \times R^3$.

Remark 5.1. The condition (B_2) is the analog of the condition (B). We could analogously formulate the generalization of the condition (B_1) in the case of equation (4.1), but we shall not do this.

Remark 5.2. It is easy to see that restrictions (4.8) are sufficient for fulfilment of (B_2).

Theorem 5.1. Let the condition (B_2) hold for (4.1), and the boundary conditions have the property

$$(-1)^i |\alpha_i| \psi_i'(\xi) \geq 0 \ .$$
(5.3)

Then such functions $\rho, \Phi(x, \xi, \eta, \lambda) \in C^2$, exist that for solution $u(x, t) \in C^{2,1}(Q_T)$, $u_{xxx} \in C(Q_T)$, $u_{xt} \in C(Q_T)$ of (4.1), (4.2) the following inequality

$$\int_0^1 \Phi(x, u, u_x, u_{xx})dx - \int_0^1 \Phi(x, u_0, u_0', u_0'')dx$$

$$\leq \int_0^t \int_0^1 \frac{\partial \Phi}{\partial u} u_t dx dt - \int_0^t \int_0^1 \rho(x, u, u_x, u_{xx}) u_{xt}^2 dx dt$$
(5.4)

holds.

Remark 5.3. Evidently, in the cases $\alpha_i = 0$ (the Dirichlet boundary condition $u|_{x=i} = A_i$) and $\psi_i(\xi) \equiv \text{const}$ (the Neuman boundary condition $u_x|_{x=i} = B_i$) inequalities (5.3) hold. They impose the restrictions on the third boundary-value problem.

Proof.   The arguments are similar to the proof of theorem 1.1 but there are some differences. We shall give this proof completely.

Assuming that all the obtained derivatives are defined, let us differentiate the left side of inequality (5.4) by t

$$\frac{d}{dt}\int_0^1 \Phi dx = \int_0^1 \left(\frac{\partial \Phi}{\partial u} u_t + \frac{\partial \Phi}{\partial u_x} u_{xt} + \frac{\partial \Phi}{\partial u_{xx}} u_{xxt} \right) dx$$

$$= \int_0^1 \frac{\partial \Phi}{\partial u} u_t dx - \int_0^1 \left(\frac{d}{dx}\frac{\partial \Phi}{\partial u_{xx}} - \frac{\partial \Phi}{\partial u_x} \right) u_{xt} dx + \frac{\partial \Phi}{\partial u_{xx}} u_{xt}|_{x=0}^1 . \qquad (5.5)$$

The smoothness assumptions of the theorem do not allow us to write down the middle part of equalities (5.5). However, if we consider a sequence of functions from $C^3(Q_T)$, which do not solve equation (4.1) but approximate the solution of (4.1), (4.2) in the space indicated under the conditions of the theorem, after integrating by t and passing to the limit, we obtain that

$$\int_0^1 \Phi dx - \int_0^1 \Phi|_{t=0} dx = \int_0^t \int_0^1 \frac{\partial \Phi}{\partial u} u_t dx dt$$

$$- \int_0^t \int_0^1 \left(\frac{d}{dx}\frac{\partial \Phi}{\partial u_{xx}} - \frac{\partial \Phi}{\partial u_x} \right) u_{xt} dx dt + \int_0^t \frac{\partial \Phi}{\partial u_{xx}} u_{xt}|_{x=0}^1 dt \qquad (5.6)$$

already for the solution of problem (4.1), (4.2).

Let us compute the derivate $u_{xt} = C_{u_{xx}} u_{xxx} + C_{u_x} u_{xx} + C_u u_x + C_x$ and set

$$f(x, \xi, \eta, \zeta) = C_\eta \zeta + C_\xi \eta + C_x(x, \xi, \eta, \zeta). \qquad (5.7)$$

By (5.6), (5,7), in order that (5.4) holds, it suffices to demand that

$$\frac{\partial^2 \Phi}{\partial u_{xx}^2} u_{xxx} + \frac{\partial^2 \Phi}{\partial u_x \partial u_{xx}} u_{xx} + \frac{\partial^2 \Phi}{\partial u \partial u_{xx}} u_x + \frac{\partial^2 \Phi}{\partial x \partial u_{xx}}$$

$$-\frac{\partial \Phi}{\partial u_x} = \rho(x, u, u_x, u_{xx})(C_{u_{xx}} u_{xxx} + f(x, u, u_x, u_{xx})), \qquad (5.8)$$

$$\frac{\partial \Phi}{\partial u_{xx}} u_{xt}\big|_{x=0}^{1} \le 0. \qquad (5.9)$$

Equate in (5.8) the coefficients of the same powers of u_{xxx} treating the arguments of the functions ρ, Φ as independent variables

$$\frac{\partial^2 \Phi}{\partial \zeta^2} = \rho C_\zeta(x, \xi, \eta, \zeta), \qquad (5.10)$$

$$\frac{\partial^2 \Phi}{\partial \eta \partial \zeta} \zeta + \frac{\partial^2 \Phi}{\partial \xi \partial \zeta} \eta + \frac{\partial^2 \Phi}{\partial x \partial \zeta} - \frac{\partial \Phi}{\partial \eta} = \rho f(x, \xi, \eta, \zeta), \qquad (5.11)$$

where, recall, the function f is defined by equality (5.7).

Using the solution g of equation (4.4), we may write the following solution of equation (5.10)

$$\Phi(x, \xi, \eta, \zeta) = \int_{g(x,\xi,\eta)}^{\zeta} (\zeta - \tau)\rho C_\tau(x, \xi, \eta, \tau) d\tau. \qquad (5.12)$$

Assume, in addition, that

$$\rho(x, \xi, \eta, g(x, \xi, \eta)) = 0, \qquad (5.13)$$

and substitute expression (5.12) for Φ into (5.11). By assumption (5.13), we have

$$\int_{g(x,\xi,\eta)}^{\zeta} [\zeta(\rho C_\tau)_\eta + \eta(\rho C_\tau)_\xi + (\rho C_\tau)_x$$

$$-(\zeta - \tau)(\rho C_\tau)_\eta] \, d\tau = \rho f(x, \xi, \eta, \zeta). \qquad (5.14)$$

After differentiating equation (5.14) by ζ and doing some transformations by ρ, we obtain the equation

$$\zeta(\rho C_\zeta)_\eta + \eta(\rho C_\zeta)_\xi + (\rho C_\zeta)_x - (\rho f)_\zeta = 0,$$

which, taking into account (5.7), may be represented as

$$- f\rho_\zeta + \zeta C_\zeta \rho_\eta + \eta C_\zeta \rho_\xi + C_\zeta \rho_x - C_\eta \rho = 0. \qquad (5.15)$$

The system of characteristics for this equation is as follows

$$\frac{dx}{C_\zeta} = \frac{d\xi}{\eta C_\zeta} = \frac{d\eta}{\zeta C_\zeta} = \frac{d\zeta}{-f} = \frac{d\rho}{C_\eta \rho}.$$

Inequality (4.3) allows us to rewrite this system in the normal form

$$\frac{d\xi}{dx} = \eta, \quad \frac{d\eta}{dx} = \zeta, \quad \frac{d\zeta}{dx} = -\frac{f}{C_\zeta}, \quad \frac{d\rho}{dx} = \frac{C_\eta}{C_\zeta}\rho. \qquad (5.16)$$

The system of the three first equations of (5.16) is equivalent to third order equation (5.2). The solution of the Cauchy problem (5.2) (determined by assumption for all $x, x_0 \in [0,1]$, $y_i \in R$) we shall denote by $\varphi(x_0, x, y_0, y_1, y_2)$.

Set

$$F(x_0, x, y_0, y_1, y_2) = \left.\frac{C_\eta(x, \xi, \eta, \zeta)}{C_\zeta(x, \xi, \eta, \zeta)}\right|_{\substack{\xi=\varphi(x_0,x,y_0,y_1,y_2), \\ \eta=\varphi_x(x_0,x,y_0,y_1,y_2), \\ \zeta=\varphi_{xx}(x_0,x,y_0,y_1,y_2).}} ,$$

and consider the solution of the last equation of system (5.16)

$$\rho = \rho_0 \exp\left\{\int_{x_0}^{x} F(x_0, \tau, y_0, y_1, y_2)d\tau\right\}.$$

Then a solution of (5.15) will be any function of the form

$$\rho(x, \xi, \eta, \zeta) = \Psi(x, \xi, \eta, \zeta)\rho_1, \qquad (5.17)$$

where Ψ is the first integral of (5.2), and ρ_1 is defined as follows

$$\rho_1(x,\xi,\eta,\zeta) = e^{\int\limits_{x_0}^{x} F d\tau} \Bigg|_{\substack{y_0 = \varphi(x,x_0,\xi,\eta,\zeta), \\ y_1 = \varphi_{x_0}(x,x_0,\xi,\eta,\zeta), \\ y_2 = \varphi_{x_0 x_0}(x,x_0,\xi,\eta,\zeta).}} \tag{5.18}$$

Set in (5.17), $\Psi = C^2(x,\xi,\eta,\zeta)\Psi_1$ (the function C is the integral of equation (5.2)). Then the function ρ by (4.4), (5.15), (5.15) will solve equation (5.14).

Thus, constructed by formulae (5.17), (5.12) the functions ρ, Φ will satisfy the equations (5.10), (5.11), where $\rho \geq 0$ if $\Psi_1 \geq 0$.

To complete the lemma proof, we need to verify equalities (5.9). We show that the following estimates

$$(-1^i)\frac{\partial\Phi}{\partial u_{xx}}\,u_{xt}\big|_{x=i} \geq 0 \tag{5.19}$$

hold. Choose one of $i = 0,1$ and demonstrate the realization of corresponding inequality from (5.19) for each type of boundary conditions.

If $\alpha_i = 0$, then, as was mentioned above, the boundary condition is $u\big|_{x=i} = A_i$. Then, taking into acount the assumptions on smoothness of a solution of (4.1), it follows that $C(x,u,u_x,u_{xx})\big|_{x=i} = 0$ or (by (4.4)) $u_{xx}\big|_{x=i} = g(x,u,u_x)\big|_{x=i}$. This means that the function Φ, constructed by formula (5.12), has the property $\frac{\partial\Phi}{\partial u_{xx}}\big|_{x=i} = 0$, i.e., the estimate (5.19) holds.

If $\alpha_i \neq 0$, then $u_{xt}\big|_{x=i} = \frac{\psi_i'(u)u_t}{\alpha_i}\big|_{x=i}$. In this case the formulae (4.1), (5.3), (5.12) yield

$$(-1^i)\frac{\partial\Phi}{\partial u_{xx}}\,u_{xt}\big|_{x=i} = (-1)^i\psi_i'(u)C\int\limits_{g(x,u,u_x)}^{u_{xx}} \rho C_\tau d\tau \geq 0,$$

since $\rho \geq 0$, $C_\tau \geq 0$. As it follows from (4.4), $C(x,u,u_x,u_{xx})(u_{xx} - g(x,u,u_x)) \geq 0$. The theorem is proved.

§2.6   *A priori* estimates of the second derivative.

As was noted by Kruzhkov (1967, 1972) for the quasilinear equation (1.1) with smooth coefficients from the known estimate of $\sup_{x,t}(|u| + |u_x|)$ for the mixed problem, it follows that the solution is bounded by the norm of the space $H^{2+\alpha}$. In the case of the more general equation (4.1) it is necessary to estimate also $\sup_{x,t}|u_{xx}(x,t)|$. After this it is easy to establish the finiteness of the Hölder norms of higher derivatives and to prove the existence theorem.

In this paragraph we shall use the set of Liapunov functionals of the second order to prove the boundedness of the second order derivative.

Theorem 6.1.   Let $u(x,t) \in C^{2,1}(Q_T)$, $u_{xxx}, u_{xt} \in C(Q_T)$ solve (4.1), (4.2); $C(x,\xi,\eta,\zeta)$, $u_0(x) \in C^3$, inequalities (4.3), (5.3) and the estimates

$$\sup_{x,t}(|u(x,t)| + |u_x(x,t)|) \leq M, \tag{6.1}$$

$$\| u_0(x) \|_2 \ \leq M_0 \tag{6.2}$$

hold. Let, besides, for equation (4.1) the condition (B_2) (see definition 5.1) and for $0 \leq x \leq 1$, $|u| + |u_x| \leq M$, $u_{xx} \in R$ the estimate

$$\frac{\partial C(x,u,u_x,u_{xx})}{\partial u} \leq N \tag{6.3}$$

hold. Then

$$\sup_{x,t}(|u_{xx}(x,t)| \leq M_1(M, M_0, N, T); \tag{6.4}$$

besides, if in (6.3) we set $N = 0$, then M_1 does not depend on T.

We preface to the proof of the theorem some auxiliary constructions. Consider a compact set $K \in R^n$ and two functions $f, \varphi(x,y) \in C(K \times R)$ with the following properties

    1) $f_y(x,y) \in C(K \times R)$, $f_y \geq \delta(y) > 0$,

2) $|f(x,y)| \to \infty$ for $|y| \to \infty$ uniformly for $x \in K$,

3) $\varphi(x,y) > 0$.

By the properties 1, 2 there exists a unique function $h(x) \in C(K)$ which satisfies the equality

$$f(x, h(x)) = 0. \tag{6.5}$$

Choosing, in accordance with property 2, the number C_0 so that for $|y| \geq C_0$, $x \in K$ we have the inequality $|f(x,y)| \geq 1$, we see that for some $\varepsilon > 0$ the estimate

$$|h(x)| < C_0 - \varepsilon_0 \tag{6.6}$$

holds. By the mean value theorem, for $|y| \geq C_0$, there exist such functions $\theta_n(x,y)$ that

$$f(x, \theta_n(x,y)) \int_{h(x)}^{y} (y - \xi) f^{n-1}(x, \xi) \varphi(x, \xi) d\xi = \int_{h(x)}^{y} (y - \xi) f^n(x, \xi) \varphi d\xi.$$

$$\tag{6.7}$$

Lemma 6.2. Under the above assumptions for each $C_1 > C_0$ we have

$$\lim_{n \to \infty} \theta_n(x,y) = y, \tag{6.8}$$

and, moreover, convergence is uniform for $x \in K$, $C_0 \leq |y| \leq C_1$.

Proof. For the case $C_0 \leq y \leq C_1$ (the case $-C_1 \leq y \leq -C_0$ is proved analogously), let us fix an arbitrary $\varepsilon \in (0, \varepsilon_0)$. The relations (6.5), (6.6) and the function f properties yield $0 < f(x, y-\varepsilon) < f(x, y - \frac{\varepsilon}{2}) < f(x,y)$ and the esimate

$$h(x) \leq \theta_n(x,y) \leq y. \tag{6.9}$$

for the function $\theta_n(x,y)$ from equalities (6.7).

Set

$$\lambda(\varepsilon) = \frac{1}{2} \inf_{x \in K, C_0 \leq y \leq C_1} \frac{f(x, y - \frac{\varepsilon}{2}) - f(x, y - \varepsilon)}{f(x, y)}.$$

By the function f properties, $\lambda(\varepsilon) > 0$. For a given λ we may choose such n_0, that for $n > n_0$, independently on $x \in K$, $C_0 \leq y \leq C_1$, the relation

$$f^{n-1}(x, y - \frac{\varepsilon}{2}) \int_{h(x)}^{y - \frac{\varepsilon}{2}} (y - \xi)\varphi d\xi \leq \lambda f^{n-1}(x, y - \frac{\varepsilon}{4}) \int_{y - \frac{\varepsilon}{4}}^{y} (y - \xi)\varphi d\xi.$$

will hold true.

Write for these values of n the following chain of inequalities.

$$\int_{h(x)}^{y - \frac{\varepsilon}{2}} (y - \xi)f^{n-1}\varphi d\xi \leq f^{n-1}(x, y - \frac{\varepsilon}{2}) \int_{h(x)}^{y - \frac{\varepsilon}{2}} (y - \xi)\varphi d\xi$$

$$\leq \lambda f^{n-1}(x, y - \frac{\varepsilon}{4}) \int_{y - \frac{\varepsilon}{4}}^{y} (y - \xi)\varphi d\xi < \frac{f(x, y - \frac{\varepsilon}{2}) - f(x, y - \varepsilon)}{f(x, y)}$$

$$\times \int_{y - \frac{\varepsilon}{2}}^{y} (y - \xi)f^{n-1}\varphi d\xi,$$

from which it follows that

$$\int_{h(x)}^{y} (y - \xi)f^{n-1}\varphi d\xi$$

$$\leq \left[\frac{f(x, y - \frac{\varepsilon}{2}) - f(x, y - \varepsilon)}{f(x, y)} + 1 \right] \int_{y - \frac{\varepsilon}{2}}^{y} (y - \xi)f^{n-1}\varphi d\xi. \qquad (6.10)$$

Monotonicity and positivity of the function $f(x, y)$ for $y > h(x)$ together with estimate (6.6) guarantee that

$$\int_{h(x)}^{y} (y - \xi) f^n \varphi d\xi \geq f(x, y - \frac{\varepsilon}{2}) \int_{y - \frac{\varepsilon}{2}}^{y} (y - \xi) f^{n-1} \varphi d\xi.$$

Using this estimate, definition (6.7) of the functions θ_n, inequality (6.10), and cancelling by a positive integral factor, we obtain

$$f(x, y - \frac{\varepsilon}{2}) \leq f(x, \theta_n(x, y)) \left[1 + \frac{f(x, y - \frac{\varepsilon}{2}) - f(x, y - \varepsilon)}{f(x, y)} \right]$$

$$\leq f(x, \theta_n) + f(x, y - \frac{\varepsilon}{2}) - f(x, y - \varepsilon). \tag{6.11}$$

In the last of the inequalities in (6.11) we have used the fact that by (6.9) $f(x, \theta_n) \leq f(x, y)$. As by (6.11) $f(x, y - \varepsilon) \leq f(x, \theta_n)$, then by the function f properties $\theta_n(x, y) \geq y - \varepsilon$, i.e., (6.8) holds, since ε is arbitrary. The lemma is proved.

Proof of theorem 6.1. Use the result of theorem 5.1. Set in (5.17) $\Psi = C^{2n}$ (as was noted, the function C is the integral of equation (5.2)) and denote the function obtained by the formulae (5.12), (5.17), (5.18) by ρ_n, Φ_n. Then, taking into account (6.2), we obtain the estimates

$$0 \leq \Phi_n(x, u_0, u_0', u_0'') \leq K_1^{2n+1}(M_0), \tag{6.12}$$

where K_1 does not depend on n. Inequality (5.4) and estimate (6.12) yield the relation

$$\int_0^1 \Phi_n dx \leq K_1^{2n+1} + \int_0^t \int_0^1 \frac{\partial \Phi_n}{\partial u} u_t dx dt. \tag{6.13}$$

Taking into account (4.4), (5.12), (5.17), (5.18), and (4.1), we may write

$$\frac{\partial \Phi_n}{\partial u} u_t = C \cdot \left[2n \int_{g(x,u,u_x)}^{u_{xx}} (u_{xx} - \tau) C^{2n-1} \frac{\partial C}{\partial u} \rho_2 d\tau \right.$$

$$+ \int_{g(x,u,u_x)}^{u_{xx}} (u_{xx} - \tau)C^{2n}\frac{\partial \rho_2}{\partial u}d\tau \Bigg], \qquad (6.14)$$

where $\rho_2(x, u, u_x, \tau) = \rho_1 C_\tau(x, u, u_x, \tau) > 0$. Now obtain the special estimate for the terms from the right side of (6.14). By inequality (6.4), we obtain

$$C(x, u, u_x, u_{xx}) \int_{g(x,u,u_x)}^{u_{xx}} (u_{xx} - \tau)C^{2n}\frac{\partial \rho_2}{\partial u}d\tau \leq K_2(M_1)\Phi_n, \qquad (6.15)$$

where $K_2 = \sup(C\frac{\partial \rho_2}{\partial u}\rho_2^{-1})$. The supremum is calculated for $0 \leq x \leq 1$ with the restrictions (6.1), (6.4) on the functions u, u_x, u_{xx}. Note, that K_2 depends on an unknown value M_1 but does not depend on n.

Consider a compact set $K = \{(x, \xi, \eta) : 0 \leq x \leq 1, \ |\xi| + |\eta| \leq M\}$ and introduce the notation

$$\Omega(t) = \{x : |C(x, u(x, t), u_x(x, t), u_{xx}(x, t))| > 1,$$

$$|u_{xx}(x, t)| > \sup(|g(x, \xi, \eta)| + 1), \ (x, \xi, \eta) \in K\}.$$

If for each $t \in [0, T]$ we have either $\Omega(t) = \emptyset$, or $\mathrm{mes}\Omega(t) = 0$, then, by the parabolicity condition (4.3) of problem (4.1), the estimate (6.4) is verified, and, therefore, the theorem is proved.

Let, for a given t, the set $\Omega(t)$ be nonempty and $\mathrm{mes}\Omega(t) \neq 0$. For $x \in \Omega(t)$ consider the expression

$$\Lambda_n = \frac{C\int_{g(x,u,u_x)}^{u_{xx}} (u_{xx} - \tau)C^{2n-1}\rho_2 d\tau}{\int_{g(x,u,u_x)}^{u_{xx}} (u_{xx} - \tau)C^{2n}\rho_2 d\tau}. \qquad (6.16)$$

Then, by the assumption (6.3) for these x the following estimate holds

$$J_n = C\int_{g(x,u,u_x)}^{u_{xx}} (u_{xx} - \tau)C^{2n-1}\frac{\partial C}{\partial u}\rho_2 d\tau \leq N\Lambda_n\Phi_n. \qquad (6.17)$$

Note, that since $\rho_2 > 0$, the relations (4.3), (4.4) yield the positiveness of Λ_n, Φ_n.

On the basis of the mean value theorem, there exist such functions $\theta_n(x, u(x,t), u_x(x,t), u_{xx}(x,t))$ that

$$C(x, u, u_x, \theta_n) \int_{g(x,u,u_x)}^{u_{xx}} (u_{xx} - \tau)C^{2n-1}\rho_2 d\tau = \int_{g(x,u,u_x)}^{u_{xx}} (u_{xx} - \tau)C^{2n}\rho_2 d\tau,$$

where the properties of C, g guarantee that $|\theta_n| \leq |u_{xx}|$.

We see that the conditions of lemma 6.2 hold, and, therefore, if (6.4) holds, then $\theta_n \to u_{xx}$ for $n \to \infty$ uniformly for $(x,t) \in \{x \in \Omega(t),$ $0 \leq t \leq T\}$. Therefore, there exists such n_0 (depending on M_1) that for $n > n_0$ we have $\Lambda_n < 2$. So, from (6.17) it follows that

$$\int_0^1 J_n dx = \int_{\Omega(t)} J_n dx + \int_{[0,1]\backslash\Omega(t)} J_n dx \leq 2N \int_0^1 \Phi_n dx + K_3^{2n+1}(M), \quad (6.18)$$

where K_3 depends on $\sup |u_{xx}|$ in $[0,1] \backslash \Omega(t)$ and does not depend on n.

Combining (6.18), (6.15), (6.13), we obtain

$$\int_0^1 \Phi_n dx \leq K_4 K_5^{2n} + (4nN + K_2) \int_0^t \int_0^1 \Phi_n dx dt, \quad (6.19)$$

where $K_5 = \max\{K_1, K_3\}$, K_2 is the constant from (6.8), K_i does not depend on n.

Applying to (6.19) the Gronuall lemma, we obtain

$$\int_0^1 \Phi_n dx \leq K_4 K_5^{2n} \left(1 + e^{(4nN+K_2)T}\right).$$

This inequality together with (6.1), (6.4), (5.12), (5.17) yield

$$\int_0^1 C^{2n+2}(x, u, u_x, u_{xx}) dx \leq (2n+2)^2 K_6(M, M_1, T) K_5^{2n} e^{4NnT}, \quad (6.20)$$

where K_6 does not depend on n. As far as the solution of (4.1), (4.2) is assumed to be sufficiently smooth, after extracting from both sides of (6.20) the root of the degree $2n + 2$ and passing to the limit $n \to \infty$, we obtain the estimate

$$\sup_{(x,t)\in Q_T} |C(x, u, u_x, u_{xx})| \leq K_5 e^{2NT}. \tag{6.21}$$

This inequality in accordance with (4.3), (6,1) denotes that the theorem is proved. Note, that if we set in estimate (6.3) $N = 0$, then $\sup |u_{xx}|$ may be estimated independently of T.

Remark 6.1. The condition (B_2) does not lose generality in the conditions of theorem 6.1. Really, the function C is the integral of equation (5.2); therefore, by (4.3), the problem (5.2) is equivalent to the following Cauchy problem

$$y'' = g_B(x, y, y'), \quad y(x_0) = y_0, \quad y'(x_0) = y_1, \tag{6.22}$$

where the functions g_B may be determined from the equations

$$C(x, y, y', g_B(x, y, y')) = B. \tag{6.23}$$

Here $B = C(x_0, y_0, y_1, y_2)$. Set $B = 0$ and determine the function g_0 so, that $g_0(x, y, y') = 0$ for $|y| + |y'| > M + 1$. The values of g_0 for $|y| + |y'| \leq M$ we shall not change. We shall change the function C so that (6.23) holds for $B = 0$, and, besides, $C(x, y, y', g_0) = g_0$ for $|y| + |y'| > M + 2$. Require also that the condition of the type (4.3) holds, i.e., $\frac{\partial}{\partial g}C(x, y, y', g) \geq \delta_0$. Then the functions g_B determined from the equations (6.23) will have the property $g_B(x, y, y') = B$ for $|y| + |y'| > M + 2$. In particular, this will denote the solvability in the whole of the Cauchy problems (6.22) and, therefore, (5.2).

As a result of the above transformations we have changed equation (4.1) for $|u| + |u_x| > M$. However, for the considered solution of the problem (4.1), (4.2), satisfying (6.1), the equation does not change.

This allows us to guarantee the validity of the proved theorem for the changed problem and for the initial object.

Remark 6.2. Condition (4.3) of the uniform parabolicity of equation (4.1) may be weakened so that $C_r(x, u, u_x, r) \geq \delta(r) > 0$, $|C| \to \infty$ for $|r| \to \infty$ uniformly for $0 \leq x \leq 1$, $|u| + |u_x| \leq M$.

Remark 6.3. Condition (6.3) may be replaced by the requirement of existence of such an integral $\Psi_0(x, y, y', y'')$ of equation (5.2), which would satisfy the lemma 6.2 conditions relative to the function f and

$$\frac{\partial \Psi_0}{\partial y} \leq N, \quad \frac{C}{\Psi_0} \leq N_1,$$

where the latter inequality is verified only for sufficiently large $|y''|$.

§2.7 Liapunov functionals in the neighborhood of a dynamic problem solution.

As was shown in §2.2, if $\frac{a}{b}(x, u, u_x)$ increases by u_x like $u_x^{2+\varepsilon}$ then the condition (B) (definition 1.2) does not hold; therefore, we can not establish the estimate $\sup_{x,t} |u_x(x, t)|$ (§2.3). It is easy to construct examples when the condition (B_1) violated and we can not construct the Liapunov functionals. Following Lavrentiev (1993), we shall construct here the local functionals, i.e., generate in them the pairs of functions $\rho, \Phi(x, \xi, \eta)$ determined not in the whole layer $(x, \xi, \eta) \in [0, 1] \times R^2$, but only in the neighborhood of a solution (x, u, u_x) trace and apply local functionals for the establishment of *a priori* solution estimates.

So, consider in the semi-strip Q equation (1.1) with the boundary conditions

$$u(0, t) = A, \quad u(1, t) = B \qquad (A, B = \text{const}) \qquad (7.1)$$

and initial data (1.3). We shall assume that $a, b \in C^3([-\varepsilon, 1 + \varepsilon] \times R^2)$ $(\varepsilon > 0)$, condition (1.4) holds and

$$b(x, \xi, 0) \equiv 0. \tag{7.2}$$

It is known (Ladyzhenskaya *et al.*, 1967) that for the above assumptions, for sufficiently small $t > 0$ there exist a smooth solution $u(x, t)$ of the problem (1.1), (7.1), (1.3) which satisfies the conditions of the "strict" maximum principle

$$\sup_{x,t} |u(x, t)| \leq \sup_{x} |u_0(x)|. \tag{7.3}$$

We shall not make any assumptions on the order of increase of $b(x, u, u_x)$ or $\frac{b}{a}(x, u, u_x)$ by the argument u_x.

As in §2.2, by (7.3), without loss of generality we shall assume $b(x, u, u_x)$ to be finite by the second argument

$$b(x, \xi, \eta) \equiv 0 \quad \text{for} \quad |\xi| > M. \tag{7.4}$$

As was remarked on in §2.2, not changing (4.1) on the considered solution of the parabolic problem, we essentially change the qualitative properties of the ordinary differential operator from the right side of the equation. As before, by $\varphi(x_0, x, y_0, y_1)$ we shall denote the Cauchy problem solution

$$y'' = -\frac{b(x, y, y')}{a(x, y, y')} \quad , \quad y(x_0) = y_0, \quad y'(x_0) = y_1. \tag{7.5}$$

From the theorem on the continuation of the solution for ordinary differential equations (Pontryagin, 1965), it follows, by (7.4), that for each $(x_0, y_0, y_1) \in [0, 1] \times R^2$ the function $\varphi(x_0, x, y_0, y_1)$ either is regular in the interval $x \in [0, 1]$, or its derivative φ_x is unbounded in a certain interval of its determination Δ containing x_0; while $|\varphi| <$ const. The problem on continuation of φ outside Δ we shall not consider here.

Definition 7.1. A noncontinuable solution $\varphi(x_0, x, y_0, y_1)$ of problem (7.5), smooth in the interval $\Delta \subset [0, 1]$, we shall call a *singular solution*, if its derivative φ_x is unbounded for $x \in \Delta$. The set of all singular solutions we shall denote by N^*.

Definition 7.2. The number

$$r_f = \inf_{z \in N^*} \inf_{x \in [0,1]} \inf_{\zeta \in I(x)} (|z(x) - f(x)| + |z'(x) - \zeta|), \qquad (7.6)$$

where N^* is the set of singular solutions of (7.5); $I(x)$ is the interval with the ends $0, f'(x)$, we shall denote as the *generalized distance from* $f(x)$ *to* N^*.

Remark 7.1. It is easy to see that the positiveness of the generalized distance from $f(x)$ to N^* denotes that for each x the pair $(f(x), \zeta)$, where $\zeta \in I(x)$, does not coincide with any pair $(z(x), z'(x))$, $z \in N^*$. In other words, either $f(x)$ does not intersect any singular solution of (7.5), or, in the point of intersection, for the derivative of this singular solution and $f'(x)$ the inequality $|z'(x)| > |f'(x)|$ holds, if these derivatives have the same sign.

Theorem 7.1. Let $u(x, t)$ be a solution of (1.1), (7.1), (1.3); and let conditions (1.4) and inequalities (7.2), (7.4) hold. If the initial data $u_0(x)$ lie at a positive generalized distance r_{u_0} from the set N^*, then the estimate

$$\sup_{x, t} |u_x(x, t)| \le M_1 \qquad (7.7)$$

holds.

Before we prove the theorem, let us preface the proof with some remarks and auxiliary constructions.

Remark 7.2. Theorem 7.1 remains valid if instead of (7.2) the equality $b(x, u, N) = 0$ $(N = \text{const})$ holds. In this case, when defining the generalized distance (7.6), we should choose $I(x)$ as the interval with the ends $N, f'(x)$.

Remark 7.3. Theorem 7.1 remains valid if we change one of the two boundary values (1.3) by $u_x(i, t) = 0$.

It is easy to see from the well-known example of the parabolic problem, which has a bounded solution with an unbounded derivative (Ladyzhenskaya *et al.*, 1967, Filippov, 1961), that conditions of theorem 7.1 do not hold. In this case the generalized distance $r_{u_0} = 0$ for any outlined initial function $u_0(x)$.

It is known that problem (1.1), (7.1), (1.3) is solved locally. This means that there exists such $t_0 > 0$ that for $t \in [0, t_0]$ there exist its solution $u(x, t) \in H^{2+\beta}$. Therefore, under the conditions of theorem 7.1, for sufficiently small t the function $u(x, t)$ will be at a positive generalized distance $r(t)$ from the set N^* of singular solutions of (7.5).

By $J^* = [0, \tau)$ we shall denote such interval of t, where a solution of (1.1), (7.1), (1.3) exists and for each $t \in J^*$ lies at a positive generalized distance from the set N^*. By the conditions of theorem 7.1 the set J^* is nonempty.

The following lemma holds:

Lemma 7.2   Under the theorem 7.1 conditions there exist the set of smooth functions $\Phi_{n\lambda}, \rho_{n\lambda}(x, \xi, \eta)$ such that for the solution $u(x, t)$ of the problem (1.1), (7.1), (1.3) for $t \in J^*$ the equality

$$\frac{d}{dt} \int_0^1 \Phi_{n\lambda}(x, u, u_x) dx = - \int_0^1 \rho_{n\lambda}(x, u, u_x) u_t^2 dx \qquad (7.8)$$

holds, where

$$\rho_{n\lambda}(x, \eta, \zeta) \geq 0. \qquad (7.9)$$

Construction of the functions $\rho_{n\lambda}, \Phi_{n\lambda}$ is just the same as in §2.1. In the case of boundary conditions (7.1), by the equality (7.2), these functions will be as follows

$$\Phi_{n\lambda}(x, \xi, \eta) = \int_0^\eta (\eta - \tau) \rho_{n\lambda} a(x, \xi, \tau) d\tau, \qquad (7.10)$$

$$\rho_{n\lambda}(x, \xi, \eta) = B_\lambda^{2n}(x, \xi, \eta) \rho_{0\lambda}(x, \xi, \eta), \qquad (7.11)$$

$$\rho_{0\lambda} = \exp\left\{ -\int_{x_0}^{x} \frac{\eta a_\xi + a_x - b_\eta}{a}\bigg|_{\substack{\xi=\varphi(x_0,x,y_0,y_1),\\ \eta=\varphi_x(x_0,x,y_0,y_1)}}\right\}\bigg|_{\substack{x_0=\lambda,\\ y_0=A_\lambda(x,\xi,\eta),\\ y_1=B_\lambda(x,\xi,\eta),}} \tag{7.12}$$

$$A_\lambda(x,\xi,\eta) = \varphi(x,\lambda,\xi,\eta), \tag{7.13}$$

$$B_\lambda(x,\xi,\eta) = \frac{\partial}{\partial\lambda}\varphi(x,\lambda,\xi,\eta). \tag{7.14}$$

Recall that $\varphi(x_0, x, y_0, y_1)$ solves the Cauchy problem (7.5), i.e., $A_\lambda, B_\lambda(x,\xi,\eta)$ are correspondingly the values in the point $x = \lambda$ of such solution of (7.5) and its derivative, that in the point x attains the value ξ with the derivative η.

The function $\varphi(x_0, x, y_0, y_1)$ is determined not for all $(x_0, x, y_0, y_1) \in [0,1]^2 \times R^2$. Therefore, not for all $\lambda, x \in [0,1]$; $(\xi,\eta) \in R^2$ the functions $A_\lambda, B_\lambda(x,\xi,\eta)$ and together with them the functions $\rho_{0\lambda}, \rho_{n\lambda}, \Phi_{n\lambda}(x,\xi,\eta)$ are determined. Domains of definitions of these functions depend on equation (7.5) and hardly may be written in the general case. We shall be interested in the points (x,ξ,η) in which the functions from formulae (7.10)-(7.14) will be determined in the interval $\lambda \in [0,1]$.

We shall denote by M^* the set of such Cauchy data (x_0, y_0, y_1) for problem (7.5) that the solutions $\varphi(x_0, x, y_0, y_1)$ with these Cauchy data would be regular in the interval $x \in [0,1]$. By theorem 13 (Pontryagin, 1965, chapter 4), the set M^*, is open; furthermore, by (7.2), $M^* \neq 0$.

We can show that if $f(x)$ lies in a positive generalized distance from N^*, then for each $x \in [0,1]$ we have $(x, f(x), f'(x)) \in M^*$. Really, taking into account (7.13), (7.14), the functions $A_\lambda, B_\lambda(x, f(x), f'(x))$ are determined and regular for $\lambda \in [0,1]$ (since $f(x)$ is not tangent to any of the singular solutions of (7.5)). The functions $\varphi, \varphi_\lambda(x, \lambda, f(x), f'(x))$ are regular in the interval $\lambda \in [0,1]$. Thus, taking into account (7.10), (7.11), it suffices to show that the function $\rho_{0\lambda}(x, f(x), f'(x))$ is determined in the interval $\lambda \in [0,1]$.

Introducing the notation

$$F(x,\xi,\eta) = \frac{\eta a_\xi + a_x - b_\eta}{a}(x,\xi,\eta),$$

rewrite equation (7.12) in the form

$$\rho_{0\lambda}(x,\xi,\eta) = e^{-\int\limits_{\lambda}^{x} F(\tau,\varphi(\lambda,\tau,A_\lambda,B_\lambda),\varphi_\tau(\lambda,\tau,A_\lambda,B_\lambda))d\tau} \qquad (7.15)$$

Consider the function $\rho_{0\lambda}(x, f(x), f'(x))$. Since $f(x)$ is not tangent to any of the functions $z \in N^*$, the pair $(A_\lambda, B_\lambda(x, f, f'))$ being the value in the point λ of the solution and its derivative to the corresponding Cauchy problem (7.5) will not coincide with any of the pairs $(z(x), z'(x))$, where $z \in N^*$. Therefore, the functions $\varphi, \varphi_\tau(\lambda, \tau, A_\lambda, B_\lambda(x, f, f'))$ will be regular for $\tau \in [0, 1]$, i.e., the functions $\rho_{0\lambda}(x, f, f')$ are determined for all $\lambda \in [0, 1]$.

Hence we obtain the following result.

For a function $f(x)$ lying in a positive generalized distance r_f from the set N^* of the singular solutions of (7.5), the following estimates

$$0 < \rho_{\min}(f) \leq a\rho_{0\lambda}(x, f(x), f'(x)) \leq \rho_{\max}(f), \qquad (7.16)$$

$$\left| \frac{\partial B_\lambda(x, f, \zeta)}{\partial \zeta} \right|_{\zeta = f'(x)} \leq M_2(f). \qquad (7.17)$$

hold uniformly by $\lambda \in [0, 1]$.

Lemma 7.3.   Under the conditions of theorem 7.1 the following estimate

$$|B_\lambda(x, u, u_x)| \leq M_1 \qquad (7.18)$$

holds for $t \in J^*$ uniformly relative to $\lambda \in [0, 1]$.

Proof.   By (7.8), (7.9), for $t \in J^*$, the following inequality

$$\int\limits_0^1 \Phi_{n\lambda}(x, u, u_x)dx \leq \int\limits_0^1 \Phi_{n\lambda}(x, u_0, u_0')dx \qquad (7.19)$$

holds. Whereas, by (7.2), we have the identity $B_\lambda(x, u, 0) \equiv 0$, then

$$\frac{1}{(2n+1)(2n+2)}B_\lambda^{2n+2}(x,u,u_x) = \int_0^{u_x}\left(\int_0^\tau B_\lambda^{2n}(x,u,\eta)\frac{\partial B_\lambda}{\partial\eta}d\eta\right)\frac{\partial B_\lambda}{\partial\eta}d\tau.$$
$$(7.20)$$

Hence the following estimate may be easily obtained (see (7.16), (7.17))

$$B_\lambda^{2n+2}(x,u,u_x) \le (2n+2)^2\frac{M_2^2(f)}{\rho_{\min}(f)}\int_0^{u_x}(u_x-\tau)B_\lambda^{2n}a\rho_{0\lambda}d\tau, \qquad (7.21)$$

which will be hold for $t \in J^*$. In this case, when t tends to the upper bound of the set J^*, it may occur that $M_2 \to \infty, \rho_{\min} \to 0$.

The following discussion presents again the schema suggested by Belonosov and Zelenyak (1975) (we have given it in §2.3). We shall give it here in order to explain the independence of the constant M_1, obtained in (7.8), on $M_2, \rho_{\max}, \rho_{\min}$. In other words, we shall explain the independence of M_1 on the generalized distance between the function $u(x,t)$ and the set N^*.

Set

$$M_1 = \sup_{\lambda\in[0,1]}\;\sup_{x\in[0,1]}\;\sup_{\zeta\in I(x)}|B_\lambda(x,u_0(x),\zeta)|, \qquad (7.22)$$

where $I(x)$ is the interval with the ends $0, u_0'(x)$ (i.e., $\zeta \in I(x)$ is equivalent to $\mathrm{sgn}u_0'(x)\zeta \in [0,|u_0'(x)|]$). By the condition of theorem 7.1 the value M_1 is finite. As follows from (7.10)-(7.14), (7.16), (7.22)

$$\int_0^1 \Phi_{n\lambda}(x,u_0,u_0')dx \le \rho_{\max}(u_0)K^2M_1^{2n}, \qquad (7.23)$$

where $K = \sup|u_0'(x)|$.

Combining estimates (7.19), (7.21), (7.23), we obtain the inequality

$$\int_0^1 B_\lambda^{2n+2}dx \le (2n+2)^2\frac{\rho_{\max}(u)}{\rho_{\min}(u)}M_2^2(u)K^2M_1^{2n}. \qquad (7.24)$$

It is well-known that

$$\lim_{n \to \infty} \| f(x) \|_{L_n(0,1)} = \underset{0<x<1}{\text{vrai sup}} |f(x)|.$$

Therefore, extracting from both sides of (7.24) the root of the degree $2n + 2$, passing to the limit for $n \to \infty$, and taking into account the smoothness of the function $B_\lambda(x, u, u_x)$, we obtain estimate (7.18) for $t \in J^*$, which was to be proved.

Corollary 7.1. Under the conditions of theorem 7.1, for $t \in J^*$, the following estimate

$$|u_x(x,t)| \leq M_1 \qquad (7.25)$$

holds, where M_1 is the constant from (7.22).

Really, inequality (7.25) follows from (7.18). since by the definition of the function $B_\lambda(x, \xi, \eta)$ (see, (7.14)), we have the identity $B_\lambda(\lambda, \xi, \eta) = \eta$.

Corollary 7.2. Under the conditions of theorem 7.1, for $t \in J^*$, the following inequality

$$\| u(x,t) \|_{2+2\beta} \leq K \qquad (7.26)$$

holds.

This statement follows from the articles of Kruzhkov (1967, 1972, 1979), from the supposed smoothness of the functions a, b, and from estimates (7.3), (7.25).

Thus, to prove theorem 7.1, it suffices to show that the interval J^* of those t that $u(x,t)$ (solution to (1.1), (7.1), (1.3)) lies in a positive such distance from the set N^*, coincides with R^+.

We can prove this by contradiction. Assume that $|J^*| < \infty$, i.e., τ is finite (see definition 7.2). By the definition of J^*, for $t < \tau$, there exists a smooth solution of (1.1), (7.1), (1.3) and estimate (7.8) holds. At the same time, for $t \to \tau$ the generalized distance from $u(x,t)$ to the set N^* tends to zero. This is so, because the distance $r(t)$ is a continuous function. Whereas, by (7.26), equation (1.1) holds up to $t = \tau$; it denotes that such $z(x) \in N^*$ and $x_1 \in [0,1]$ exist that

$$\lim_{t \to \tau} u(x_1, t) = z(x_1), \qquad \lim_{t \to \tau} u_x(x_1, t) = z'(x_1), \qquad (7.27)$$

where the value $z'(x_1)$ is finite (see (7.25)).

In the strict sense, for $r(\tau) = 0$, we may only state that

$$\operatorname{sgn}(u_x(x_1, \tau)) z'(x_1) \in [0, |u_x(x_1, \tau)|].$$

However, the relation $z'(x_1) \neq u_x(x_1, \tau)$ denotes that for $t < \tau$ and sufficiently close to τ, we have $r(t) = 0$. This contradicts the definition of τ.

By the function $z(x)$, which satisfies property (7.27), we may choose such $x_i \in [0,1]$ that $z(x)$ will be regular in the interval I_1 with the ends x_1, x_2, and the inequality (see (7.18))

$$|z'(x_2)| > M_1 + 2 \qquad (7.28)$$

will hold.

Taking into account theorem 13 (Pontryagin, 1965, chapter 4) on continuous dependence of a solution of the Cauchy problem for an arbitrary differential equation on the parameters, it follows from (7.28), that for a certain $\varepsilon > 0$ all solutions $y(x)$ of (7.5) with the Cauchy data $y(x_1) = y^0, y'(x_1) = y^1$ are determined and regular in the interval I_1, and satisfy the inequality

$$|y^0 - z(x_1)| + |y^1 - z'(x_1)| < \varepsilon_1, \qquad (7.29)$$

where

$$|y'(x_2)| > M_1 + 1. \qquad (7.30)$$

By (7.27), there exists such $t_1 \in (0, \tau)$ that for $t \in (t_1, \tau)$ the functions $y^0(t) = u(x_1, t)$ and $y^1(t) = u(x_1, t)$ satisfy estimate (7.29). It denotes (taking into account (7.14) and (7.30)) that

$$|B_{x_2}(x_1, u(x_1, t), u_x(x_1, t))| \geq M_1 + 1.$$

This inequality contradicts the proved estimate (7.18). Thus, the assumption that the set J^* is bounded is false. The theorem is proved.

As it was noted, the estimate (7.7) of the derivative of the bounded solution to quasilinear equation (1.1) is the basic estimate. This means that if we apply the standard methods, we may prove the existence theorem for sufficiently smooth coefficients without any additional restrictions on the nonlinearities. Without detailing, formulate again theorem 7.1 and its corollaries which follow from the results of Ladyzhenskaya *et al.* (1967), Kruzhkov (1967, 1972), Belonosov and Zelenyak (1975) as the sequence of statements.

Corollary 7.3. Let the assumptions (1.4) hold, the initial data lie in a positive generalized distance $r_{u_0} > 0$ from the set of singular solutions of (7.5) (see definitions 7.1, 7.2), and $b(x, \xi, 0) \equiv 0$.

Then the problem (1.1), (7.1), (1.3) has the following properties.

1). Functions (7.14) are uniformly bounded for the solution of considered problem in the whole existence interval

$$|B_\lambda(x, u, u_x)| \leq M_1.$$

2). The solution derivative is uniformly bounded

$$|u_x(x, t)| \leq M_1.$$

3). The classical solution of the problem $u(x, t) \in H^{2+2\alpha}(Q)$ exists for all $t > 0$.

4). There exists a stationary solution $v(x)$, i.e., the solution of the problem

$$a(x, v, v')v'' + b(x, v, v') = 0, \qquad v(0) = A, \qquad v(1) = B$$

such that

$$\lim_{t \to \infty} \| u(x,t) - v(x) \|_{2+\alpha} = 0.$$

Chapter 3

The behavior of solutions of one-dimensional nonlinear problems over extended time.

In this chapter we shall investigate the asymptotic behavior of solutions of quasilinear parabolic equations with one spatial variable. Now, there are many works where the qualitative properties of quasilinear parabolic equations with one spatial variable have been investigated. Here we mainly give the results of the authors of this book. The bibliography lists other publications that are closest to our research.

Note, also, that a great many of the techniques of this chapter fail for quasilinear parabolic equations with many spatial variables. There are other results that hold even for parabolic equations with many spatial variables and these will be considered further. The methods for solving these statements for quasilinear parabolic equations with one spatial variable and for quasilinear parabolic equations with many spatial variables differ essentially from each other. Therefore, we shall give both of the methods, the more so, as for the parabolic equation with one spatial variable we have evolved a refined method demanding fewer assumptions about the problem.

The examples in this chapter are of great importance, since they allow us to clarify the results and to understand which restrictions are essential.

§3.1 Liapunov's functionals and asymptotic behavior of solutions for extended time.

We shall consider the quasilinear problem

$$u_t = a(x, u, u_x)u_{xx} + b(x, u, u_x), \qquad (1.1)$$

$$\alpha_i u_x(i, t) + \varphi_i(u(i, t)) = 0, \quad i = 0, 1, \quad t > 0, \qquad (1.2)$$

$$u(x, 0) = u_0(x), \qquad (1.3)$$

in the semi-strip $(x, t) \in Q = [0, 1] \times [0, +\infty)$, or in the rectangle $(x, t) \in Q_T = [0, 1] \times [0, T]$. Further, we shall assume that functions a, b, φ_i are from C^3 by their arguments. This assumption may be weakened, but it is not of paramount importance when considering these questions. We shall assume that

$$a(x, \xi, \eta) \geq a_0 > 0, \qquad (x, \xi, \eta) \in [0, 1] \times R^2 \qquad (1.4)$$

$$\alpha_i^2 + \varphi_i'^2(\xi) \geq a_1 > 0, \quad \xi \in R, \quad \alpha_i = \text{const}$$

$$u_0(x) \in E = \left\{ u_0(x) \in C^1[0, 1], \ \alpha_i u_0'(i) + \varphi_i(u_0(i)) = 0, \ i = 0, 1 \right\}.$$

In the first chapter we have established the existence theorem locally for t and the uniqueness theorem for problem (1.1)-(1.3) with initial data from E. We shall denote this solution by $u(x, t; u_0)$. It was shown in chapter 1 that the height T of Q_T, where the solution $u(x, t; u_0)$ is determined, depends only on $\| u_0 \|_1$. This allows us to establish the following statements (see lemma 5.3, chapter 1).

Lemma 1.1. For the above assumptions, for each $R > 0$ there exists such $T(R) > 0$ that if $\| u_0 \|_1 \leq R$, then the solution of (1.1)-(1.3) exists in the interval $t \in (0, T(R))$. Moreover, for each ε there exists such $C(\varepsilon)$ that

$$\| u(x, t; u_0) \|_{3+\alpha}^{Q_T^\varepsilon} \leq C(\varepsilon) \| u_0 \|_1^{[0,1]}, \tag{1.5}$$

where $Q_T^\varepsilon = [0,1] \times [\varepsilon, T)$, $\alpha > 0$; the constant $C(\varepsilon)$ depends on the nonlinearities and grows for $\varepsilon \to 0$.

Lemma 1.2. Under the lemma 1.1 assumptions, if

$$\lim_{t \to \infty} \| u(x, t; u_0) \|_1^{[0,1]} < \infty,$$

then

$$\lim_{t \to \infty} \| u(x, t; u_0) \|_{2+\alpha}^{[0,1]} < \infty.$$

Proof of lemma 1.2. Let

$$\lim_{t \to \infty} \| u(x, t; u_0) \|_1^{[0,1]} < \infty;$$

then there exists a bounded sequence $u(x, t; u_0)$ for $t_i \to +\infty$. By lemma 1.1, there exists such $T > 0$ that

$$\| u(x, t_i + t; u_0) \|_{1,3+\alpha}^{Q_T} \leq C;$$

therefore, if $0 < \varepsilon < T$, then

$$\| u(x, t_i + t; u_0) \|_{3+\alpha}^{Q_T^\varepsilon} \leq C_1.$$

This means that, in particular,

$$\| u(x, t_i + T; u_0) \|_{3+\alpha}^{[0,1]} \leq C_1.$$

This proves the lemma.

Here we shall consider such initial data u_0, that the solution $u(x, t; u_0)$ may be extended for all $t \in (0, +\infty)$. In this case two situations are possible.

1) The value $\| u(x, t; u_0) \|_1$ is uniformly bounded for $t \geq 0$.

2) The value $\| u(x, t; u_0) \|_1$ is not bounded for $t \geq 0$.

In this paragraph we shall consider the first situation leaving the second for the next paragraphs.

By analogy with the theory of ordinary differential equations, consider the following notions.

Definition 1.1.   The set of partial limits for $t \to \infty$ in the norm $C^1[0, 1]$ of the solution $u(x, t; u_0)$ we shall call the ω-*limit set* and denote it by $\omega(u_0)$.

Note that

1) by lemma 1.2, if the lower limit of the solution norm in C^1 for $t \to \infty$ is bounded, then $\omega(u_0)$ is not empty and contains the smooth (from the space $C^{2+\alpha}$) elements.

2) By lemma 1.1, if the solution $u(x, t; u_0)$ is uniformly by $t > 0$ bounded in the norm of $C^1[0, 1]$, then for $t \geq t_0 > 0$, this solution is bounded in the norm of $C^{3+\alpha}[0, 1]$ also (Vishnevskii, 1987).

Definition 1.2.   We shall say that the solution $u(x, t; u_0)$ is *quasistabilized*, if $\omega(u_0)$ consists only of stationary (not depending on t) solutions of the problem.

Definition 1.3.   We shall say that the solution $u(x, t; u_0)$ is *stabilized*, if $\omega(u_0)$ consists only of a unique stationary solution.

The following statement holds (Zelenyak, 1968, 1977; Belonosov and Zelenyak, 1975):

Theorem 1.1.   Let $u(x, t; u_0)$ solve problem (1.1)-(1.3) and be uniformly by $t \geq 0$ bounded in $C^1[0, 1]$. Then the solution $u(x, t; u_0)$ is stabilized.

Remark 1.1. As it was noted, for the above assumptions, $u(x, t; u_0) \in C^{3+\alpha}[0, 1]$, and its limit for $t \to \infty$ may be taken in the norm of this space.

First, let us prove some auxiliary statements.

Lemma 1.3. Let $a(x), b(x), f(x)$ be continuous for $x \in [0, 1]$; and $\alpha, \beta, \gamma, \delta, \theta_1, \theta_2$ be constants, $\alpha^2 + \beta^2 = \gamma^2 + \delta^2 = 1$. Let $u(x)$ solve the problem

$$Lu = u_{xx} + au_x + bu = f,$$

$$z_1 u = \alpha u_x + \beta u|_{x=0} = \theta_1,$$

$$z_2 u = \gamma u_x + \delta u|_{x=1} = \theta_2,$$

and $s(x) \neq 0$ be such that

$$Ls = z_1 s = z_2 s = 0, \quad \int_0^1 u \cdot s \, dx = 0.$$

Then there exists a constant K depending only on $\sup |a|$, $\sup |b|$, α, β, γ, δ such that

$$\| u \|_{W_2^2} \leq K \left\{ \| f \|_{L_2} + |\theta_1| + |\theta_2| \right\},$$

$$K = K \left(\frac{\beta}{\alpha}, \frac{\delta}{\gamma}, \sup |a|, \sup |b| \right). \tag{1.6}$$

Proof. Set

$$u = \theta_1 [\alpha x + \beta](2x^3 - 3x^2 + 1) + v(x) + \theta_2 [(x-1)\gamma + \delta] (-2x^3 + 3x^2).$$

It is easy to show that

$$Lv = f_1, \qquad z_i v = 0,$$

where

$$\| f_1 \|_{L_2} \leq K_1 \{|\theta_1| + |\theta_2|\} + \| f \|_{L_2},$$

and K_1 depends only on a, b. Consider the solution of the Cauchy problem

$$Lv_1 = f_1, \quad v_1(0) = v_1'(0) = 0.$$

Then

$$\| v_1 \|_{W_2^2} \leq K_2 \{\| f \|_{L_2} + |\theta_1| + |\theta_2|\},$$

where K_2 depends only on $\sup |a|$, $\sup |b|$. We have $z_1 v_1 = 0$.

Then $v - v_1 = w$ satisfies the equation

$$Lw = 0, \qquad z_1 w = 0.$$

As $s \neq 0$, then w and s are linearly dependent (otherwise, all solutions of the homogeneous equation would satisfy the condition $z_1 w = 0$): $w = cs$, where c is a constant.

Finally, for the function u we obtain

$$v = v_1 + cs, \quad u = (u - v) + v_1 + cs = v_1 + v_2 + cs,$$

where v_2 is a fourth degree polynomial in x with the coefficients linearly dependent on θ_i.

As, by the lemma supposition $\int_0^1 u s dx = 0$, then

$$\int_0^1 (v_1 + v_2) s dx = -c \int_0^1 s^2 dx.$$

Considering $\int_0^1 s^2 dx = 1$, we obtain

$$|c| \leq K_3 \{|\theta_1| + |\theta_2| + \| f \|_{L_2}\}.$$

Rewrite the equation for s as follows

$$(ps_x)_x + qs = 0, \quad s_x + \beta_1 s|_{x=0} = 0; \quad s|_{x=1} = 0$$

taking the most typical boundary conditions in the ends of the interval, i.e. $\alpha \neq 0$, $\gamma = 0$, $\delta = 1$, $\beta_1 = \frac{\beta}{\alpha}$.

Set

$$p = e^{\int_0^x a(\xi)d\xi}, \quad q = pb.$$

Then

$$\int_0^1 ps_x^2 dx = \int_0^1 qs^2 dx - p(0)\beta_1 s^2(0).$$

But

$$-s^2(0) = 2\int_0^1 s(x)s'(x)dx,$$

whence

$$s^2(0) \leq \varepsilon \int_0^1 s'^2(x)dx + \frac{1}{\varepsilon}\int_0^1 s^2(x)dx.$$

Thus,

$$\| s \|_{W_2^1} \leq K_4(\beta_1, \sup |a|, \sup |b|)$$

and, finally, from the equation we obtain

$$\| s \|_{W_2^2} \leq K_5(\| \beta_1 \|, \sup |a|, \sup |b|).$$

Taking into account the representation for u, and estimates for v_i, c, s, we obtain estimate (1.6), where

$$K = K\left(\left|\frac{\beta}{\alpha}\right|, \left|\frac{1}{\gamma}\right|, \sup |a|, \sup |b|\right)$$

does not depend on the first (second) argument if $\alpha = 0(\delta = 0)$, i.e., in the case of the boundary conditions $u|_{x=i} = \theta_{i+1}$. The lemma is proved.

Lemma 1.4. Let for $0 \leq t \leq T$ the inequality

$$\int\limits_{t}^{\infty} \left(\int\limits_{0}^{1} u_t^2 dx \right) dt \leq K e^{-\lambda t}$$

hold and let $0 \leq t \leq \tau \leq T$. Then

$$\int\limits_{0}^{1} |u(x,t) - u(x,\tau)| dx \leq \sqrt{K} \frac{e^{\frac{\lambda}{2}}}{e^{\frac{\lambda}{2}} - 1} e^{-\frac{\lambda t}{2}}.$$

Proof. Let $0 \leq t \leq \tau \leq 1$. Then

$$\int\limits_{0}^{1} |u(x,t) - u(x,\tau)| dx = \int\limits_{0}^{1} \left| \int\limits_{t}^{\tau} \frac{\partial u}{\partial t} dt \right| dx$$

$$\leq \sqrt{\tau - t} \int\limits_{0}^{1} \left(\int\limits_{t}^{\tau} u_t^2 dt \right)^{\frac{1}{2}} dx \leq \int\limits_{0}^{1} \left(\int\limits_{t}^{\tau} u_t^2 dt \right)^{\frac{1}{2}} dx$$

$$\leq \left(\int\limits_{0}^{1} \int\limits_{t}^{\tau} u_t^2 dt dx \right)^{\frac{1}{2}} \leq \sqrt{K} e^{-\frac{\lambda t}{2}}.$$

Let now $\tau - t > 1$, where $\tau = t + N + \delta$, $0 \leq \delta < 1$. Then

$$\int\limits_{0}^{1} |u(x,t) - u(x,\tau)| dx = \int\limits_{0}^{1} \left| \sum_{0}^{N-1} \int\limits_{t+i}^{t+i+1} u_t dt + \int\limits_{t+N}^{t+N+\delta} u_t dt \right| dx$$

$$\leq \sum_{0}^{N-1} \int\limits_{0}^{1} \left[\int\limits_{t+i}^{t+i+1} u_t^2 dt \right]^{\frac{1}{2}} dx + \int\limits_{0}^{1} \left[\int\limits_{t+N}^{t+N+\delta} u_t^2 dt \right]^{\frac{1}{2}} dx$$

$$\leq \sum_{i=0}^{N} \sqrt{K} e^{-\frac{t+i}{2}\lambda} \leq \sqrt{K} e^{-\frac{\lambda t}{2}} \cdot \frac{e^{\lambda/2}}{e^{\lambda/2} - 1},$$

which was to be proved.

Proof of theorem 1.1. By lemma 8.2 (Belonosov and Zelenyak, 1975) each partial limit of $u(x,t)$ is a stationary solution. Further, (see theorem 1.3) we shall give the proof of this result for the more general case. To simplify calculations, we shall take the boundary conditions with $\alpha_0 \neq 0$. Consider the problem

$$v'' = -\frac{b(x,v,v')}{a(x,v,v')}, \quad v\Big|_{x=0} = \mu, \quad v_x\Big|_{x=0} = -\frac{1}{\alpha_0}\varphi_0(\mu). \tag{1.7}$$

Let

$$\mu_1 = \varliminf_{t\to\infty} u(0,t), \quad \mu_2 = \varlimsup_{t\to\infty} u(0,t).$$

If $\mu_1 = \mu_2$, then all partial limits of $u(x,t)$ coincide, since they all solve (1.7) and $\mu_1 = \mu_2 = \mu$. The functions $v(x,\mu)$ for $\mu_1 \leq \mu \leq \mu_2$ satisfy also the boundary condition

$$\alpha_1 v_x + \varphi_1(v)\big|_{x=1} = 0.$$

Really, let

$$u(0,t_i)\underset{t_i\to\infty}{\longrightarrow}\mu.$$

As $|u(x,t)|_{2+\alpha}^{[0,1]}$ is bounded, we may choose such a subsequence t_{i_k} that $u(x,t_{i_k})$ converges in $C^{2+\beta}$ with $\beta < \alpha$, where the limit is a stationary solution, i.e., it satisfies (1.7) and the boundary condition for $x = 1$.

The function $v(x,\mu)$, evidently, is differentiable by the parameter, where

$$\frac{\partial^2 v}{\partial x \partial \mu} = -\frac{\partial}{\partial v}\frac{a}{b}\cdot\frac{\partial v}{\partial \mu} - \frac{\partial}{\partial v_x}\frac{b}{a}\cdot\frac{\partial^2 v}{\partial x \partial \mu}, \tag{1.8}$$

$$\alpha_0\frac{\partial^2 v}{\partial x \partial \mu} + \varphi_0'(v)\frac{\partial v}{\partial \mu}\bigg|_{x=0} = \alpha_1\frac{\partial^2 v}{\partial x \partial \mu} + \varphi_1'(v)\frac{\partial v}{\partial \mu}\bigg|_{x=1} = 0. \tag{1.9}$$

Set $w(x,t) = u(x,t) - v(x,\mu)$. Using the Taylor formula, we obtain

$$\frac{1}{a(x,u,u_x)}\cdot\frac{\partial u}{\partial t} = w_{xx} + w\frac{\partial}{\partial u}\frac{b}{a}\bigg|_{\substack{u=v \\ u_x=x_x}} \tag{1.10}$$

$$+ w_x \frac{\partial}{\partial u_x} \frac{b}{a}\bigg|_{\substack{u=v \\ u_x=x_x}} + R = N_\mu w + R,$$

where

$$|R| \le K(r)(w^2 + w_x^2) \quad \text{for} \quad w^2 + w_x^2 \le r^2.$$

The boundary conditions for w have the form

$$\alpha_0 w_x + \varphi_0'(v)w|_{x=0} = R_1(t),$$

$$\alpha_1 w_x + \varphi_1'(v)w|_{x=1} = R_2(t), \qquad (1.11)$$

where

$$|R_i| \le K_i |w(i-1,t)|^2, \quad i = 1,2.$$

Consider

$$\inf_\mu \int_0^t w^2 dx = z_1(t).$$

If this infimum is achieved in an interior point of the interval $[\mu_2, \mu_1]$, then

$$\frac{\partial}{\partial \mu} \int_0^1 w^2 dx = 2 \int_0^1 (v-u)\frac{\partial v}{\partial \mu} dx = 0,$$

i.e. the function w is orthogonal to $\partial v / \partial \mu$.

Let $\inf_\mu \int_0^1 w^2 dx$ be achieved for $\mu = \mu(t_0) \in (\mu_2, \mu_1)$. For certain $\mu(t) \in (\mu_2, \mu_1)$ the same property has $z_1(t)$ with $t_0 \le t < t_1$. Moreover, for $t = t_1$ infimum is achieved either for $\mu = \mu_2$, or for $\mu = \mu_1$. The set of such t we shall denote by $\Pi(t_0, t_1)$.

Consider now equation (1.10)

$$N_\mu w = R - \frac{1}{a(x, u, u_x)} u_t \qquad (1.12)$$

with boundary conditions (1.11). For each $t \in \Pi(t_0, t_1)$ after the natural normalization connected with the boundary conditions, we may use lemma 1.3. For all $t \in \Pi(t_0, t_1)$ the equation $N_{\mu(t)}w = 0$ with the homogeneous boundary conditions (1.11) has a nontrivial solution $u_\mu|_{\mu=\mu(t)}$, where the arguments of K entering in (1.6) are bounded, since they are bounded in the norms

$$\| v(x, \mu) \|_{2+\alpha}^{[0,1]}, \qquad \| u(x, t) \|_{2+\alpha}^{[0,1]} .$$

Therefore,

$$\| w \|_{W_2^2} \leq \frac{K}{a_0} \left\{ \| u_t \|_{L_2} + \| R \|_{L_2} + \sup_t (|R_1| + |R_2|) \right\}. \qquad (1.13)$$

The constant K may be chosen not depending on t. We may also take that

$$|R| \leq K_3 \left(w^2 + w_x^2 \right), \quad |R_i| \leq K_3 \left(w^2(0,t) + w^2(1,t) \right) \qquad (1.14)$$

with the constant K_3 not depending on t. Further, for $\varepsilon_0 > 0$ we have

$$\int_0^1 w_x^2 dx = ww_x|_0^1 - \int_0^1 ww_{xx} dx \qquad (1.15)$$

$$\leq K_4 \left[w^2(0,t) + w^2(1,t) + \varepsilon_0 \| w_{xx} \|_{L_2}^2 + \frac{1}{\varepsilon_0} \| w \|_{L_2}^2 \right]$$

and, obviously,

$$w^2(i,t) \leq 2 \int_0^1 |ww_x| dx + w^2(x,t).$$

Integrating the latter inequality with respect to x and using the triangle inequality, we obtain

$$w^2(i,t) \leq \varepsilon_0 \| w_x \|_{L_2}^2 + \frac{1}{\varepsilon_0} \| w \|_{L_2}^2 + \| w \|_{L_2}^2 . \qquad (1.16)$$

Choosing ε_0 sufficiently small, from (1.15), (1.16) we obtain the estimates with an arbitrary $\varepsilon > 0$:

$$\int_0^1 w_x^2 dx \leq \varepsilon \parallel w \parallel_{W_2^2}^2 + \frac{K_5}{\varepsilon} \parallel w \parallel_{L_2}^2, \qquad (1.17)$$

$$w^2(0,t) + w^2(1,t) \leq \varepsilon \parallel w \parallel_{W_2^2}^2 + \frac{K_5}{\varepsilon} \parallel w \parallel_{L_2}^2. \qquad (1.18)$$

Further, we obtain the following inequalities

$$\int_0^1 R^2 dx \leq K_3^2 \int_0^1 (w^2 + w_x^2)^2 dx \leq 2K_3^2 \int_0^1 (w^4 + w_x^4) dx$$

$$\leq 2K_3^2 \int_0^1 w^4 dx + 4K_3^2 \left(\int_0^1 w^4 dx\right)^{\frac{1}{2}} \left(\int_0^1 w_x^4 dx\right)^{\frac{1}{2}}$$

$$+ 2K_3^2 \int_0^1 w_x^4 dx \leq 2K_3^2 \left[\left(\int_0^1 w^4 dx\right)^{\frac{1}{2}} + \left(\int_0^1 w_x^4 dx\right)^{\frac{1}{2}}\right]^2.$$

Thus,

$$\parallel R \parallel_{L_2} \leq \sqrt{2}K_3 \left(\sup |w| + \sup |w_x|\right) \left[\parallel w \parallel_{L_2} + \parallel w_x \parallel_{L_2}\right]$$

$$\leq K_6 \parallel w \parallel_{W_2^2} \left[\parallel w \parallel_{L_2} + \parallel w_x \parallel_{L_2}\right]. \qquad (1.19)$$

The constants K_i may be chosen not depending on t. Taking into account inequality (1.17), we obtain

$$\left(\int_0^1 R^2 dx\right)^{\frac{1}{2}} \leq K_6 \parallel w \parallel_{W_2^2} \qquad (1.20)$$

$$\times \left[\parallel w \parallel_{L_2} + \left(\varepsilon \parallel w \parallel_{W_2^2}^2 + \frac{K_5}{\varepsilon} \parallel w \parallel_{L_2}^2\right)^{\frac{1}{2}}\right]$$

$$\leq K_6 \parallel w \parallel_{W_2^2} \left[\parallel w \parallel_{L_2} + \sqrt{\varepsilon} \parallel w \parallel_{W_2^2} + \sqrt{\frac{K_5}{\varepsilon}} \parallel w \parallel_{L_2}\right]$$

$$\le K_7 \parallel w \parallel_{W_2^2} \left[\frac{1}{\sqrt{\varepsilon}} \parallel w \parallel_{L_2} + \sqrt{\varepsilon} \parallel w \parallel_{W_2^2} \right].$$

Analogously,

$$|R_i| \le K_8 \left(\sqrt{\varepsilon} \parallel w \parallel_{W_2^2}^2 + \frac{1}{\sqrt{\varepsilon}} \parallel w \parallel_{L_2}^2 \right). \tag{1.21}$$

Substituting (1.21), (1.20) into (1.19), we obtain the estimate

$$\parallel w \parallel_{W_2^2} \le \frac{K}{a_0} \left\{ \parallel u_t \parallel_{L_2} + \parallel w \parallel_{W_2^2} \left[\frac{K_9}{\sqrt{\varepsilon}} \parallel w \parallel_{L_2} + K_9 \sqrt{\varepsilon} \right] \right\}.$$

Choose $\varepsilon > 0$ so that $\frac{K K_9}{a_0} \sqrt{\varepsilon} \le \frac{1}{2}$. Then

$$\parallel w \parallel_{W_2^2} \le \frac{2K}{a_0} \parallel u_t \parallel_{L_2} + \frac{K_{10}}{\sqrt{\varepsilon}} \parallel w \parallel_{L_2} \parallel w \parallel_{W_2^2}. \tag{1.22}$$

We were choosing $\mu(t)$ so that $\parallel w \parallel_{L_2}$ for $\mu = \mu(t)$ to be minimal. Thus,

$$\parallel w \parallel_{L_2} \underset{t \to \infty}{\longrightarrow} 0$$

and from (1.22), for sufficiently large t_0, we obtain

$$\parallel w \parallel_{W_2^2} \le K_{11} \parallel u_t \parallel_{L_2}, \tag{1.23}$$

where K_{11} does not depend on t, and (1.23) holds for $t \subset \Pi(t_0, t_1)$ if t_0 is large enough.

By theorem 1.1 and remark 1.5 from §2.1 (see also corollary 1 from theorem 7.1, Belonosov and Zelenyak, 1975), we have a pair of functions Φ and $\rho \ge \rho_0 > 0$ generating the Liapunov functional (see definition 1.1 from §2.1) and three times continuously differentiable.

Then, for our solution

$$\frac{\partial}{\partial t} \int_0^1 \Phi(x, u, u_x) dx = - \int_0^1 \rho u_t^2 dx; \tag{1.24}$$

therefore, the function $\int\limits_0^1 \Phi(x, u, u_x)dx$ has a limit for $t \to \infty$. Denote this limit by B. Consider the difference,

$$J = \int\limits_0^1 \Phi(x, u, u_x)dx - B \equiv \int\limits_0^1 \Phi(x, u, u_x)dx$$

$$- \int\limits_0^1 \Phi(x, v(x, \mu), v_x(x, \mu))dx.$$

This identity, by our assumption, holds for all $\mu \in [\mu_2, \mu_1]$. Using the Taylor formula and the equality

$$\int\limits_0^1 \left[\frac{\partial \Phi}{\partial u}\bigg|_{u \equiv v} (u - v) + \frac{\partial \Phi}{\partial u_x}\bigg|_{u \equiv v} (u - v)_x \right] dx = \frac{\partial \Phi}{\partial u_x}\bigg|_{u \equiv v} (u - v)\bigg|_0^1$$

$$+ \int\limits_0^1 \left(\frac{\partial \Phi}{\partial u} - \frac{d}{dx}\frac{\partial \Phi}{\partial u_x} \right)\bigg|_{u \equiv v} (u - v)dx = 0, \qquad (1.25)$$

we may obtain the estimate

$$|J| \leq \int\limits_0^1 \left| A_1(u - v)^2 + 2A_2(u - v)(u - v)_x + A_3(u - v)_x^2 \right| dx$$

$$\leq A \parallel w \parallel_{W_2^2}^2, \qquad (1.26)$$

where A is a constant,

$$A_1 = \frac{\partial^2 \Phi}{\partial u_x^2}\bigg|_{\substack{u=\theta_1 \\ u_x=\theta_2}}, \qquad A_2 = \frac{\partial^2 \Phi}{\partial u \partial u_x}\bigg|_{\substack{u=\theta_1 \\ u_x=\theta_2}},$$

$$A_3 = \left. \frac{\partial^2 \Phi}{\partial u_x^2} \right|_{\substack{u=\theta_1 \\ u_x=\theta_2}} ,$$

θ_1, θ_2 are certain intermediate points. The equality (1.25) follows from the fact that $v(x, \mu)$ is a stationary solution of the problem, i.e., solves the following equations

$$0 = \rho(a v_{xx} + b) = \left. \Phi_u - \frac{d}{dx} \Phi_{u_x} \right|_{u \equiv v} ,$$

where v satisfies the boundary conditions. By theorem 7.1 (Belonosov and Zelenyak, 1975) the integrated term vanishes for corresponding Φ and ρ. Estimates (1.23), (1.26) for $t \in \Pi(t_0, t_1)$ and for sufficiently large t_0 yield

$$\frac{dJ}{dt} = -\int_0^1 \rho u_t^2 dx \leq -\rho_0 \int_0^1 u_t^2 dx \leq -\frac{\rho_0}{K_{11}} \parallel w \parallel_{w_2^2}^2 \leq -K_{12} J.$$

The function $J(t)$ monotonically decreases, $\lim_{t \to \infty} J(t) = 0$; therefore, $J(t) \geq 0$. Hence follows that for $t \in \Pi(t_0, t_1)$

$$J(t) \leq J(t_0) e^{-K_{12}(t-t_0)} \tag{1.27}$$

if t_0 is sufficiently large. We must now show that we may choose t_0 so that (1.27) would hold for all $t \geq t_0$. Let

$$t_i \to \infty, \quad |u(x, t_i) - v(x, \bar{\mu})|_{2+\alpha}^{[0,1]} \to 0 \quad \text{for } i \to \infty,$$

where $\mu_2 < \bar{\mu} < \mu_1$. For sufficiently large i, find $\bar{\tau}_i > t_i$ such that for $\tau \in [t_i, \bar{\tau}_i]$, we have

$$\int_0^1 |u(x, \tau) - v(x, \bar{\mu})|^2 dx < \frac{1}{4} \min \left\{ \int_0^1 |v(x, \mu_2) - v(x, \bar{\mu})|^2 dx , \right.$$

$$\left. \int\limits_0^1 |v(x,\mu_1) - v(x,\bar{\mu})|^2 \, dx \right\} = \delta^2$$

Then,

$$2\int\limits_0^1 |u(x,\tau) - v(x,\mu_k)|^2 \, dx \geq \int\limits_0^1 |v(x,\bar{\mu}) - v(x,\mu_k)|^2 \, dx$$

$$-2\int\limits_0^1 |u(x,\tau) - v(x,\bar{\mu})|^2 \, dx \geq \delta^2 \cdot 2.$$

Therefore,

$$\min_{\mu} |u(x,t) - v(x,\mu)|^2_{L_2}$$

is achieved in the interior point of the interval $[\mu_2, \mu_1]$, since the value of this function in the point $\bar{\mu}$ is less than in the ends of the interval. Therefore, for sufficiently large i $\Pi(t_i, \tau_i)$ are not empty and (1.27) holds. Next, we must show that for sufficiently large i, $\tau_i = \infty$, i.e., (1.27) holds for all $t \geq t_i$. Really,

$$J(t) = \int\limits_\tau^\infty d\tau \int\limits_0^1 \rho u_t^2 dx \leq J(t_i) e^{-K_{12}(t-t_i)},$$

and, by lemma 1.4,

$$\int\limits_0^1 |u(x,\tau) - u(x,t_i)|^2 \, dx \leq 2K \int\limits_0^1 |u(x,\tau) - u(x,t_i)|^2 \, dx$$

$$\leq 2K\sqrt{J(t_i)}e^{\frac{K_{12}}{2}} \frac{1}{e^{\frac{K_{12}}{2}}-1} = K_{13}\sqrt{J(t_i)} \tag{1.28}$$

for sufficiently large i and $\tau \in \Pi(t_i, \tau_i)$. If τ_i is finite for all i, we have a contradiction. Really, choose t_i so that

$$16K_{13}\sqrt{J(t_i)} < \delta^2, \quad z(t_i) < \frac{\delta^2}{16}.$$

Then

$$\| u(x,\tau_i) - u(x,t_i) \|_{L_2} < \frac{\delta}{4},$$

$$\| u(x, \tau_i) - v(x, \bar{\mu}) \|_{L_2} \leq \| u(x, \tau_i) - u(x, t_i) \|_{L_2}$$

$$+ \| u(x, t_i) - v(x, \bar{\mu}) \|_{L_2} \leq \frac{1}{2}\delta,$$

$$\| u(x, \tau_i) - v(x, \mu_k) \|_{L_2} \geq \| v(x, \mu_k) - v(x, \bar{\mu}) \|_{L_2}$$

$$- \| u(x, \tau_i) - v(x, \bar{\mu}) \|_{L_2} > \delta.$$

From these inequalities it follows that

$$\min_{\mu} \| u(x, \tau_i) - v(x, \mu) \|_{L_2}$$

is not achieved for $\mu = \mu_k$, $k = 1, 2$. Therefore, $\tau_i = \infty$. In this case

$$\| u(x, \tau) - v(x, \mu_k) \| > \frac{\delta}{2}$$

for all τ; consequently, $v(x, \mu_k)$ are not partial limits of the solution of problem (1.1)-(1.3). Compactness of the solution in $C^{2+\beta}$ and its convergence in L_2 complete the proof of the theorem.

Theorem 1.1 reflects one of the basic qualitative properties of a problem (1.1)-(1.3) solution - stabilization over extended time.

However, for an equation with many spatial variables this theorem becomes false. In later chapter we shall consider examples which show that it is so.

Now we shall point out the basic properties of the problem which were used when we proved theorem 1.1.

1. The problem has Liapunov functional. This allows us to establish the quasistabilization of bounded in C^1 solution. Analogous results are known in the theory of dynamic systems of the form $x_t = \nabla f(x)$. Quasistabilization of solution, bounded in a corresponding norm, as will be shown below, is the corollary of the existence of the Liapunov functional in wide class of mixed problems for multidimensional parabolic equations and systems.

2. When we have the continuum set of stationary solutions, these solutions compose of a one-parameter set. The problem linearized on

the functions of this set has the trivial eigenvalue of unit multiplicity. The proof may be extended for the following case. The stationary solutions of the problem form a k-parameter smooth manifold in a Banach space, where the linearization of the infinitesimal operator of the corresponding semi-group on these solutions leads to the selfadjoint elliptic operator (since the regular Liapunov functional exists). This operator has the zero as the spectrum point of multiplicity k.

The method which is based on these properties allows us to consider the problem of which data depend on time explicitly and converge to their limits for $t \to \infty$ with a certain speed.

We shall give without proof, the theorem on stabilization of the nonautonomous problem (Provorova, 1969, 1973; Zelenyak, 1972).

Consider the problem

$$\rho(x, u, u_x, t) u_t = \Phi_u - \frac{d}{dx} \Phi_{u_x}, \tag{1.29}$$

$$u_x \big|_{x=0} = \varphi(u), \quad u \big|_{x=1} = 0, \tag{1.30}$$

where Φ is four times continuously differentiable; u_0 is three times continuously differentiable, $\rho \in C^2$,

$$\rho \geq \rho_0 > 0, \quad -\frac{\partial^2 \Phi}{\partial u_x^2} \geq a_0 > 0.$$

Here ρ_0, a_0 are constants, φ is two times continuously differentiable. Let

$$\left\| \Phi(x, u, u_x, t) - F(x, u, u_x) \right\|_2 \leq \frac{K}{t^{n+\varepsilon}}, \tag{1.31}$$

$$\left\| \rho(x, u, u_x, t) - \rho(x, u, u_x) \right\|_2 \leq \frac{K}{t^{n+\varepsilon}}, \quad \left\| \frac{\partial \Phi}{\partial t} \right\|_2 \leq \frac{K}{t^{n+1+\varepsilon}}$$

for each of the arguments u, u_x, $0 < x < 1$, $\varepsilon > 0$.

Theorem 1.2. (Provorova, 1973) Let condition (1.31) hold for $n = 0$; and F, ρ be smooth functions. Then each partial limit of bounded by the norm $C^{2+\alpha}$ solution of problem (1.29)-(1.30) solves the equation

$$\frac{\partial F}{\partial u} - \frac{d}{dx}\frac{\partial F}{\partial u_x} = 0$$

and satisfies conditions (1.30). If (1.31) holds for $n = 1$, then there exists the limit of this solution for $t \to \infty$.

The proof of this theorem is analogous to the proof of theorem 1.1. In this case, the following modification of lemma 1.4 is used (it is derived from Fokin M.V.).

Lemma 1.5. Let for $t > 0$

$$\int\limits_{t}^{\infty}\int\limits_{0}^{1} u_t^2 \, dx \, dt \leq \frac{K}{t^{n+\varepsilon}}, \quad 0 < \xi < \eta.$$

Then

$$\int\limits_{0}^{1} |u(x,\xi) - u(x,\eta)| \, dx \leq 2 \sup_{t \geq 0} |u| \sqrt{2} K M(\varepsilon) \frac{1}{\xi^{\varepsilon/2} + 1}.$$

Instead of conditions (1.30) we may take more general conditions. Note, that this theorem is unimprovable.

Example 1.1. The solution of the problem

$$u_t = u_{xx} + u + \frac{\cos \ln \ln t}{t \ln t} \sin x$$

$$u\big|_{t=e} = u_x\big|_{x=0} = u_x\big|_{x=\pi} = 0$$

is the function $u = \sin x \sin \ln \ln t$. Each partial limit of this function is a stationary solution; however, $\lim\limits_{t \to \infty} u$ does not exist.

Now we shall prove the important corollary of the existence of the Liapunov functional of the evolutionary (not necessarily normally parabolic) problem (definition 1.1, §2.1).

Consider in a domain $\Omega \subset R^n$ with a smooth boundary a vector-function $u(x,t) = (u_1, \ldots, u_m)$ determined in the set $\Omega \times [0, \infty)$. Let this function solve the system

$$\rho_j \frac{\partial u_j}{\partial t} = \frac{\partial F}{\partial u_j} - \sum_{i=1}^{n} \frac{d}{dx_i} \frac{\partial F}{\partial u_{jx_i}}, \quad j = 1, \ldots, m \qquad (1.32)$$

$$u(x, 0) = u_0(x)$$

in the classic sense, with one of the following boundary conditions

$$u_j(x,t)|_{x \in \partial\Omega} = 0, \qquad j = 1, \ldots, m;$$

or

$$\sum_{i=1}^{n} \frac{\partial F}{\partial u_{jx_i}} \cos(n, x_i)|_{x \in \partial\Omega} = 0, \qquad j = 1, \ldots, m.$$

Here $\rho_j, F(x, u, \nabla u)$ are smooth functions ($\rho_j > 0$, $j = 1, \ldots, m$); n is a unit exterior normal vector to $\partial\Omega$; $\nabla u = (u_{1x_1}, \ldots, u_{mx_n})$.

Evidently, problem (1.32) has the Liapunov functional of the form

$$\frac{\partial}{\partial t} \int_\Omega F(x, u, \nabla u) dx = \sum_{j=1}^{m} \int_\Omega \rho_j(x, u, \nabla u) u_{jt}^2 dx dt. \qquad (1.33)$$

Theorem 1.3. Let $u(x,t) \in H^{2+\alpha}(Q)$ solve (1.32) and be uniformly bounded in this norm. Then the solution $u(x,t)$ is quasistabilized.

Proof. Let $v_0(x)$ belong to the ω-limit set of $u(x,t)$, i.e., there exists such a sequence $t_n \to \infty$ that $u(x, t_n) \to v_0(x)$ in $C^{2+\alpha}(\Omega)$. Prove that $v_0(x)$ is a stationary solution of (1.32). Consider the sequence

$$w_n(x,t) = u(x, t_n + t) \quad \text{for} \quad (x,t) \in K_\tau = \Omega \times [0, \tau].$$

This sequence is bounded in $H^{2+\alpha}(K_\tau)$, and therefore, is compact in $H^{2+\alpha'}(K_\tau)$ ($\alpha' < \alpha$). Taking a converging subsequence (saving the notations), we obtain that the limit function $v(x,t) \in H^{2+\alpha}(K_\tau)$; moreover,

$$\| u(x, t_n + t) - v(x, t) \|_{2+\alpha'} \to 0, \quad n \to \infty.$$

As $u(x, t_n) \to v_0(x)$, then $v(x, 0) = v_0(x)$.

Passing to the limit in (1.32) written for the function $u(x, t_n + t)$, we obtain that the function $v(x, t)$ solves problem (1.32) with the initial data $v_0(x)$.

Since, by (1.33)

$$P(t) \overset{\text{def}}{=} \sum_{j=1}^m \int_0^t \int_\Omega \rho_j u_{jt}^2 \, dx \, dt = \int_\Omega F(x, u, \nabla u) dx \Big|_{t=0}^t,$$

and the function $P(t)$ is monotone, if $P(t_n)$ has a finite limit, then $P(t)$ has a finite limit for $t \to \infty$, and therefore, is bounded.

Consequently,

$$\sum_{j=1}^m \int_{t_n}^t \int_\Omega \rho_j(x, u, \nabla u) u_{jt}^2 \, dx \, dt \to 0$$

for $n \to \infty$, $t \in [t_n, t_n + \tau]$. Going to the limit, we obtain

$$\sum_{j=1}^m \int_0^t \int_\Omega \rho_j(x, v, \nabla v) v_{jt}^2 \, dx \, dt = 0, \quad t \in [0, \tau].$$

This is possible only when $v_{jt}^2(x, t) \equiv 0$ $(j = 1, \ldots, m)$, i.e., $v(x, t) \equiv v_0(x)$ and is a stationary solution of the problem. The theorem is proved.

Remark 1.2. This proof does not require the local solvability of the problem. In particular, the theorem holds for equations which are not normally parabolic, for example, for degenerate equations and for the equations with inverse time (the equations where the leading coefficients are negative $a \le a_0 < 0$).

The following corollary of theorem 1.3 holds.

Theorem 1.4. Let $u(x,t) \in H^{2+\alpha}(Q)$ be a solution of problem (1.1)-(1.3), uniformly by $t > 0$ bounded in $C^{2+\alpha}[0,1]$, which has the Liapunov functional (definition 1.1, §2.1), and the coefficient a of equation optionally satisfied (1.4).

Then the solution $u(x,t)$ is quasistabilized.

If, in addition, one of the two following conditions hold:

1) the problem has no more than a countable number of stationary solutions;

2) the function a is separated from zero;

then the solution $u(x,t)$ is stabilized.

Stabilization in the case of a countable set of stationary solutions is the corollary of connectedness of the partial limits set. This follows from the theorem conditions (from the supposed smoothness of the solution).

Stabilization for nondegenerate function a (as it may be negative) is established similar to the theorem 1.1 proof.

Corollary 1.1. Under the theorem 1.4 conditions for $a \neq 0$ the bounded for $t \in (-\infty, +\infty)$ solution of (1.1)-(1.3) may be only a connected orbit (the definition of the connected orbit see in Brunovsky and Fiedler, 1988, 1989).

To illustrate the proved theorem let us give some examples.

Example 1.2. Consider the problem

$$u_t = \frac{\partial}{\partial x} a'(u_x) + f(u), \tag{1.34}$$

$$u(0,t) = u(1,t) = 0, \qquad u(x,0) = u_0(x).$$

The function $a''(\eta) \in C^1$ is assumed to be alternating, asymptotically positive for large $|\eta|$. Not repeating the results of chapter 2, we refer to this possibility of application of theorems 1.3, 1.4.

It is easy to show (Lar'kin et al., 1983) that the sufficiently smooth solution of (1.34) satisfies the identity

$$\int\limits_{t_1}^{t_2}\int\limits_0^1 u_t^2\,dx\,dt = \int\limits_0^1 \Big(-a(u_x)+F(u)\Big)dx\Big|_{t=t_1}^{t_2},$$

where F is a primitive for f.

By theorem 1.3, the smooth solution of this problem is quasistabilized.

Example 1.3. Here we shall consider the so-called Cahn-Hilliard equation which is not of the form (1.1)-(1.3). However, the existence of the Liapunov functional allows us to repeat the proof of theorem 1.3. Let us formulate the results obtained. Consider the problem

$$u_t = -\gamma u_{xxxx} + \big(a'(u_x)\big)_x, \tag{1.35}$$

$$\gamma u_{xxx}(i,t) - a'(u_x(i,t)) = u_x(i,t) = 0, \qquad i = 0,1;$$

$$u(x,0) = u_0(x),$$

where the function a is taken from example 1.2. If we denote $v(x,t) = u_x(x,t)$, then this problem is easily reduced to the Cahn-Hilliard problem (Elliot and French, 1989) suggested by Cahn and Hilliard (1958) to describe the process of the phase separation in a cooling binary metal solution. The solution of this problem, the existence of which is established in the standard way, has some interesting qualitative properties. They have been obtained both numerically and theoretically (Alikakos *et al.*, 1991; Bates and Zheng Songmu, 1992; Furihata *et al.*, 1993). Note that the existence theorem for this problem considered as the regularization of (1.34) was proved by Zelenyak *et al.*(1974).

We confine ourself to formulate the analog of theorem 1.3.

It is easy to see that problem (1.35) has a Liapunov functional of the form

$$\int\limits_{t_1}^{t_2}\int\limits_0^1 u_t^2\,dx\,dt = -\int\limits_0^1 \Big(\gamma\frac{u_{xx}^2}{2}+a(u_x)\Big)dx\Big|_{t=t_1}^{t_2}.$$

Therefore, the following theorem holds.

Theorem 1.5 Let $u(x,t)$ be uniformly by $t > 0$ bounded in the $C^{4+\alpha}[0,1]$ norm and solve (1.35). Then $u(x,t)$ is quasistabilized.

If, problem (1.35) has no more than a countable set of stationary solutions, then $u(x,t)$ is stabilized.

Thus, the existence of the Liapunov functional, allows us to investigate the behavior of solution for extended time.

§3.2   The discrete Liapunov functional.

Theorem 1.1 on stabilization of (1.1)-(1.3) solutions was proved by Zelenyak (1968) by means of the Liapunov functionals (1.6). Begining with Matano (1978), mathematicians began to use the so-called discrete Liapunov functionals. This method makes use of the maximum principle and of the fact that the problem is one-dimensional.

It turns out (Matano, 1978, 1979, 1982; Angenent, 1988) that if $u(x,t;u_0)$ is a sufficiently smooth solution of (1.1)-(1.3), then for each $t > 0$ the number of intersections $z_y(t)$ of its chart with the chart of each stationary solution $y(x)$ is finite and does not increase when t grows. So the number of intersections is a nonnegative function with respect to time attaining for $t > 0$ the natural values. This function is called a discrete Liapunov functional.

The number of sign changes for a function $\varphi(x)$ we shall denote by $z(\varphi)$.

For the discrete Liapunov functional, the following statements hold.

Let $v(x,t)$ solve the boundary-value problem for the linear parabolic equation with one spatial variable

$$v_t = d(x,t)v_{xx} + g(x,t)v_x + c(x,t)v, \qquad (2.1)$$

$$\alpha_i v_x(i,t) - \beta_i(t)v(i,t) = 0, \quad i = 1,0,$$

$$v(x,0) = v_0(x),$$

where

$$v_0(x) \in W_2^1[0,1], \quad d(x,t) \in C^2(\bar{Q}), \quad d(x,t) \geq d_0 > 0,$$

$$g(x,t), \; g_t(x,t), \; g_x(x,t), \; c(x,t) \in L^\infty_{\text{loc}}(Q).$$

(The definition of $L^\infty_{\text{loc}}(Q)$ may be taken, for example, from Ladyzhen-skaya *et al.*(1967)).

Theorem 2.1. (Angenent, 1988). Let the above assumptions hold. Then the number of sign changes of $z(v(x,t))$ possesses the following properties.

1) $z(v(x,t))$ is finite for each $t > 0$ (even if the number of sign changes $z(v(x,t))$ possesses the following properties $z(v_0(x)) = \infty$).

2) $z(x(x,t))$ is strictly decreasing in the point $t = t_0 > 0$ if and only if $v(x,t_0) \not\equiv 0$, $x \to v(x,t_0)$ has a multiple zero in a certain point $x_0 \in (0,1)$. This denotes that $v(x_0,t_0) = v_x(x_0,t_0) = 0$.

3) If $v(x,t_0) \equiv 0$, then $v(x,t) \equiv 0$ for each t.

Now we consider equation (1.1)-(1.3) which was considered in §1 with all the suppositions which were made there. Then, the following theorem holds (it is the corollary of theorem 2.1).

Theorem 2.2. Let $u(x,t;u_0^1)$, $u(x,t;u_0^2)$, $u_0^1(x) \not\equiv u_0^2(x)$ be two solutions of problem (1.1)-(1.3). Then

1) $z\left(u(x,t;u_0^1) - u(x,t;u_0^2)\right)$ is finite for $t > 0$;

2) $z\left(u(x,t;u_0^1) - u(x,t;u_0^2)\right)$ does not increase for $t \geq 0$;

3) there exists $t_0 > 0$ such that for $t \geq t_0$ $z\left(u(x,t;u_0^1) - u(x,t;u_0^2)\right)$ is constant;

4) the properties 1) - 3) hold true if instead of $u(x,t;u_0^1) - u(x,t;u_0^2)$ we consider $u_t(x,t;u_0)$, $u_t \not\equiv 0$ (in this case the nonlinear terms have to be sufficiently smooth).

The existence of the discrete Liapunov functional for the parabolic problem (like the existence of the Liapunov functionals set) is the basic property of the problem. These properties allow us to make some proofs clearer. We shall point out those problems which become more evident from the point of view of these concepts.

The following theorem holds (Belonosov and Zelenyak, 1975; Provorova, 1969).

Theorem 2.3. Let for problem (1.1)-(1.3) condition (B) hold (see definition 1.2 from §2.1) $u, u_0 \in C^1[0,1]$.

Then there exists such T (finite or infinite), that the solution $u(x,t)$ belongs to $H^{2+\alpha}(Q_T)$, and one of two properties holds

i) $\lim\limits_{t \to T} \sup\limits_{0 \leq x \leq 1} |u(x,t)| = \infty$.

ii) The solution $u(x,t)$ is determined for all $t > 0$ and is stabilized. This means that such a stationary solution $v(x)$ exists that

$$\| u(x,t) - v(x) \|_{2+\alpha}^{[0,1]} \to 0 \quad \text{for} \quad t \to \infty.$$

Remark 2.1. This statement is the corollary of theorem 3.1 (chapter 2) and theorem 1.1 of this chapter.

The proof of the theorem requires the derivatives (we know, only, that $|u|$ is bounded) and estimates of this could not be realized by means of the discrete functional only. When we have to estimate the rate of convergence of a solution, or when we consider problems with time dependent coefficients (Belonosov and Zelenyak, 1975; Provorova, 1969), the discrete functional method is useless.

It seems to be impossible to apply the discrete functionals method for the problems with one spatial variable, when the maximum principle does not hold (for example, theorem 1.3 and the Cahn-Hilliard equation, example 1.3), or to the equations with many spatial variables. For the problems with many spatial variables, similarly (Matano, 1978), we may define the number of sign changes for solutions only in particular cases. But simple geometric considerations show that to establish the monotonicity of this value, we have essential difficulties.

It should be noted that the first results obtained by this method were weaker than in theorem 1.1 (Matano, 1978, 1979). Furthermore, the cases considered were different from those in theorem 1.1: the stabilization of the solution of the periodically time-dependent problem to the periodic solution with the same period (Brunovskii et al., 1992); or with the periodic boundary conditions (Fiedler and Mallet-Paret, 1989).

The following theorem gives the answer to the problem posed by Zelenyak (1972) and is the corollary of the statement formulated by Angenent (1988) and by Brunovskii et al.(1992). Similar results may be obtained by means of functional (1.5), chapter 2.

Suppose, for simplicity, $a, b, \varphi_i \in C^4$ by all their arguments.

Theorem 2.4.   For the solution $u(x, t; u_0)$ of (1.1)-(1.3) one of the following statements holds.

i) There exists such $T > 0$ (finite or infinite) that

$$\lim_{t \to T} \| u(x, t; u_0) \|_1 = \infty;$$

ii) The solution $u(x, t; u_0)$ is uniformly bounded by $t > 0$ in the space $C^{2+\alpha}[0, 1]$ and is stabilized for $t \to \infty$ (has a limit for $t \to \infty$ which is a stationary solution of the problem).

Remark 2.2.   Theorem 2.4 becomes a trivial corollary of theorem 1.1 if i) is replaced as follows

i') there exists such $T > 0$ (finite or infinite) that

$$\overline{\lim_{t \to T}} \| u(x, t; u_0) \|_1 = \infty.$$

Remark 2.3.   The theorem conditions do not include condition (B) (definition 1.2, §2.1).

Before we prove the theorem we must formulate a corollary.

Corollary 2.1.   Let $u(x, t; u_0)$ be a classic solution of (1.1)-(1.3) determined for all $t \geq 0$ and

$$\lim_{t \to T} \| u(x, t; u_0) \|_1 < \infty.$$

Then the solution $u(x, t; u_0)$ is stabilized by the norm $C^{2+\alpha}[0, 1]$.

Corollary 2.2.   Let all the suppositions of theorem 2.4 hold and $u(x, t; u_0)$ be a bounded in $C^1(Q)$ solution of (1.1)-(1.3). Then $u(x, t; u_0)$ is stabilized for $t \to +\infty$.

Remark 2.4.   Note, that when we prove theorem 2.4 or corollary 2.2 we do not need the existence of the Liapunov functional. Therefore, it is not necessary that the condition (B) holds.

In the proof of theorem 2.4 we shall use the following result (Angenent, 1988; Brunovskii *et al.*, 1992).

Theorem 2.5. Let functions d, d_t, d_x, d_{xx}, g, g_t, g_x, $c(x,t)$ be continuous in $Q = [0,1] \times [0, +\infty)$; $d(x,t) \geq d_0 > 0$ and either for $i = 0$ or for $i = 1$ the boundary condition has the form

$$v(i,t) = 0 \quad (\alpha_i = 0, \quad \beta_i(t) \equiv 1).$$

Then, either $v(x,t) \equiv 0$, or there exists such t^* that for $t > t^*$ the solution $v(x,t)$ of the problem (2.1) for given i satisfies the inequality

$$v_x(i,t) \neq 0.$$

Remark 2.5. In the papers of Angenent (1988) and Brunovskii et al.(1992) this result is formulated also for arbitrary boundary conditions from (2.1). Then, starting with a certain t^* in the corresponding part of the boundary the inequality $v(i,t) \neq 0$ holds. The conditions imposed by the authors do not require the uniform boundedness by t of equation coefficients and their derivatives. Then, for the case of the third boundary conditions $\alpha_i \neq 0$, $\beta_i \neq 0$ (Brunovskii, et al.1992) the uniform boundedness of $\beta_i(t)$ is required in the space $C^1[0, \infty)$. This restriction may easily be overcome by the change

$$w(x,t) = v(x,t)e^{(i-x)\beta_i}.$$

This change, saving the solution value in the considered part of the boundary turns the condition $v_x(i,t) = \beta_i(t)v(i,t)$ into the homogeneous condition $w_x(i,t) = 0$.

Proof of theorem 2.4. We shall give the proof for the case of the first boundary-value problem, and point out the differences for the other boundary conditions.

By lemma 1.1, if the solution exists only for a finite time $t \in (0, T)$, then

$$\varlimsup_{t \to T} \| u(x,t;u_0) \|_1 = \infty.$$

Hence, by lemma 1.1, we have

$$\lim_{t \to T} \| u(x,t;u_0) \|_1 = \infty.$$

This proves the statement i) of the theorem.

So, let the solution $u(x,t;u_0)$ be determined for all $t \geq 0$ and

$$\overline{\lim_{t \to \infty}} \| u(x,t;u_0) \|_1 = \infty.$$

(In the case when the upper limit is bounded the stabilization follows from theorem 1.1). Taking into account the additional smoothness of coefficients, we see that $u(x,t;u_0) \in C^{4+\alpha}[0,1]$ for $t > 0$ (Belonosov and Zelenyak, 1975; Ladyzhenskaya and Ural'tseva, 1986). As we are interested in the behavior of solutions for extended time, without loss of generality, we shall consider that $u \in C^{4+\alpha}[0,1]$ for $t \geq 0$.

By the theorem supposition, there exists a finite partial limit $v_0(x) = \lim_{t \to \infty} u(x,t_i;u_0)$, where the sequence of norms $\| u(x,t^i;u_0) \|_1$ is uniformly bounded. Then, by lemma 1.2, there exists a partial limit $v_0(x) \in C^{3+\alpha}[0,1]$, where

$$\| u(x,t_i;u_0) - v_0(x) \|_{2+\alpha} \to 0. \tag{2.2}$$

By the change of the desired function $w = u - v_0$ this element of the ω-limit set turns into zero. Further, we shall assume that $0 \in \omega(u_0)$.

The derivative $u_t(x,t;u_0)$, evidently, solves the problem (2.1) with

$$d(x,t) = a\Big(x, u(x,t;u_0), u_x(x,t;u_0)\Big),$$

$$g(x,t) = \frac{\partial}{\partial u_x}\Big(a(x,u,u_x)u_{xx} + b(x,u,u_x)\Big), \tag{2.3}$$

$$c(x,t) = \frac{\partial}{\partial u}\Big(a(x,u,u_x)u_{xx} + b(x,u,u_x)\Big).$$

Therefore, by theorem 2.5, for sufficiently large $t > t^*$, $u_{xt}(0,t;u_0) \neq 0$. This means that $u_x(0,t;u_0)$ is monotone (in the cases of another boundary conditions, the function $u(0,t;u_0)$ will be monotone). Then, from the existence of the trivial partial limit, it follows that

$$\lim_{t\to\infty} u_x(0,t) = 0. \tag{2.4}$$

In particular, it follows that $\omega(u_0)$ consists of the functions $\varphi(x)$ such that

$$\varphi(0) = \varphi'(0) = 0. \tag{2.5}$$

For a fixed positive τ consider in the rectangle $Q_\tau = [0,1] \times [0,\tau]$ a functional sequence $\{u(x, t_i + t; u_0)\}$. By the theorem on continuous dependence of solution on initial data in Q_τ, this sequence converges in the norm of $C^{2+\alpha}$ to a solution of (1.1)-(1.3) with the initial data $u(x, t; 0)$ (see (2.2)). Moreover, by (2.5), we have the additional boundary condition

$$u(0, t; 0) = 0, \qquad u_x(0, t; 0) = 0.$$

Consider again the derivative $w(x, t) = u_t(x, t; 0)$ determined in Q_τ, which solves the linear problem (2.1) with the additional boundary condition $w_x(x, t) = 0$. As was shown by Angenent (1988), it is possible only in the case when

$$w(x, t) = u_t(x, t; 0) \equiv 0.$$

Thus, taking into account the boundary conditions, we have

$$u(x, t; 0) \equiv 0 \quad \text{in} \quad Q_\tau,$$

i.e., the function $v_0(x)$ (see (2.3)) is a stationary solution of initial problem (1.1)-(1.3).

Analogously we may prove that each element of the ω-limit set is a stationary solution. As the Cauchy problem for the ordinary differential equation has a unique solution, it denotes (take into account (2.5)) that $\omega(u_0)$ consists of a unique trivial element, i.e., the solution $u(x, t; u_0)$ is stabilized.

We prove now that $u(x, t; u_0)$ is bounded in $C^1[0, 1]$ uniformly by t, and hence it will follow that $u(x, t; u_0)$ is bounded in the $C^{3+\alpha}[0, 1]$ norm. Consider the sequence

$$u(x, t_n; u_0) \to 0 \quad \text{for} \quad t_n \to \infty$$

and the intervals $[t_{n-1}, t_n]$. If the solution (smooth in each rectangle of a finite height T) does not tend to 0 in the $C^1[0, 1]$ norm for $t \to \infty$, then such $\varepsilon > 0$, $\tau_n \in [t_{n-1}, t_n]$ exist that

$$\| u(x, \tau_n; u_0) \|_1 = \varepsilon, \tag{2.6}$$

$$\| u(x, t; u_0) \|_1 < \varepsilon \quad \text{for} \quad t \in [t_{n-1}, \tau_n).$$

Then, by lemma 1.1, in $P_n = [0, 1] \times [t_{n-1}, \tau_n]$ the inequalities

$$\| u(x, t; u_0) \|_{2+\alpha}^{P_n} \leq C\varepsilon$$

hold. Therefore, we may choose from $\{u(x, \tau_n; u_0)\}$ a converging subsequence. This subsequence must converge to 0 (as it was shown above, $\omega(u_0) = \{0\}$), which contradicts (2.6). Thus, $u(x, t; u_0)$ is uniformly bounded by t in the space $C^1[0, 1]$. The theorem is proved.

§3.3 Qualitative properties of mixed problem solutions for nonlinear parabolic equations.

In this paragraph we shall state the behavior of mixed problem global solutions, i.e., we shall describe the attraction domains of stationary solutions, and the properties of solutions near the boundary of its domain of definition. We shall formulate some statements about the number of stationary solutions of mixed problems.

We shall assume that the coefficients are smooth (see §3.1), and also that the condition (B) holds (definition 1.2, §2.1).

Consider the equation

$$u_t = a(x, u, u_x)u_{xx} + b(x, u, u_x);$$ (3.1)

one of the boundary conditions

$$u\big|_{x=0} = u\big|_{x=1} = 0,$$ (3.2)

$$u_x\big|_{x=0} = \varphi(u)\big|_{x=0}, \quad u\big|_{x=1} = 0,$$ (3.3)

$$u_x\big|_{x=0} = \varphi(u)\big|_{x=0}, \quad u_x\big|_{x=1} = \psi(u)\big|_{x=1},$$ (3.4)

and, also, the initial condition

$$u\big|_{t=0} = u_0(x).$$ (3.5)

Let $v(x, x_0, y_0, y_1)$ be such that

$$v\big|_{x=x_0} = y_0, \quad v_x\big|_{x=x_0} = y_1$$

and

$$a(x, v, v_x)v_{xx} + b(x, v, v_x) = 0.$$ (3.6)

Set

$$y_1(x, \alpha) = v(x, 0, 0, \alpha;), \quad y_2(x, \alpha) = v(x, 0, \alpha, \varphi(\alpha)).$$

The following lemma holds.

Lemma 3.1. (Akramov and Zelenyak; 1975) Let the zero be the asymptotically stable solution of each of the problems in question. Let

$$y_1(\xi_i, \alpha_1) = 0, \quad y_2(\eta_k, \alpha_2) = 0 \quad \text{for}$$

$$0 < \xi_1 < \ldots < \xi_n \le 1, \quad 0 < \eta_1 < \ldots < \eta_m \le 1, \quad \alpha_k > 0.$$

Then there exist not less than $n + 1$ solutions of (3.6), (3.2); not less than $m + 1$ solutions of (3.6), (3.3); and not less than m solutions of (3.6), (3.4).

Before we prove the lemma we need one auxiliary statement. This statement deals with the behavior of Liapunov functionals in the neighborhoods of stationary solutions; namely, it is the criterion of asymptotic stability of solutions to the following problem.

Let

$$u_t = p(x)u_{xx} + q(x)u_x + r(x)u = Lu, \tag{3.7}$$

$$u\Big|_{t=0} = u_0(x), \tag{3.8}$$

be the problem, and one of the following boundary conditions hold:

$$u\Big|_{x=0} = u\Big|_{x=1} = 0, \tag{3.9}$$

$$u_x + \alpha u\Big|_{x=0} = u\Big|_{x=1} = 0, \tag{3.10}$$

$$u_x + \alpha u\Big|_{x=0} = u_x + \beta u\Big|_{x=1} = 0. \tag{3.11}$$

Let $p \geq p_0 > 0$, q, r, p' be continuous functions; p_0, α, β be constants. Setting $u = \rho w$, where

$$\rho = e^{-\frac{\beta - \alpha - 2}{2}x^2 - (1+\alpha)x}$$

we obtain for w that conditions (3.9)-(3.11) transform into conditions of the same form with $\alpha = -1$, $\beta = 1$. Therefore, without loss of generality we may take $\alpha = -1$, $\beta = 1$, since the property of asymptotic stability of the trivial solution is saved under this change.

Theorem 3.1. (Zelenyak, 1977). The asymptotic stability of solutions to the above problems in C^1 is equivalent to each of the conditions below.

1. The operator L spectrum determined on the functions which satisfy the corresponding boundary condition is negative.

2. In the case of conditions (3.9) there exists a solution $\rho > 0$ of the equation $L\rho = 0$. For conditions (3.10), we have, in addition, $\rho_x + \alpha\rho|_{x=0} < 0$. For conditions (3.11), this solution satisfies the inequalities

$$\rho_x + \alpha\rho\Big|_{x=0} < 0, \qquad \rho_x + \beta\rho\Big|_{x=1} > 0.$$

3. There exist functions $\mu_i > 0$, $\rho > 0$ and constants $\alpha_1, \beta_1 \geq 0$ such that the corresponding problem has the Liapunov functional in the integral form. This means that for each solution we have

$$\frac{\partial}{\partial t} \int_0^1 \mu_1 u^2 dx = -\alpha_1 u^2(0, t) - \beta_1 u^2(1, t) - \int_0^1 \mu_2(\rho u)_x^2 dx.$$

This theorem will be considered in the more general form in chapter 4; therefore, its proof will be omitted now.

Proof of lemma 3.1. The lemma suppositions and theorem 3.1 yield that $y_{k\alpha}(x, \alpha) \geq 0$ for $x > 0$, and for sufficiently small α. Therefore, $y_k(x, \alpha) > 0$ for sufficiently small $\alpha > 0$. If $y_k(1, \alpha) > 0$ for all α, $0 \leq \alpha \leq \alpha_k$, then $y_k(x, \alpha)$ has no zeros. Really, if α_0 is such that $y_k(x, \alpha) \geq 0$ for $0 \leq \alpha \leq \alpha_0$, $y_k(\xi, \alpha_0) = 0$, then by the continuity of y_k by its arguments, for the interior point ξ, we have

$$y'_{k\xi}(\xi, \alpha_0) = 0,$$

i.e., $y_k(x, \alpha_0) \equiv 0$, as a solution of (3.11) satisfying the trivial initial data in the point ξ. By assumption, $y \equiv 0$ also solves this equation; therefore, we obtain a contradiction since

$$y'_1(0, \alpha_0) = \alpha_0, \quad y_2(0, \alpha_0) = \alpha_0, \quad \alpha_0 \neq 0.$$

The function $y_k(x, \alpha)$ in a neighborhood of its zero, which is an interior point of the interval $[0, 1]$ changes the sign. By its continuity, for α' sufficiently close to α, the number of sign changes of the function $y_k(x, \alpha')$ may differ from the number of changes of the function $y_k(x, \alpha)$ only if $y_k(1, \alpha) = 0$. As $y_k(x, \alpha) > 0$ for small α, it follows that if α

changes from 0 up to α_k, the function $y_1(1, \alpha)$ has not less than $n + 1$ zeros, and the function $y_2(1, \alpha)$ has not less than $(m + 1)$ zeros. So, the statement of the lemma relative to problems (3.1), (3.2) and (3.1), (3.3) is proved.

Consider the function (see condition (3.4))

$$\theta(\alpha) = y_2'(1, \alpha) - \psi(y_2(1, \alpha)).$$

Theorem 3.1 yields $y_{2\alpha}|_{\alpha=0} > 0$, i.e., y_2 is positive for small positive α. Therefore, the function $\theta(\alpha)$ for α, which changes from 0 up to α_2, becomes zero not less than m times. Really, we have proved that $y_2(1, \beta_i) = 0$ for $0 < \beta_1 < \ldots < \beta_m < \alpha_2$. Let for $\alpha < \beta_k$ and α sufficiently close to β_k we have

$$y_2(1, \alpha) < 0, \qquad y_2'(1, \beta_k) < 0,$$

and for $\alpha > \beta_k$ and small $|\alpha - \beta_k|$ we have $y_2(1, \alpha) > 0$. Then the number of zeros of the function $y_2(x, \alpha)$ for $\alpha > \beta_k$, $|\alpha - \beta_k|$ being small, is less than the number of zeros of the function $y_2(x, \beta_k)$.

Really, the number of zeros of the function $y_2(x, \alpha)$ is not less than the number of zeros being inside the interval $[0, 1]$ of the function $y_2(x, \beta_k)$. We have $y_2(1, \alpha) > 0$ for $\alpha > \beta_k$ and small $|\alpha - \beta_k|$, and since

$$y_2(x, \alpha) - y_2(1, \alpha) = y_2'(\theta, \alpha)(x - 1) > 0,$$

then, near the point $x = 1$ for α close to β_k, we have $y_2(x, \alpha) > 0$. Consequently, the number of zeros of the function $y_2(x, \alpha)$ for $\alpha > \beta_k$ and $\alpha - \beta_k$ being small is less on the unit than the number of zeros of the function $y_2(x, \beta_k)$. The analogous situation holds if

$$y_2(1, \alpha) > 0, \qquad y_2'(1, \beta_j) > 0$$

for small $\alpha - \beta_j < 0$: the number of zeros of the function $y_2(1, \beta_j)$ is greater on the unit than the number of zeros of the function $y_2(x, \alpha)$ for $\alpha > \beta_j$, if for $\alpha > \beta_j$ and small $\alpha - \beta_j$ we have $y_2(1, \alpha) < 0$.

Let β_1 be such that

$$y_2(1, \beta_1) = 0, \qquad y_2(1, \alpha) > 0$$

for $0 < \alpha < \beta_1$. Then, evidently, $y_2'(1, \beta_1) < 0$, since in the point $x = 1$ the function y_2 obtains its minimal value. Let, further, $\beta_1, \ldots, \beta_j$ be such that

$$y_2(1, \beta_j) = 0, \qquad \mathrm{sgn} y_2'(1, \beta_j) = (-1)^j$$

and the functions $y_2(x, \beta_j)$ have exactly j zeros. Then, if for each $\alpha_0 > \beta_j$ for which $y_2(1, \alpha_0) = 0$ we have $\mathrm{sgn} y_2'(1, \alpha_0) = (-1)^j$, then for each $\alpha > \beta_j$ the function $y_2(x, \alpha)$ may have no more than j zeros. Therefore, without loss of generality we may consider that

$$\mathrm{sgn} y_2'(1, \beta_j) = (-1)^j, \quad y_2(1, \beta_j) = 0, \quad j < m.$$

But then $\mathrm{sgn}\theta(\beta_j) = (-1)^j$ for $j = 1, \ldots, m - 1$. By the lemma supposition $\theta'(0) > 0$ (see theorem 3.1), therefore, the function $\theta(\alpha)$ has no less than m zeros. The lemma is proved.

Corollary 3.1. When we prove the lemma, we have established, in addition, that either the zero is the unique stationary solution, or there exists such $\beta_1 > 0$ that $y_k(x, \beta_1) \geq 0$, where y_k satisfies the corresponding boundary conditions. By analogous considerations, we obtain that for $\beta < 0$ there exists such $\beta_2 < 0$ that $y_k(x, \beta_2) \leq 0$ satisfies the boundary conditions. If such β_2 does not exist, then $y_k(x, \beta) \leq 0$ for all negative β.

Lemma 3.2. Let the zero be an asymptotically stable solution of the problem in question. Suppose we know that there are no nonnegative solutions which satisfy the boundary conditions. Then for each function $u_0 \in C^1([0, 1])$ satisfying one of the boundary conditions (3.2), (3.3), (3.4), there exists $y_k(x, \alpha)$ satisfying for $x = 0$ the same condition as u_0 and such that

$$y_k(x, \alpha) > u_0(x) \quad \text{for} \quad x > 0. \tag{3.12}$$

Proof. Firstly, consider the first boundary-value problem. By the lemma supposition

$$y_1(1, |\alpha|) > 0 = u_0(1).$$

Really, the trivial solution is asymptotically stable, and, therefore, $y_{1\alpha}(x, 0) \geq 0$. There are no other solutions which are nonnegative and satisfy the boundary conditions. Therefore, $y_1(1, |\alpha|) > 0$. Suppose, there is no α such that (3.12) holds. Then, for each $\alpha > 0$ the function

$$v(x, \alpha) = y_1(x, \alpha) - u_0(x)$$

is such that

$$v(0, \alpha) = 0, \quad v(1, \alpha) > 0, \quad v(\xi, \alpha) < 0$$

for a certain $0 < \xi < 1$.
Consequently, there exists such $\eta(\alpha)$ that

$$v_x \Big|_{x=\eta} = 0, \quad v(\eta(\alpha), \alpha) \leq 0.$$

From the solvability "as a whole" of the Cauchy problem for (3.6) it follows that

$$|y_1'| + y_1 \to \infty, \quad \text{for} \quad \alpha \to \infty$$

uniformly by x. In the points $\eta(\alpha)$ we have

$$|y_1'| + y_1 \leq |v'| + |u_0'| + v + u_0,$$

i.e., $(|y_1'| + y_1)|_{x=\eta}$ does not tend to ∞ for $\alpha \to \infty$.
This contradiction proves the lemma in the case of the first boundary-value problem.

Let u_0 satisfy condition (3.3). For sufficiently large α we have

$$v(x, \alpha) = y_2(x, \alpha) - u_0(x), \quad v(0, \alpha) = \alpha - u_0(0) > 0,$$

$$v(1, \alpha) = y_2(1, \alpha) > 0.$$

Therefore, either $v(x, \alpha) > 0$ for all x, or there exists such a point $\eta(\alpha)$ that

$$v'(\eta, \alpha) = 0, \quad v(\eta, \alpha) \leq 0.$$

This means that $(|y_2'| + y_2)\,|_{x=\eta}$ does not tend to infinity for $\alpha \to \infty$.

In the case of the problem (3.6), (3.4) the condition of stability of the trivial solution leads to the inequality (by theorem 3.1)

$$\theta(\alpha) > 0, \quad \alpha > 0, \quad \theta = y_2' - \psi(y_2)\Big|_{x_4=1}.$$

If we have no positive stationary solutions, then $\theta > 0$ for all $\alpha > 0$. Let $u_0 \in C^1$ satisfy boundary condition (3.4). Consider the function

$$v(x, \alpha) = y_2(x, \alpha) - u_0(x).$$

For large values of α we have

$$v(0, \alpha) = \alpha - u_0(0) > 0.$$

We must show that

$$\lim_{\alpha \to \infty} y_2(1, \alpha) = \infty.$$

Assume the contrary. Then there exists such a sequence α_i, that

$$y_2(1, \alpha_i) \leq M, \quad \alpha_i \to \infty, \quad y_2'(1, \alpha_i) \to -\infty.$$

Really, if it is not so, then

$$y_2'(1, \alpha_i) \to +\infty, \quad \text{i.e.,} \quad \min y_2(1, \alpha_i) \leq M$$

is attained in the interior point ξ, since for large α_i we have

$$y_2(0, \alpha_i) = \alpha_i > M.$$

Consequently, there exists such $\xi(\alpha_i)$ that

$$v_x\big|_\xi = 0, \qquad v \le M,$$

and $(|y_2'| + y_2)\,|_{\xi(\alpha)}$ does not tend to infinity if $\alpha_i \to \infty$. We have $\theta(\alpha_i) < 0$ for sufficiently large α_i, $\theta(0) = 0$, $\theta(1/\alpha_i) > 0$, i.e., θ attains, also, positive values. This means that such α_0 exists that $\theta(\alpha_0) = 0$. This α_0 corresponds to a positive stationary solution, which contradicts the lemma supposition. The lemma is proved, since from

$$\lim_{\alpha \to \infty} y_2(1, \alpha) \to \infty$$

it follows that for large α we have

$$v(0, \alpha) > 0 \qquad v(1, \alpha) > 0;$$

and if $v(x, \alpha)$ is not positive for each α, then its minimum is nonpositive. Therefore, for each α there exists such $\xi(\alpha)$ that $v_x|_\xi = 0$, $v|_\xi \le 0$ and

$$|y_2'| + y_2 \le |u_0'| + u_0,$$

which contradicts the solvability condition "as a whole" for the Cauchy problem.

Remark 3.1. Without essential change we may obtain the analog of lemma 3.1 for negative α. More precisely, one of the following cases is possible.

1). For each $u_0(x) \in C^1$, which satisfies the boundary condition, there exists such $\alpha < 0$, that for the corresponding $y_k(x, \alpha)$ we have $y_k(x, \alpha) < u_0(x)$.

2). There exists such $\alpha_0 < 0$ that $y_k(x, \alpha_0) \leq 0$ and $y_k(x, \alpha_0)$ satisfies the corresponding boundary condition in the point $x = 1$, i.e., there exists a nonpositive nontrivial stationary solution.

Definition 3.1. The set $M(v)$ is called by *the attraction domain* of the stationary solution of one of the problems in question if for each $u_0 \in M(v)$ there exists a solution $u(x, t)$ of this problem such that

$$u(x, 0) = u_0(x), \quad \lim_{t \to \infty} \| u(x, t) - v(x) \|_2^{(0,1)} = 0.$$

In contrast to dynamic systems in R^n the problem of describing $M(v)$ is rather difficult. We shall prove the theorems which characterize the essential part of $M(v)$, namely, $M(v) \cap C^1$. The following theorem was established in Akramov and Zelenyak (1975) for the first boundary-value problem.

Recall, that we assume the condition (B) from chapter 2 still holds.

Theorem 3.2. Let $v(x)$ be an asymptotically stable stationary solution of the problem (1.1)-(1.3). Then one of the following statements holds.

a). There exist functions $v_k(x)$, which are stationary solutions of the problem, such that

$$T_a = \Big\{ u_0, \ v_1 \leq u_0 \leq v_2 \ \text{ for } \ 0 \leq x \leq 1 \Big\} \cap \overset{\circ}{C^1} \subset M(v).$$

b) There exists a stationary solution v_1 such that

$$T_b = \Big\{ u_0, \ v_1 \leq u_0 \ \text{ for } \ 0 \leq x \leq 1 \Big\} \cap \overset{\circ}{C^1} \subset M(v).$$

c) There exists a stationary solution v_2 such that

$$T_c = \Big\{ u_0, \ u_0 \leq v_2 \ \text{ for } \ 0 \leq x \leq 1 \Big\} \cap \overset{\circ}{C^1} \subset M(v).$$

d)

$$T_d = \Big\{ u_0, \ u_0 \in \overset{\circ}{C^1} [0, 1] \Big\} \subset M(v).$$

The sign of equality in a), b), c) may be only in those points $x = i$ of the boundary, where the conditions $u(i, t) = 0$ hold. In this case we assume that $u_0'(i) \neq v_k'(i)$. Here $\overset{0}{C}{}^1$ is the space of continuously differentiable functions satisfying the corresponding boundary conditions.

The theorem analogous to theorem 3.2 for somewhat different suppositions will be proved in chapter 5 for the general quasilinear parabolic equation with many spatial variables.

To prove theorem 3.2 we shall need the following comparison theorem (Peterson and Maple, 1966).

Theorem 3.3. Let u, v be continuously differentiable functions by their arguments for $0 \leq x \leq 1$, $0 \leq t \leq T$; $F(x, t, u, u_x, u_{xx})$ be continuously differentiable, $F_{u_{xx}} \geq \delta > 0$ with a certain δ. Let

$$u(x, 0) \geq v(x, 0) \quad \text{for} \quad 0 \leq x \leq 1,$$

$$u_x(0, t) = f_1(u(0, t)), \qquad u_x(1, t) = f_2(u(1, t)),$$

$$v_x(0, t) = g_1(v(0, t)), \qquad v_x(1, t) = g_2(v(1, t)),$$

where f_i, g_i are continuously differentiable and

$$f_1(u) \leq g_1(u), \qquad f_2(u) \geq g_2(u).$$

If for all (x, t)

$$F(x, t, v, v_x, v_{xx}) - v_t \geq F(x, t, u, u_x, u_{xx}) - u_t,$$

then for these (x, t) we have

$$v(x, t) \leq u(x, t).$$

The inequality will remain, if one or both boundary conditions are replaced by the inequality $u - v|_\Gamma \geq 0$.

Remark.   Note, that in Peterson and Maple (1966) in the boundary and for $t = 0$ the inequalities must be rigorous. This is unessential here. Really, taking into account the Taylor formula, we obtain

$$(u - v)_t - a_1(u - v)_{xx} - b_1(u - v)_x - c_1(u - v) \geq 0,$$

where $a_1 \geq \delta$, a_1, b_1, c_1 may be taken continuous by x, t. Further,

$$(u - v)_x - \frac{f_1(u) - f_1(v)}{u - v}(u - v) - f_1(v) + g_1(v)\Big|_{x=0} = 0,$$

$$(u - v)_x - \frac{f_2(u) - f_2(v)}{u - v}(u - v) - f_2(v) + g_2(v)\Big|_{x=1} = 0.$$

As f_i, u, v are differentiable, the functions

$$\theta_i(x, t) = \frac{f_2(u) - f_2(v)}{u - v}$$

are bounded. Let $|\theta_i| \leq M$. Set

$$\rho = e^{(M+1)(x^2 - x)}, \qquad u - v = \rho w.$$

Then

$$w_t - a_2 w_{xx} - b_2 w_x - c_2 w \geq 0,$$

$$w_x - [1 + M + \theta_1]w\Big|_{x=0} \leq 0, \qquad w_x + [1 + M + \theta_2]w\Big|_{x=1} \geq 0.$$

The function w, evidently, does not achieve in the boundary points the negative minimum, since $1 + M \pm \theta_2 \geq 1$. Setting $w = e^{\lambda t}z$, where $\lambda > |c_2| + 1$ for all x and t, we see that z, and therefore, w does not achieve the negative minimum in the domain

$$0 < x < 1, \quad 0 < t \leq T.$$

As w is nonnegative for $t = 0$, then w, and, therefore, $u - v$ could not attain negative values. If we replace the inequality in the boundary by the inequality $u - v|_{x=i} \geq 0$, then the theorem, evidently, will remain valid.

Consider the problem (1.1)-(1.3). Assume that there exists a stationary solution $v(x)$. Consider the first boundary-value problem. Evidently, $v(x) \equiv y(x, \alpha_0)$, where α_0 is one of the roots of equation

$$a(x, y, y')y'' + b(x, y, y') = 0, \qquad y(1, \alpha_0) = 0$$

$$y(0, \alpha) = 0, \qquad y'(0, \alpha) = \alpha.$$

By the theorem on continuous dependence of the Cauchy problem solution on initial data (we do not assume its solvability "as a whole") for α, sufficiently close to α_0, the function $y(x, \alpha)$ is defined and is twice continuously differentiable by x in the interval $[0, 1]$, where $y_\alpha, y_{x\alpha}, y_{xx\alpha}$ are continuous. Without loss of generality we may assume that $v(x) \equiv 0$, $\alpha_0 = 0$. Let now $y_\alpha > 0$ for $0 < x \leq 1$, $|\alpha| < \varepsilon$. For x, α_i we have

$$y(x, \alpha_1) < 0 < y(x, \alpha_2) \quad \text{if} \quad \alpha_1 < 0, \quad \alpha_2 > 0.$$

From theorem 3.3, if

$$y(x, \alpha_1) \leq u_0(x) \leq y(x, \alpha_2)$$

holds, then

$$y(x, \alpha_1) \leq u(x, t) \leq y(x, \alpha_2).$$

Now, consider the problem

$$u_t = a_K(x, u, u_x)u_{xx} + b_K(x, u, u_x),$$

$$u\big|_{t=0} = u_0(x), \qquad u\big|_{x=0} = u\big|_{x=1} = 0, \qquad (3.13)$$

where a_K, b_K are chosen as in theorem 1.1, chapter 2, where

$$K = \max_i \| y(x, \alpha_i) \|_1^{(0,1)} + 1.$$

From theorem 1.1, chapter 2 for the solution of our problem, we have the estimate

$$\| u_K \|_1^{(0,1)} \leq C \left(R_1 + \| u_K \|_0^{(0,1)} \right), \quad R_1 = \| u_0 \|_1^{(0,1)}. \qquad (3.14)$$

Taking $\| u_0 \|_1^{(0,1)}$ sufficiently small we may deduce that solution of (3.13) would be a solution of (1.1)-(1.3)

$$u_K(x,t) \equiv u(x,t).$$

Thus, the trivial solution is stable by Liapunov: by each $\varepsilon > 0$, we may find such $\delta(\varepsilon) > 0$ that if

$$\| u_0 \|_1^{(0,1)} \leq \delta(\varepsilon), \quad \text{then} \quad \| u \|_1^{(0,1)} \leq \varepsilon.$$

Asymptotic stability of this solution follows from the results of chapter 1. Really, $y_\alpha |_{\alpha=0}$ solves the equation

$$Lw = w'' + \frac{\partial}{\partial y} \frac{b}{a} \Big|_{y \equiv 0} w + \frac{\partial}{\partial y'} \frac{b}{a} \Big|_{y \equiv 0} w' = 0,$$

and is not equal to zero for $0 < x \leq 1$, where $y_\alpha |_{\substack{\alpha=0 \\ x=0}} = 0$. The same property has the solution of the equation

$$Lw = 0, \quad w \Big|_{x=0} = \varepsilon, \quad w_x \Big|_{x=0} = y_{\alpha x} \Big|_{\substack{\alpha=0 \\ x=0}}$$

for sufficiently small positive ε because of the continuous dependence of a solution on initial data. Then, by theorem 3.1, the spectrum of L is negative. Further,

$$Lw = w'' + \frac{1}{a(x,0,0)} \frac{\partial b}{\partial y} \Big|_{y \equiv 0} w + \frac{1}{a(x,0,0)} \frac{\partial b}{\partial y'} \Big|_{y \equiv 0} w', \qquad (3.15)$$

since $b(x, 0, 0) \equiv 0$. Therefore, the linearized problem (1.1)-(1.3) has the form

$$w_t = a(x, 0, 0) Lw, \qquad w\big|_{x=0} = w\big|_{x=1} = 0.$$

As the spectrum of L is negative, then the trivial solution of (1.1)-(1.3) is asymptotically stable. If $y_\alpha|_{\alpha=0}$ changes sign, then there exists such x_1 that $y_\alpha|_{x_1} = 0$, and, by the Sturm theorem, each solution of the equation $Lw = 0$ vanishes. Hence follows the existence of a positive eigenvalue of L, and also of the problem

$$a(x, 0, 0) Lw = \lambda w, \qquad w\big|_{x=0} = w\big|_{x=1} = 0.$$

From the results of chapter 1 it follows that the trivial solution is unstable.

To investigate the critical cases we need the following lemma, which immediately follows from theorem 1.3.

Lemma 3.3. Let $u, u_x, u_t, u_{xx}, u_{xt}, u_{xxt}$, be continuous for $0 \leq x \leq 1$, $0 \leq t < \infty$, where

$$\left\| u \right\|_{2+\alpha}^{(0,1)} \leq K.$$

Let u solve equation (1.1) and one of the conditions (1.2) hold. Then each partial limit

$$\lim_{t_i \to \infty} u(x, t_i)$$

is a stationary solution of the corresponding mixed problem.

Suppose for $0 < \alpha < \varepsilon$, $0 < x \leq 1$ we have $y_\alpha > 0$. Then the trivial solution is semi-stable: if $\| u_0 \|_1^{(0,1)}$ is sufficiently small, $0 \leq u_0 \leq y(x, \varepsilon)$, then $\| u(x, t) \|_1^{(0,1)}$ is small for all $t > 0$, and it is easy to show that $\| u(x, t) \|_3^{(0,1)}$ is bounded for $t > 0$. Then, by lemma 3.3, $u(x, t)$ is stabilized. By the maximum principle

$$0 \le u(x,t) \le y(x,\varepsilon)$$

and each partial limit is a stationary solution. But the unique stationary solution satisfying the boundary conditions and being between 0 and $y(x,\varepsilon)$ is zero.

Now, for $\varepsilon > \alpha > 0$, let $y_\alpha \ge 0$. Then for such α we have $y(x,\alpha) > 0$. But $y(x,\alpha)$ for positive α, $\alpha < \varepsilon$ could not be zero in an interior point, since then, it will be the point of minimum, $y'_x = 0$, and by the uniqueness theorem, $y \equiv 0$. Therefore, it is only possible that such $\alpha_i \to 0$, $\alpha_i > 0$ exist that $y(1,\alpha_i) = 0$. In this case the trivial solution, evidently, could not be asymptotically stable, since in each of its neighborhoods there exist other stationary solutions. But this is semi-stable, since from

$$0 \le u_0 \le y(x,\alpha_i)$$

it follows that

$$0 \le u(x,t) \le y(x,\alpha_i).$$

If y_α changes sign for each $\alpha > 0$, $\alpha < \varepsilon$ for a certain ε, then the trivial solution is unstable. Really, then, there exists such $\alpha_i \to 0$ that

$$y(x,\alpha_i) \ge 0 \quad \text{for} \quad 0 \le x \le \xi_i,$$

$$y(x,\alpha_i) \le 0 \quad \text{for} \quad \xi_i \le x < \eta_i.$$

Let $u_0^i(x)$ be such that $u_0^i(x) \ge 0$ and

$$u_0^i(x) \ge y(x,\alpha_i) \quad \text{for} \quad 0 \le x \le \xi_i,$$

where

$$\lim_{t \to \infty} \| u_0^i \|_2^{(0,1)} = 0.$$

Corresponding solutions to these initial data could not converge to the zero since

$$u^i(x,t) \ge 0, \qquad u^i(x,t) \ge y(x,\alpha_i).$$

All $u^i(x,t)$ are going from the sufficiently small zero neighborhood, otherwise, they all would be stabilized to the close to zero nonnegative stationary solutions which satisfy the boundary conditions. But in a sufficiently small zero neighborhood there are no such stationary solutions. Repeating the same considerations for $\alpha < 0$, we obtain the analogous result.

Now consider the boundary conditions

$$u_x - \varphi(u)\big|_{x=0} = 0; \qquad u_x - \psi(u)\big|_{x=1} = 0.$$

Let $y(x,\alpha)$ be such that

$$ay'' + b = 0, \quad y(0,\alpha) = \alpha, \quad y'(0,\alpha) = \varphi(\alpha).$$

Consider the equation

$$y'(1,\alpha) = \psi(y(1,\alpha)).$$

As earlier, assume that $y(x,0) \equiv 0$. Then $\psi(0) = 0$ and $y(x,0) \equiv 0$. Let

$$y_\alpha \geq 0, \qquad \frac{\partial^2 y}{\partial x \partial \alpha} - \psi'(0)\frac{\partial y}{\partial \alpha}\bigg|_{x=1} \geq 0, \qquad |\alpha| < \varepsilon.$$

There exist such $\alpha_1 > 0$, $\alpha_2 < 0$ that

$$y(x,\alpha_2) < 0 < y(x,\alpha_1),$$

where

$$\theta(\alpha) = y'(1,\alpha) - \psi(y(1,\alpha))$$

is such that $\theta(\alpha_1) \geq 0$, $\theta(\alpha_2) \leq 0$. Really, if $y(\xi,\alpha_2) = y(\xi_1,\alpha_1) = 0$ in the interior point then

$$\frac{\partial y}{\partial \xi}(\xi,\alpha_2) = \frac{\partial y}{\partial \xi}(\xi_1,\alpha_1) = 0$$

and
$$y(x, \alpha_2) \equiv y(x, \alpha_1) \equiv 0.$$

But
$$y(1, \alpha_2) < y(1, \alpha_1), \quad y(0, \alpha_2) < y(0, \alpha_1).$$

Now, apply theorem 3.3

$$a(x, u, u_x)u_{xx} + b(x, u, u_x) - u_t = a(x, y, y')y''$$

$$+b(x, y, y') - \frac{\partial y}{\partial t}\bigg|_{\alpha=\alpha_i} \qquad i = 1, 2,$$

$$y(x, \alpha_2) \le u_0 \le y(x, \alpha_1),$$

$$y_x - \varphi(y)\bigg|_{\substack{x=0 \\ \alpha=\alpha_i}} = u_x - \varphi(u)\bigg|_{x=0} = 0; \quad \theta(\alpha_i) = y_x - \psi(y)\bigg|_{\substack{x=1 \\ \alpha=\alpha_i}},$$

$$0 = u_x - \psi(u)\bigg|_{x=1} \ge \theta(\alpha_2), \qquad 0 = u_x - \psi(u)\bigg|_{x=1} \le \theta(\alpha_1).$$

For the solution $u(x, t)$ we obtain the estimates

$$y(x, \alpha_2) \le u(x, t) \le y(x, \alpha_1).$$

If we assume that in a certain neighborhood of zero there are no stationary solutions satisfying the boundary conditions, then the asymptotic stability of solution $u \equiv 0$ is obtained analogously. This holds also if, for α sufficiently small, but not equal to zero, we have $\theta(\alpha) \ne 0$. If there exists a subsequence α_n such that $\theta(\alpha_n) = 0$, $\alpha_n \to 0$ then the solution is stable but not asymptotically stable.

Now let $y_\alpha \ge 0$ and such $\alpha_n \to 0$ exist that $\theta'(\alpha_n) < 0$. We shall assume that $\alpha_n > 0$ (the case $\alpha_n < 0$ may be considered analogously) for all n. There exists such $\beta_n \to 0$ that $\theta(\beta_n) < 0$, $\theta'(\beta_n) \le 0$, where $\beta_n > 0$ (otherwise, $\theta(\beta) \geq 0$ for small positive β). Consider the sequence of smooth functions $u_0^n(x)$ such that

$$\lim_{n \to \infty} \| u_0^n \|_1^{(0,1)} = 0, \quad u_0^n \geq 0, \quad u_0^n \geq y(x, \beta_n)$$

and use theorem 3.3 for the corresponding solution $u_n(x,t)$ where u_0^n satisfies the boundary conditions

$$0 = \frac{\partial u_n}{\partial x} - \varphi(u_n) = \left. \frac{\partial y}{\partial x} - \varphi(y) \right|_{\substack{x=0 \\ \alpha = \beta_n}},$$

$$\left. \frac{\partial u_n}{\partial x} - \varphi(u_n) \right|_{x=1} = 0 > \theta(\beta_n).$$

We obtain

$$u_n(x,t) \geq y(x, \beta_n), \qquad u_n(x,t) \geq 0,$$

i.e., the trivial solution could not be asymptotically stable. If $\theta'(\beta) < 0$ for $0 < \beta < \varepsilon$, then this solution is unstable, since in the sufficiently small neighborhood of zero there are no positive stationary solutions of the problem. For semi-stability in this case it is necessary and sufficient that there exist positive stationary solutions with an arbitrary small norm. But then $\theta(0) = 0$, $\theta'(0) = 0$ and, evidently, $y_\alpha|_{\substack{x=0 \\ \alpha = 0}}$ could not vanish, i.e., $y_\alpha > 0$. It remains to consider the case when y_α changes sign. Linearizing the problem in the neighborhood of zero, we obtain the equation

$$w_t = a(x, 0, 0)Lw,$$

where w satisfies the conditions

$$\left. w - \varphi'(0)w \right|_{x=0} = \left. w' - \psi'(0)w \right|_{x=1} = 0,$$

and L is determined by (3.15).

For solution $w(x,t)$ of this problem constructed by the initial data $w(x,0) = w_0(x)$ the following relations are possible

$$|w|\underset{t\to\infty}{\longrightarrow} 0, \qquad |w|\underset{t\to\infty}{\longrightarrow} \infty,$$

or $w(x,t)$ converges to a nontrivial stationary solution. If $\theta'(0) = 0$, then there exists at least one positive eigenvalue of the operator aL, since the solution constructed by positive initial data

$$w_0(x) \geq y_\alpha\big|_{\alpha=0}$$

could not converge by the maximum principle to a stationary solution. Therefore, it converges to infinity for $t \to \infty$. In this case, w_0 may be chosen satisfying the boundary conditions since y_α satisfies these conditions.

If $\theta'(0) \neq 0$ then the unique stationary solution satisfying the boundary conditions is the trivial solution. If positive eigenvalues are absent, then, by theorem 3.1, there exists such $\rho(x)$ that

$$\rho(x) > 0, \quad L\rho = 0, \quad \rho' - \varphi'(0)\rho\big|_{x=0} = -1, \quad \rho' - \psi'(0)\rho\big|_{x=1} = 1.$$

Setting $w = \rho z$, we obtain

$$z_t = a_1(x)z_{xx} + b_1(x)z_x = L_1 z$$

$$z_x - z\big|_{x=0} = z_x + z\big|_{x=1} = 0.$$

Evidently, $y_\alpha = C\rho z_1$, where

$$L_1 z_1 = 0, \quad z_1\big|_{x=0} = 1, \quad z_{1x}\big|_{x=0} = 1,$$

C is a certain constant. By immediate computation we obtain

$$z_1(x) = 1 + \int_0^x e^{-\int_0^\xi \frac{b_1(\eta)}{a_1(\eta)}d\eta} d\xi \neq 0,$$

i.e., $y_\alpha \neq 0$. We have established that if y_α changes sign, then there exists at least one positive eigenvalue. So, we have proved the following theorem.

Theorem 3.4. Let

$$y_1(x,0) \equiv 0, \quad ay_1'' + b = 0, \quad y_1(0,\alpha) = 0, \quad y_1'(0,\alpha) = \alpha.$$

1) In order that the trivial solution of (1.1)-(1.2) with conditions $u|_{x=0} = u|_{x=1} = 0$ be asymptotically stable in C^1 it is necessary and sufficient that

$$y_{1\alpha}\big|_{\alpha=0} \geq 0 \quad \text{and} \quad y_1(1,\alpha) \neq 0$$

for $\alpha \neq 0$ sufficiently small.
    Let

$$ay_2'' + b = 0, \quad y_2(0,\alpha) = \alpha, \quad y_2'(0,\alpha) = \varphi(\alpha),$$

$$\theta(\alpha) = y_2 - \psi(y_2)\big|_{x=1}.$$

2) In order that the trivial solution of (1.1)-(1.2) with the condition

$$u_x - \varphi(u)\big|_{x=0} = 0; \quad u\big|_{x=1} = 0;$$

be asymptotically stable in C^1 it is necessary and sufficient that

$$y_{2\alpha}\big|_{\alpha=0} \geq 0, \quad (-1)^k y_2\big(1, (-1)^k |\alpha|\big) > 0$$

for small $\alpha \neq 0$.
    3) In order that the trivial solution of (1.1)-(1.2) with the condition

$$u_x - \varphi(u)\big|_{x=0} = 0; \quad u_x - \psi(u)\big|_{x=1} = 0$$

be asymptotically stable in C^1 it is necessary and sufficient that

$$y_{2\alpha}\Big|_{\alpha=0} > 0, \quad (-1)^k \theta((-1)^k|\alpha|) > 0$$

for small α.

4) In order that the trivial solution of the corresponding problem be stable but not asymptotically stable it is necessary and sufficient that for each $\varepsilon > 0$ there exist $v_1(x), v_2(x)$ satisfying the equation and the boundary conditions so that $v_1 \geq 0; v_2 \leq 0; |v_i| \leq \varepsilon$.

Theorem 3.4 and theorem 3.1 chapter 2 yield the following statement.

Theorem 3.5.   Let the condition (B) (definition 1.2, §2.1) hold. Then the conditions of theorem 3.4 are necessary and sufficient for stability in the following sense: for each ε there exists such $\delta(\varepsilon)$ that if

$$u_0 \in C^1, \qquad \| u_0 \|_0^{(0,1)} \leq \delta(\varepsilon),$$

then

$$\| u(x,t) \|_0^{(0,1)} \leq \varepsilon, \qquad u(x,t) \in C_{xt}^{2,1}, \quad t > 0,$$

where $u(x,t)$ solves the problem and $u(x,0) = u_0(x)$.

If conditions of theorem 3.4, which are related to asymptotical stability, hold, then, in addition,

$$u(x,t) \to 0, \qquad t \to \infty$$

by the C^2 norm.

Proof.   As when proving theorem 3.4 we may estimate the solution. From theorem 1.1 chapter 2, it follows that the smooth solution is bounded in C^1. It suffices to show the existence "as a whole" of the mixed problem solution by the initial data from C^1 if we know the *a priori* estimate. Approximate u_0 by a sequence of smooth solutions choosing them so that they lie between stationary solutions of the problem. As follows from the theorem 3.4 proof, this is possible.

Then for corresponding solutions we have the estimate in C^1 uniform by t. Moreover, for each $\delta > 0$ the sequence of solutions is bounded in C^3 for $t \geq \delta$ and, therefore, it converges in C^2 to the solution of our problem (since from the maximum principle, the uniqueness of this solution follows). Thus, theorem 3.5 is the corollary of theorem 3.4, theorem 1.1 chapter 2, and of the local solvability of the corresponding problems from smooth initial data.

Consider the example which illustrates the theorems 3.4, 3.5.

Example 3.1. (Zelenyak and Slin'ko, 1977a, b). The following equation

$$u_t = u_{xx} + \lambda f(u), \quad u_x\big|_{x=0} = u\big|_{x=1} = 0, \quad u\big|_{t=0} = u_0(x)$$

is often met in chemical technology. Here, λ is a parameter, $f(u) \geq 0$. Set

$$v_x\big|_{x=0} = 0, \quad v\big|_{x=0} = v_0$$

and consider the solution of equation

$$v_{xx} + \lambda f(v) = 0$$

with these initial conditions. Denote it by $v(x, v_0, \lambda)$. As the function $f(v)$ is supposed to be nonnegative for $\lambda > 0$, $v_0 < 0$ there are no solutions such that $v(1, v_0, \lambda) = 0$. Let $v_0 > 0$. Then $v(x, v_0, \lambda)$ monotonically decreases and for a certain $\xi(\lambda)$ we have $v(\xi, v_0, \lambda) = 0$. Consequently, the function $v(x\xi, v_0, \lambda_1/\xi^2)$ is a stationary solution of our problem for $\lambda = \lambda_1$. Set $v_x = \varphi(v)$ (recall that the function v is monotone). Then, for a fixed λ we have

$$v_{xx} = \varphi\varphi', \quad \frac{1}{2}(\varphi^2)' = -\lambda f(v), \quad \varphi^2(v) = -2\lambda \int_{v_0}^{v} f(\xi)d\xi,$$

and since $\varphi \leq 0$, then

$$\varphi(v) = -\sqrt{2\lambda} \left(\int_v^{v_0} f(\xi)d\xi \right)^{1/2}.$$

Further

$$G(v, v_0) = \int_v^{v_0} \frac{dz}{\left(\int_z^{v_0} f(\xi)d\xi \right)^{1/2}} = \sqrt{2\lambda}x. \qquad (3.16)$$

Thus, the initial data for the stationary solution may be determined from the equation

$$F(v_0) = G(0, v_0) = \sqrt{2\lambda}.$$

Let $f(v) > 0$ for $0 \le v < V$, $f(V) = 0$. Then the equation $F(v_0) = \sqrt{2\lambda}$ has at least one solution for each λ: for $v_0 \in [0, V)$ the function $F(v_0)$ is continuous, $F(0) = 0$, $F(v_0) \underset{v_0 \to V}{\longrightarrow} \infty$. Let $f(\xi)$ be differentiable. Then integrating by parts we obtain

$$G(v, v_0) = \frac{2}{f(v)} \left(\int_v^{v_0} f(\xi)d\xi \right)^{1/2} - 2 \int_v^{v_0} \frac{f'(z)}{f^2(z)} \left(\int_z^{v_0} f(\xi)d\xi \right)^{1/2} dz,$$

whence

$$F'(v_0) = \frac{\partial G(0, v_0)}{\partial v_0} = \frac{f(v_0)}{f(0) \left[\int_0^{v_0} f(\xi)d\xi \right]^{1/2}} - \int_0^{v_0} \frac{f'(z)}{f^2(z)} \frac{f(v_0)dz}{\left[\int_z^{v_0} f(\xi)d\xi \right]^{1/2}},$$

$$\frac{\partial v}{\partial v_0} = \left\{ \frac{f(v_0)}{f(v) \left(\int_v^{v_0} f(\xi)d\xi \right)^{1/2}} - \int_v^{v_0} \frac{f'(z)}{f^2(z)} \frac{f(v_0)dz}{\left(\int_z^{v_0} f(\xi)d\xi \right)^{1/2}} \right\}$$

$$\times \frac{1}{\left(\int_v^{v_0} f(\xi)d\xi \right)^{1/2}}.$$

Thus,

$$\frac{\partial v}{\partial v_0} = K(v, v_0) \frac{1}{\left(\int_v^{v_0} f(\xi)d\xi \right)^{1/2}},$$

where $F'(v_0) = K(0, v_0)$. Evidently,

$$\frac{\partial v}{\partial v_0}\bigg|_{v=0} = \frac{\partial v}{\partial v_0}\bigg|_{x=1}$$

and this function differs from $K(0, v_0) = F'(v_0)$ by the positive multiplier. As

$$\frac{\partial K}{\partial v} = \frac{1}{2} \frac{f(v_0)}{\left(\int\limits_{v}^{v_0} f(\xi) d\xi\right)^{3/2}} > 0, \quad 0 \le v_0 < V,$$

then $\partial v/\partial v_0$ has no more that one zero. Hence follows that the corresponding stationary solution to v_0 is asymptotically stable if $F'(v_0) > 0$ and is unstable if $F'(v_0) < 0$. The critical case is $F'(v_0) = 0$. Now we may apply theorem 3.5, since from the maximum principle it follows that if $0 \le u_0 \le V$ then for the solution of the nonstationary problem we have $0 \le u \le V$. If all zeros of the equation $F(v_0) = \sqrt{2\lambda}$ are isolated then the stability conditions coincide with the conditions of asymptotic stability. Therefore, the necessary and sufficient condition of stability in this case is the condition

$$F'(v_0) \ge 0, \quad F(\alpha) > 0 > F(-\alpha)$$

for $\alpha > 0$ sufficiently small. Except for this case the stability is possible if only there exist such sequences

$$\alpha_n \to v_0, \quad \beta_n \to v_0, \quad \alpha_n > v_0, \quad \beta_n < v_0,$$

that

$$F(\alpha_n) = F(v_0), \quad F(\beta_n) = F(v_0) = \sqrt{2\lambda}.$$

The analogous examples for stability criterious are in Zelenyak (1972). The possibility of lowering the order of the stationary equation in question allows us to solve the problem on stability of the stationary solution by simply investigating the function, the roots of which determine the initial data corresponding to the stationary solution.

In particular, in example 3.1, it turns out well if we reduce the problem of determining these data to solving the equation $F(v_0) = \sqrt{2\lambda}$, i.e., to resolve the corresponding functional equation relative to the parameter.

Investigating the graph of this function we obtain the information on stability of stationary solutions corresponding to different parameter v_0 values. In applications there arises the problem: of finding a stationary solution optimal by a certain criterion. As the example 3.1 shows, this problem may be essentially simplified if we investigate the properties of $F(v_0)$, construction of which does not require the stationary solutions for all x.

Proof of theorem 3.2. Without loss of generality we may set $v \equiv 0$. Let there exist such stationary solutions v_k of the problems such that $v_1 < 0$, $v_2 > 0$ for $0 < x < 1$, and each stationary solution w of the problem, such that $v_1 < w < v_2$ is trivial. The maximum principle (theorem 3.3) applied to $v_1, v_2, u(x, t)$, solutions of the problem such that $u(x, 0) = u_0(x)$, $v_1 \leq u_0 \leq v_2$ yields that for $u(x, t)$ we have $v_1 \leq u \leq v_2$. So, we have obtained an *a priori* estimate for solutions constructed by initial data from T_a. Theorem 1.1, chapter 2 allows us to obtain an *a priori* estimate in C^1. As was done in the proof of theorem 3.5, we may establish that there exists a twice continuously differentiable solution with initial data from T_a which is uniformly by $t \geq \delta$ bounded with its derivatives. By theorem 1.3 this solution is stabilized; therefore, its limit is either v_1, or v_2, or zero. Consider conditions (3.2). By the definition of T_a, we have

$$u_0(x) < v_2(x), \qquad 0 < x < 1,$$

$$u_0(i) = v_2(i) = 0, \qquad i = 0, 1$$

$$u_0'(0) < v_2'(0),$$

$$|u_0'(1)| < -v_2'(1).$$

Let

$$y_1(x, \alpha) > 0, \quad \text{for} \quad 0 < \alpha < \alpha_0, \quad 0 < x \leq 1,$$

and

$$\lim_{\alpha \to \alpha_0} y_1(x, \alpha) = v_2(x).$$

There exists such $\alpha' > 0$, $\alpha' < \alpha_0$ that

$$u_0(x) < y_1(x, \alpha'), \qquad x \geq 0.$$

As

$$y_1(0, \alpha') = 0, \qquad y_1'(0, \alpha') = \alpha' < \alpha_0,$$

then in a certain neighborhood of zero we have

$$y_1(x, \alpha') < v_2(x), \qquad x \neq 0.$$

But $y_1(1, \alpha') > 0$ for $\alpha' < \alpha_0$; therefore, in the neighborhood of the unit we have

$$y_1(x, \alpha') > v_2(x).$$

From the maximum principle it follows that

$$u(x, t) \leq v_2(x), \qquad u(x, t) \leq y_1(x, \alpha'),$$

since y_1 solves the equation and

$$u(x, 0) \leq y_1(x, \alpha'), \quad u\big|_{x=0} = y_1(0, \alpha') = 0, \quad u\big|_{x=1} = 0 < y_1(1, \alpha').$$

Therefore, in the neighborhood of $x = 0$ we have

$$\lim_{t \to \infty} u(x, t) \leq y_1(x, \alpha') < v_2(x);$$

consequently u does not converge to v_2. Analogously we may obtain that u does not converge to v_1. In the case of condition (3.3) we have

$$u_0(x) < v_2(x), \qquad 0 \leq x < 1$$

and, therefore, there exists such α' that

$$u_0(x) < y_2(x, \alpha'), \quad \alpha' < \alpha_0, \quad v_2(0) = \alpha_0$$

in a certain neighborhood of zero $y_2(x, \alpha') < v_2(x)$. It follows from theorem 3.3 that $u(x,t) < y_2(x, \alpha')$ and v_2 can not be the limit for $u(x,t)$. In the case of conditions (3.4) from $u_0 \in T_a$ it follows that

$$u_0 < v_2(x) = y_2(x, \alpha_0)$$

for all x, $0 \le x \le 1$. As the trivial solution is asymptotically stable, then for $\alpha > 0$, $\alpha < \alpha_0$ we have

$$\theta(\alpha) = y_2 - \psi(y_2)\big|_{x=1} > 0.$$

Apply to solutions u, $y_2(x, \alpha')$ of (3.1) theorem 3.3

$$0 = u_x - \varphi(u)\big|_{x=0} = y_2' - \varphi(y_2)\big|_{x=0},$$

$$0 = u_x - \psi(u)\big|_{x=1} < y_2' - \psi(y_2)\big|_{x=1} = \theta(\alpha'),$$

$$u(x,0) - y_2(x, \alpha') < 0$$

for $\alpha' < \alpha_0$ sufficiently close to α_0. Therefore,

$$u(x,t) \le y_2(x, \alpha'), \quad u(0,t) \le \alpha' < \alpha_0.$$

Hence follows that $u(x,t)$ can not converge to v_2. Analogously we may prove that v_1 is not the limit of $u(x,t)$. Therefore,

$$u(x,t) \to 0, \quad t \to \infty.$$

Let there be no stationary solution $v_2(x)$ of the problem such that $v_2(x) > 0$ for $0 < x < 1$. From corollary 3.1 of lemma 3.1 it follows that we may apply lemma 3.2. Consequently, for each $u_0(x)$ there exists such $y_k(x, \alpha)$ that $y_k(x, \alpha) \geq u_0(x)$. Apply to the functions $y_k(x, \alpha), u(x, t)$ the maximum principle. Really, in the case of problem (3.1), (3.4) we have

$$u_0(x) \leq y_2(x, \alpha), \quad \theta(\alpha) = y_2' - \psi(y_2)\big|_{x=1} > 0 = u_x - \psi(u)\big|_{x=1},$$

$$y_2' - \varphi(y_2)\big|_{x=0} = u_x - \varphi(u)\big|_{x=0} = 0.$$

The functions y_2 and u satisfy equation (3.1); therefore, $u(x, t) \leq y_2(x, \alpha)$. If there exists the function v_1 - a stationary solution of the problem such that $v_1 < 0$ for $0 < x < 1$, then for $u_0 \in T_\delta$, by analogous considerations we may prove that there exists the solution converging to zero for $t \to \infty$. If there exists $v_2(x) > 0$ and no stationary solutions which attain negative values, then each solution $u(x, t)$ such that $u(x, 0) \in T_c$ converges to zero for $t \to \infty$.

Finally, if the stationary solution of the problem is unique, i.e., $y_k(x, \alpha)$ satisfies both boundary conditions only for $\alpha = 0$, then for the arbitrary initial data from C^1 satisfying the boundary conditions we may use lemma 3.2. As a result we may obtain $\alpha_1 > 0$, $\alpha_2 < 0$ such that

$$y_k(x, \alpha_2) < u_0 < y_k(x, \alpha_1)$$

and, in the case of condition $u|_{x=1} = 0$,

$$y_k(1, \alpha_2) < 0, \qquad y_k(1, \alpha_1) > 0$$

y_k satisfies the problem condition for $x = 0$. When we consider problem (3.1), (3.4) we have $\theta(\alpha_2) < 0$, $\theta(\alpha_1) > 0$, and, by theorem 3.3, we have

$$y_k(x, \alpha_2) < u(x, t) < y_k(x, \alpha_1).$$

Hence follows, as earlier, that $u(x, t)$ is stabilized to the trivial solution (the problem has no other stationary solutions in this case). The theorem is proved.

When proving theorem 3.2 we have essentially used that there exists at least one asymptotically stable stationary solution. The following theorem proved by Provorova (1973) for initial data from C^1 in the case of the first boundary-value problem (3.1), (3.2), (3.5) gives the description of the qualitative properties of solution "as a whole" without this assumption.

Theorem 3.6. Let $u_0(x)$ belong to $C^{2+\alpha}[0,1]$ and satisfy one of the boundary conditions in question. Then for the solution $u(x,t)$ of the corresponding boundary-value problem one of the following properties holds

1) $u(x,t)$ exists for all $t \geq 0$, $t < t_0$, where

$$\lim_{t \to t_0} \sup_x |u| = \infty.$$

Recall that we assume the condition (B) from chapter 2 still holds.

2) $u(x,t)$ is determined for all t, and there exists such stationary solution $v(x)$ that

$$\lim_{t \to \infty} \| u(x,t) - v(x) \|_2^{(0,1)} = 0.$$

Proof. The local solvability by smooth initial data we have proved before. From the estimates obtained by Kruzhkov (1967, 1969a, 1972) it follows that if $\| u \|_1^{(0,1)}$ is bounded for $0 \leq t \leq T \leq \infty$, then $\| u(x,t) \|_{2+\alpha}^{(0,1)}$ is bounded for the same T. Therefore, $\| u \|_{2+\alpha}^{(0,1)}$ is bounded for all t. Really, if it is not so, then such $t_0 \leq \infty$ exists that

$$\overline{\lim_{t \to t_0}} \| u \|_1^{(0,1)} = \infty.$$

In the first case the stabilization of solution (the property 2)) follows from theorem 1.3. In the second case we have the property 1). Really, in theorem 3.1 chapter 2, we have shown that if $\| u \|_1^{(0,1)}$ is unbounded in the neighborhood of finite or infinite t_0, then

$$\lim_{t \to t_0} \| u \|_0^{(0,1)} = \infty.$$

The theorem is proved.

To prove the theorem in the case when $u_0 \in C^1$, it sufficies to prove the theorem on local solvability of the mixed problem by initial data from C^1. For this purpose it sufficies to obtain , as we have done earlier, an *a priori* estimate of $\sup |u|$ in a certain fixed time interval. In the case of the first boundary-value problem such estimate follows from the maximum principle. In the case of conditions (3.3) and (3.4), the difficulty is in growth of the functions φ and ψ for large values of u. If we suppose that φ and ψ are growing not faster than a linear function, the theorem condition may be weakened by considering $u_0 \in C^1$.

Now, we shall illustrate the results obtained.

Firstly, we show that from the asymptotic stability in the first approximation, generally speaking, does not follow the asymptotic stability in L_p even in the case of the first boundary-value problem for an almost linear normally parabolic equation.

Example 3.2.   Consider the equation

$$u_t = u_{xx} + u^3 e^{u^2} \tag{3.17}$$

for $t \geq 0$, $-1 \leq x \leq 1$ with the boundary conditions

$$u\Big|_{x=1} = u\Big|_{x=-1} = 0 \tag{3.18}$$

and with the initial condition $u|_{t=0} = u_0(x)$. Let us show that the Cauchy problem with arbitrary data in x_0, $-1 \leq x_0 \leq 1$ for equation

$$y'' + y^3 e^{y^2} = 0$$

is solvable "as a whole" for $-1 \leq x \leq 1$ (which means that the condition (B) from definition 1.2, §2.1 holds).

Really, the relation

$$\varphi = y'^2 + (y^2 - 1)e^{y^2} = C$$

gives the first integral of the equation under consideration

$$\frac{d\varphi}{dx} = 2y'\left(y'' + y^3 e^{y^2}\right) = 0$$

for each solution. Therefore, in each point x of this interval one of the following two conditions holds

$$y^2 \le 1, \quad y'^2 \le y'^2(x_0) + (1 - y^2(x_0))e^{y^2(x_0)} + 1 \le K$$

(K is a constant), or

$$y^2 \ge 1, \quad y'^2 \le \varphi(x_0) = C, \quad (y^2 - 1)e^{y^2} \le C.$$

As

$$(y^2 - 1)e^{y^2} \to \infty \quad \text{for} \quad y^2 \to \infty,$$

then $y^2(x) \le K$, where K is a certain constant depending on C. Thus, for each solution, $|y| + |y'|$ is uniformly bounded, and the Cauchy problem is solvable "as a whole". The trivial solution of (3.17), (3.18) is asymptotically stable in C^0, as follows from the chapter 1 results.

Now we show that there exists a sequence of solutions of this problem so that

$$\int\limits_{-1}^{1} |u_n(x,0)|^p dx \xrightarrow[n \to \infty]{} 0$$

for each $p \ge 1$, but for each $t_0 > 0$

$$\lim_{n \to \infty} \int\limits_{0}^{t_0} \int\limits_{-1}^{1} |u_n|^p dx dt = \infty.$$

Thus, the solution of the problem does not continuously depend on initial data in L_p, $p \ge 1$. We have asymptotical stability by the C^0 norm, and the asymptotic stability of the trivial solution by the L_p norm for $p \ge 1$, but the trivial solution is not stable in $L_p, p \ge 1$.

Set

$$u(x,t,\varepsilon)\Big|_{t=0} = \frac{1}{\varepsilon^\lambda}\left[e^{-\frac{x^2}{\varepsilon}} - e^{-\frac{1}{\varepsilon}}\right]^3; \quad \text{with} \quad \varepsilon, \lambda > 0.$$

Then $u(x,0,\varepsilon) = u_0 \geq 0$ for $|x| \leq 1$. Further, we obtain

$$u_0' = \frac{3}{\varepsilon^\lambda}\left[e^{-\frac{x^2}{\varepsilon}} - e^{-\frac{1}{\varepsilon}}\right]^2 \cdot \left(-\frac{2x}{\varepsilon}\right),$$

$$u_0'' = \frac{6}{\varepsilon^{1+\lambda}}\left[e^{-\frac{x^2}{\varepsilon}} - e^{-\frac{1}{\varepsilon}}\right]\left\{-e^{-\frac{x^2}{\varepsilon}} + e^{-\frac{1}{\varepsilon}} + \frac{4x^2}{\varepsilon}\right\}.$$

Denoting

$$\psi = -e^{-\frac{x^2}{\varepsilon}} + e^{-\frac{1}{\varepsilon}} + \frac{4x^2}{\varepsilon}$$

we have

$$\frac{d\psi}{dx} = \frac{2x}{\varepsilon}e^{-\frac{x^2}{\varepsilon}} + \frac{8x}{\varepsilon} > 0 \quad \text{for} \quad x > 0,$$

$$\psi(0) < 0, \qquad \psi(\sqrt{\varepsilon}) = 4 - e^{-1} + e^{-\frac{1}{\varepsilon}} > 0.$$

Therefore, if $1 \geq x^2 \geq \varepsilon$, then $u_0'' \geq 0$. For $x^2 \leq \varepsilon < 1/2$ we have

$$\left|u_0''\right| \leq \frac{K}{\varepsilon^{1+\lambda}},$$

where K is an absolute constant.

For each $\lambda > 0$ there exists $\varepsilon_0(\lambda)$ such that for $|x| \leq 1$, $\varepsilon < \varepsilon_0(\lambda)$ we have

$$z_0 = \frac{d^2 u_0}{dx^2} + u_0^3 e^{u_0^2} \geq 0.$$

Really, for $x^2 \geq \varepsilon$ it is evident, since $u_0 \geq 0$, $u_0'' \geq 0$. For $x^2 \leq \varepsilon$ we obtain

$$z_0 \geq K_1 \varepsilon^{-3\lambda} e^{K_2 \varepsilon^{-2\lambda}} - \frac{K}{\varepsilon^{1+\lambda}} \geq 0,$$

if $\varepsilon_0(\lambda)$ is chosen sufficiently small and $\varepsilon \leq \varepsilon_0(\lambda)$. The constant entering into this inequality does not depend on λ, ε.

Set

$$\Psi_1(u) = \left(3u^2 + 2u^4\right)e^{u^2}, \quad w = u_t.$$

Then, (3.17), (3.18) yield

$$w_t = w_{xx} + \Psi_1(u)w, \quad w\Big|_{x=-1} = w\Big|_{x=1} = 0; \quad w\Big|_{t=0} = z_0 \geq 0$$

for $\varepsilon < \varepsilon_0(\lambda)$. From the maximum principle it follows that $w \geq 0$ for all t for which the solution exists, i.e., $u(x,t,\varepsilon) \geq u(x,0,\varepsilon)$. Moreover, for a certain $\varepsilon_1(\lambda)$, $x^2 \leq \varepsilon < \varepsilon_1(\lambda)$ we shall have

$$u_t\Big|_{t=0} = w\Big|_{t=0} = z_0 \geq K_3 \exp\left(\frac{K_2}{\varepsilon^{2\lambda}}\right) \geq (\varepsilon - x^2)K_3 \exp\left(\frac{K_2}{\varepsilon^{2\lambda}}\right) = v_0,$$

$$\Psi_1(u) \geq K_3\varepsilon^{-2\lambda}e^{K_2\varepsilon^{-2\lambda}} = \Psi_2(\varepsilon, \lambda),$$

where K_3 depends on $\varepsilon_1(\lambda)$ and does not depend on ε if $0 < \varepsilon < \varepsilon_1(\lambda)$. Now consider the problem

$$v_t = v_{xx} + \Psi_2 v, \quad v\Big|_{t=0} = v_0, \quad v\Big|_{x^2=\varepsilon} = 0. \qquad (3.19)$$

For the difference $u_t - v$ we have

$$(u_t - v)_t = (u_t - v)_{xx} + \Psi_2(u_t - v) + (\Psi_1 - \Psi_2)u_t$$

$$\geq (u_t - v)_{xx} + \Psi_2(u_t - v),$$

$$(u_t - v)\Big|_{t=0} \geq 0 \quad \text{for} \quad x^2 \leq \varepsilon < \varepsilon_1, \quad (u_t - v)\Big|_{x^2=\varepsilon} \geq 0.$$

By the maximum principle $u_t - v \geq 0$ for $x^2 \leq \varepsilon$ and for all t for which the solution is regular. In (3.19) set

$$\tau = \frac{\pi^2}{4\varepsilon}t, \quad \xi = \frac{\pi}{2\sqrt{\varepsilon}}\left(x + \sqrt{\varepsilon}\right),$$

$$v = \frac{4\varepsilon}{\pi^2} K_3 e^{K_2 \varepsilon^{-2\lambda} + t\Psi_2} \cdot v_1 = R(t)v_1.$$

We obtain

$$\frac{\partial v}{\partial t} = \Psi_2 R v_1 + R \frac{\partial v_1}{\partial \tau} \cdot \frac{\pi^2}{4\varepsilon}, \qquad \frac{\partial^2 v}{\partial x^2} = R \frac{\partial^2 v_1}{\partial \xi^2} \cdot \frac{\pi^2}{4\varepsilon},$$

i.e.,

$$\frac{\partial v_1}{\partial \tau} = \frac{\partial^2 v_1}{\partial \xi^2}, \qquad v_1\Big|_{\tau=0} = R^{-1}(0)v\Big|_{t=0} = (\varepsilon - x^2)\frac{\pi^2}{4\varepsilon} = \xi(\pi - \xi),$$

$$v_1\Big|_{\xi=0} = v_1\Big|_{\xi=\pi} = 0.$$

But, for $0 \le \xi \le \pi$, we have

$$\xi(\pi - \xi) \ge \frac{2}{\pi} \sin \xi.$$

From the maximum principle, it follows that

$$v_1(\xi, \tau) \ge \frac{2}{\pi} \sin \xi \cdot e^{-\tau}.$$

Therefore, for $x^2 \le \varepsilon$, $\varepsilon < \varepsilon_1(\lambda)$ we have $u_t \ge R(t)v$. But for $\varepsilon \le \xi \le \pi - \varepsilon$ the inequality $\sin \xi \ge \varepsilon$ holds. Consiquently, for these ξ we have

$$\inf_{\varepsilon \le \xi \le \pi - \varepsilon} v \ge R(t)\frac{2\varepsilon}{\pi} e^{-\frac{\pi^2 t}{4\varepsilon}}. \tag{3.20}$$

Now fix a sufficiently small positive t_0. For each $n \ge 1$ we may find such positive λ_n and $\varepsilon_n < \varepsilon_1(\lambda)$ that for all $\varepsilon < \varepsilon_n$ the inequality

$$\| u(x, 0, \varepsilon) \|_{L_n} \le \frac{1}{n}$$

holds. Really,

$$u_0(x) \le \frac{8}{\varepsilon^\lambda} e^{-\frac{x^2}{\varepsilon}}, \quad \| u_0 \|_{L_n} \le \frac{8}{\varepsilon^\lambda} \left(\frac{\varepsilon}{n}\right)^{\frac{1}{2n}} \left(\sqrt{\frac{n}{\varepsilon}} \int_{-\infty}^{+\infty} e^{-\frac{x^2 n}{\varepsilon}} dx\right)^{\frac{1}{n}} \le K \varepsilon^{\frac{1}{2n} - \lambda},$$

where K is an absolute constant, and it is sufficient to choose

$$\lambda_n = \frac{1}{4n}, \quad \varepsilon_n \le \varepsilon_1\left(\frac{1}{4n}\right), \quad K\varepsilon_n^{\frac{1}{4n}} < \frac{1}{n}.$$

Further, for $\varepsilon \le \xi \le \pi - \varepsilon$ we have

$$u(x,t,\varepsilon) \ge u(x,0,\varepsilon) + t \cdot \inf_{\varepsilon \le \xi \le \pi-\varepsilon} u_t,$$

i.e., for these ξ and $t \le 4\varepsilon/\pi^2$ the inequality

$$u(x,t,\varepsilon) \ge K_4 t e^{K_5\varepsilon^{-\lambda}}$$

holds. The constants K_4, K_5 may be chosen independent on ε for $\varepsilon \le \varepsilon_2(\lambda)$ and small $\varepsilon_2(\lambda)$. Consequently,

$$J(t_0) = \left(\int_0^{t_0}\int_{-1}^{1} |u|^n \, dx \, dt\right)^{1/n} \ge K_6 \cdot \frac{1}{(n+1)^{1/n}} \varepsilon^{\frac{n+1}{n}+\frac{1}{2}} e^{K_5\varepsilon^{-\lambda}} \xrightarrow[\varepsilon \to 0]{} \infty.$$

By the diagonal method, we may choose such sequences $\lambda_n, \bar{\varepsilon}_n$ that

$$\| u(x,0,\bar{\varepsilon}_n) \|_{L_n} < \frac{1}{n}$$

and $J(t_0) > n$, $J(\bar{\varepsilon}_n) > n$, which was to be proved. So, we have proved that some solutions by initial data from C^0 (and even from C^∞) do not converge to zero. From theorem 3.2 it follows that there exist such stationary solutions $v_i(x)$, $v_1(x) = -v_2(x)$ such that each solution $u(x,t)$ constructed by initial data from C^1 such that

$$v_1 < u_0 < v_2$$

for

$$-1 < x < 1, \quad u_0'(1) \ne v_i'(1), \quad u_0'(-1) \ne v_i'(-1)$$

converges to zero for $t \to \infty$.

The example shows that the norm restrictions in our theorems could not be weakened, generally speaking, without additional restrictions on the coefficient growth for $|u| + |u_x| \to \infty$.

§3.4 Some examples.

In this paragraph we shall show how the obtained results may be applied for description of the structure of the stationary solutions of the problem and its attraction domains. Also we shall give some examples which will illustrate the possible situations. Except for the examples considered below we shall assume that for the problem (1.1)-(1.3) the condition (B) holds (see definition 1.2, §2.1).

Definition 4.1. A stationary solution of (1.1)-(1.3) $v(x)$ we shall call *stable from above (from below)* if for each $\varepsilon > 0$ there exists such $\delta > 0$ that for $u_0 \in E$ from

$$u_0 \geq v \ (u_0 \leq v), \quad \| u_0(x) - v(x) \|_1 < \delta$$

it follows that the solution $u(x, t; u_0)$ is determined for all $t \geq 0$, where

$$\| u(x, t; u_0) - v(x) \|_1 < \varepsilon.$$

If, in addition, we know that

$$\lim_{t \to +\infty} \| u(x, t; u_0) - v(x) \|_1 = 0,$$

then $v(x)$ will be called *asymptotically stable from above (from below)*.
As usual, a solution $v(x)$ we shall call *stable*, if for each $\varepsilon > 0$ there exists such $\delta > 0$ that for $u_0 \in E$, from

$$\| u_0(x) - v(x) \|_1 < \delta$$

it follows that the solution $u(x, t; u_0)$ is determined for all $t \geq 0$ and

$$\| u(x, t; u_0) - v(x) \|_1 < \varepsilon.$$

If

$$\lim_{t \to +\infty} \| u(x, t; u_0) - v(x) \|_1 = 0,$$

then we shall call $v(x)$ *asymptotically stable*.

Applying the comparison theorem, it is easy to prove that if $v(x)$ is stable from above and from below, then it is stable. Analogously, if $v(x)$ is asymptotically stable from above and from below, then it is asymptotically stable.

The following examples clarify the role of condition (B) (see definition 1.2 chapter 2) involved in the statements of the theorems in the previous and current sections. This condition generalizes the well-known Bernstein-type constraints (see for instance Ladyzhenskaya and Ural'tseva, 1986) that require, roughly speaking, at most quadratic growth of the ratio b/a in the gradient variable and allow obtaining the estimate for the $C^1[0,1]$-norm of a bounded solution.

In example 4.1, condition (B) is violated, although all solutions are smooth and stabilize to zero as $t \to \infty$. At the same time, for an isolated solution $u(x,t)$, we can make condition (B) hold by extending the right-hand side of the equation to the exterior of the range of $u(x,t)$ so as to have a compact support for the extention. In the following example the problem has unbounded solutions which cannot stabilize disregarding the fact that condition (B) is valid (as well as the conditions on the growth in u_x). This relates to unboundedness of the solution itself.

Observe that consition (B) enables us to interpret the right-hand side of the equation as the Euler operator of some variational problem. When the corresponding Lagrangian is strictly convex in u_x and demonstrates that the violations of one-sided boundedness of the Lagrangian of a variational problem may lead to absence of stabilization for a solution of the corresponding dynamical problem.

Example 4.3 shows that the presence of unbounded solutions relates both to the form of the nonlinearity in the equation and to the boundary conditions.

Consider the examples.

Example 4.1. It is easy to see that the trivial solution of the problem

$$u_t = u_{xx} - u^3, \quad u(\pm 1, t) = 0, \quad u(x,0) = u_0(x) \qquad (4.1)$$

is asymptotically stable in the space $C^{2+\alpha}$. Let $v(x)$ be a stationary solution, then

$$0 = \int_0^1 \left(v_{xx} - v^3\right)vdx = -\int_0^1 \left(v_x^2 + v^4\right)dx,$$

whence follows that $v(x)$ may be only the trivial solution.

Evidently, if

$$u_0(x) \in E = \{u_0 \in C^1[0,1] : u_0(\pm1) = 0\},$$

then the solution $u(x,t;u_0)$ is uniformly bounded for $t \geq 0$ in the norm $C^1[0,1]$. Therefore, (see theorem 1.1) it is stabilized (tends to zero) for $t \to \infty$.

Further, calculations show that for each smooth positive function $\psi(\tau)$, the pair

$$\Phi(x,\xi,\eta) = \int_0^\eta (\eta - \zeta)\psi(\zeta^2 - \xi^4/2)d\zeta + \frac{1}{2}\int_0^{\xi^4/2} \psi(-\tau)d\tau,$$

$$\rho(x,\xi,\eta) = \psi(\eta^2 - \xi^4/2)$$

generates the Liapunov functional on the solution of problem (4.1) (see definition 1.1, §2.1).

Really, we have

$$\frac{d}{dt}\int_{-1}^1 \Phi(u,u_x)dx = \int_{-1}^1 \left[-2u^3 \int_0^{u_x} (u_x - \eta)\psi'd\eta \right.$$

$$+u^3\psi\left(-\frac{u^4}{2}\right)\bigg]u_tdx + \int_{-1}^1 \int_0^{u_x} \psi\left(\eta^2 - \frac{u^4}{2}\right)d\eta u_{xt}dx$$

$$= \int_{-1}^1 \left\{-2u^3 \left[u_x \int_0^{u_x} \psi'd\eta - \frac{1}{2}\int_0^{u_x} \frac{d}{d\eta}\psi\left(\eta^2 - \frac{u^4}{2}\right)d\eta\right]\right.$$

$$+u^3\psi\left(-\frac{u^4}{2}\right)\Big\}u_t dx - \int\limits_{-1}^{1}\left[\psi\left(u_x^2 - \frac{u^4}{2}\right)u_{xx} - 2u^3 u_x \int\limits_{0}^{u_x}\psi' d\eta\right]u_t dx$$

$$= -\int\limits_{-1}^{1}\left(\psi u_{xx} - \psi u^3\right)u_t dx = -\int\limits_{-1}^{1}\rho(u, u_x)u_t^2 dx.$$

At the same time, since the equation $y'' = y^3$ has the set of solutions $y(x) = \sqrt{2}(x - c)^{-1}$ destroyed for $x = c$, the condition (B) for this problem fails.

Thus, if (B) fails, all the solutions of the problem may stabilize for $t \to \infty$.

Example 4.2. (See also, Bernstein, 1912). As earlier, the trivial solution of the problem

$$u_t = u_{xx} + u^3, \quad u(\pm 1, t) = 0, \quad u(x, 0) = u_0(x) \tag{4.2}$$

is asymptotically stable (the linearized equation coincides with the heat conductivity equation and the problem has a negative spectrum). The equation which determines the stationary solutions

$$y'' + y^3 = 0 \tag{4.3}$$

has the first integral

$$y'^2 + \frac{1}{2}y^4 = C. \tag{4.4}$$

Therefore, each solution $\varphi(x_0, x, y_0, y_1)$ of the Cauchy problem

$$y(x_0) = y_0, \qquad y'(x_0) = y_1$$

for equation (4.3) may be extended in the whole real axis, where

$$|\varphi_x(x_0, x, y_0, y_1)| \leq C^{1/2}, \quad |\varphi(x_0, x, y_0, y_1)| \leq (2C)^{1/4}.$$

Therefore, for problem (4.2) the condition (B) holds, i.e., as was proved above, there exists a set of Liapunov functionals and each bounded solution in $C[-1,1]$ norm is bounded in $C^1[-1,1]$ and is stabilized for $t \to \infty$ (theorem 1.1).

Consider in detail the structure of the stationary solutions set for problem (4.2). By (4.4) in the neighborhood of the point x_0 we have

$$x - x_0 = F(y) = \int_{y_0}^{y} \frac{dv}{(C - v^4/2)^{1/2}}, \quad C = y_1^2 + \frac{1}{2}y_0''.$$

In the points $x_i = x_0 + F(\pm(2C)^{1/4})$, evidently, we have the equalities

$$y(x_i) = \pm(2C)^{1/4}, \qquad y'(x_i) = 0 \quad (i = 1, 2).$$

In this case, in the point x_1 the function $y(x)$ has the positive maximum, and in the point x_2 the function $y(x)$ has the negative minimum. By symmetry (see (4.3)), it follows that

$$\varphi\left(x_0, \frac{x_1 + x_2}{2}, y_0, y_1\right) = 0,$$

and the function φ is odd relative to the point $(x_1 + x_2)/2$.

Thus, the solution of equation (4.3) is periodic with the period $2T$.

$$2T = 2 \int_{-(2C)^{1/4}}^{(2C)^{1/4}} \frac{dv}{(C - v^4/2)^{1/2}} = 2\frac{(2C)^{1/4}}{C^{1/2}} \int_{-1}^{1} \frac{dt}{(1 - t^4)^{1/2}}.$$

Besides, each solution of (4.3) has the following properties

$$\varphi(x_0, x, y_0, y_1) = \varphi(0, x - x_0, y_0, y_1) = \lambda^{-1}\varphi\left(\frac{x_0}{\mu}, \frac{x}{\mu}, \lambda y_0, \lambda\mu y_1\right),$$

where $\lambda = \pm\mu \neq 0$ is arbitrary.

Let us construct the stationary solutions of problem (4.2). Consider the function $\varphi(-1, x, 0, \alpha)$. By the periodicity, $\varphi(-1, -1 + nT, 0, y_1) = 0$. Set

$$\theta(x,n) = \varphi\left(-1, \frac{nT}{2}(x+1) - 1, 0, y_1\right)\frac{nT}{2} = \varphi\left(-1, x, 0, y_1\frac{n^2T^2}{4}\right).$$

As in this case $C = y_1^2$, then

$$y_1\frac{n^2T^2}{4} = \sqrt{2}\frac{y_1|y_1|}{4y_1^2}n^2\left(\int_{-1}^{1}\frac{dt}{(1-t^4)^{1/2}}\right)^2$$

$$\sqrt{2}n^2\left(\int_{0}^{1}\frac{dt}{(1-t^4)^{1/2}}\right)^2.$$

It means that all the stationary solutions of problem are as follows

$$\theta_\pm(x,n) = \varphi\left(-1, x, 0, \pm\sqrt{2}n^2\left(\int_{0}^{1}\frac{dt}{(1-t^4)^{1/2}}\right)^2\right).$$

As follows from Zelenyak (1977) and Vishnevskii (1984c) the functions $\theta_\pm(x,1)$ lie in the boundary of the attraction domain of the stationary solution. Evidently, the oscillations of $\theta_\pm(x,n)$ increase in proportion to n, and their zeros, by the periodicity, are disposed uniformly. All the solutions $\theta_\pm(x,n)$ for $n \neq 0$ are unstable (Belonosov and Zelenyak, 1975).

Now, we show the existence of solutions of (4.2) which are not stabilized. Choose the parameter $\alpha \in (\theta'_+(-1,1), \theta'_+(-1,2))$ and consider such initial data $u_0 \in E$ that

$$u_0(x) > \psi(x) = \varphi(-1, x, 0, \alpha), \quad u_0(x) > \theta_+(x,1)$$

for $x \in (-1,1)$. As when α grows, the distance between the zeros of $\psi(x)$ is diminished, then the graphs of the functions $\psi(x)$ and $\theta_+(x,1)$ intersect. Moreover, near the point $x = -1$ we have $\psi(x) > \theta_+(x,1)$ and $\psi(1) < 0$. From the maximum principle it follows that $u(x,t) \geq \psi(x)$; therefore, $u(x,t)$ can not tend to $\theta_+(x,1)$ for $t \to \infty$. Each of the functions $\theta_+(x,n)$ $(n \neq 1)$ has at least one zero inside the interval (-1,1), i.e., attains the values less than $\theta_+(x,1)$. Consequently, $\theta_\pm(x,n)$ also can not be limits for $u(x,t)$.

Hence follows (see theorem 1.1) that the solution $u(x,t)$ can not be stabilized; therefore, there exists finite or infinite T such that

$$\lim_{t \to T} \sup_{-1 \le x \le 1} |u(x,t)| = \infty.$$

Remark 4.1. Example 4.2 shows that the result of lemma 3.1 on the number of stationary solutions is exact in the class of solutions with the bounded by the number α_p derivative $u_x(0,t)$.

Example 4.3. (Vishnevskii *et al.*, 1995). Consider the problem

$$u_t = u_{xx} - x u_x^3, \tag{4.5}$$

$$u(0,t) = 0, \quad u(1,t) = 2\pi, \quad u(x,0) = u_0(x). \tag{4.6}$$

It is easy to verify that the equation

$$y'' - x y'^3 = 0, \tag{4.7}$$

which determine stationary solutions of this problem has the first integrals $A = $const, $B = $const, where

$$A = \frac{y_x}{(1 + x^2 y_x^2)^{1/2}}; \quad B = \frac{1}{(1 + x^2 y_x^2)^{1/2}} \sin y - \frac{x y_x}{(1 + x^2 y_x^2)^{1/2}} \cos y$$

Its general solution is as follows

$$y = C_1 + \sin^{-1}(C_2 x). \tag{4.8}$$

Setting (Belonosov and Zelenyak, 1975)

$$\rho(x, y, y') = \frac{1}{(1 + x^2 y'^2)^{3/2}} \Psi\Big(A(x, y, y'), B(x, y, y')\Big),$$

$$\Phi(x, y, y') = \int_0^{y'} (y' - \eta) \rho(x, y, \eta) d\eta$$

we see that for each smooth nonnegative function Ψ the pair ρ, Φ generates the Liapunov functional (definition 1.1, §2.1) on the solutions of (4.5), (4.6). In other words, equation (4.5) may be rewritten as follows

$$\rho(x, u, u_x)u_t = \frac{d}{dx}\Phi_{u_x} - \Phi_u.$$

In particular, if $\Psi \equiv 1$, equation (4.5) is equivalent to the equation

$$\frac{1}{(1 + x^2 u_x^2)^{3/2}}u_t = \frac{d}{dx}\frac{u_x}{(1 + x^2 u_x^2)^{1/2}}.$$

For problem (4.5)-(4.6) the maximum principle holds

$$\sup |u(x, t)| \leq \sup |u_0(x)| + 2\pi.$$

Therefore (see theorem 1.1), if u_x is uniformly bounded for $t \geq \delta > 0$, then u is bounded in the $H^{2+\alpha}$ norm and is stabilized to the smooth stationary solution of the problem. But, as follows from the form (4.8) of the general solution of (4.7), the oscillation of such solution does not exceed π. Therefore, taking into account boundary conditions (4.6), there are no smooth stationary solutions.

Therefore, for a certain T (finite, or infinite) we have

$$\lim_{t \to T} \sup_{-1 \leq x \leq 1} |u_x(x, t)| = \infty.$$

The function $w = u_t$ satisfies the problem

$$w_t = w_{xx} - 3xu_x^2 w_x,$$

$$w(0, t) = w(1, t) = 0, \qquad w(x, 0) = w_0(x) = u_0'' - xu_0'^3.$$

Thus, if $u_0 \in C^4$ and the compatibility conditions hold, then we may apply to the function w the maximum principle and obtain

$$|w| = |u_t| \leq K \quad \text{for} \quad 0 \leq t < T.$$

Set $u_0(x) = a \sin \beta x$, where $a = 2\pi / \sin \beta$, $\beta \in (\pi/2, \pi)$. It is easy to see that in the point $x = 0$ the compatibility conditions hold and $u_0(1) = 2\pi$. The calculations show that such β^* exists that

$$u_0'' - x u_0'^3 \le 0 \qquad x \in (0,1),$$

$$u_0''(1) - u_0'^3(1) = 0.$$

Then, evidently, $u_t \le 0$. So, the bounded function $u(x,t)$ monotonically decreases when t grows and, therefore, has a finite limit for $t \to T$ in each point x.

For equation (4.5) consider the problem

$$u(0,t) = u_x(1,t) - u(1,t) = 0, \quad u(x,0) = \varphi(x),$$

where $\varphi(x)$ is a smooth function satisfying the compatibility conditions. Denote the solution to equation (4.5) with these initial boundary conditions and $\varphi(x) = N_x$ by $u^N(x,t)$. To establish the existence "as a whole" with respect to time, it suffices to obtain the uniform by t estimate for $|u^N(x,t)| + |u_x^N(x,t)|$.

Evidently, $w(x,t) = u_t^N(x,t)$ solves the problem

$$w_t = w_{xx} - 3x \left(u_x^N \right)^2 w_x, \quad w(0,t) = w_x(1,t) - w(1,t) = 0,$$

$$w(x,0) = \varphi''(x) - x\varphi'^3(x) = -N^3 x.$$

As initial data satisfy the compatibility conditions with boundary values, we may apply to the function w the maximum principle, i.e., $w(x,t) \le 0$; therefore, $u_t^N(x,t) \le Nx$. Hence, from the maximum principle it follows that $u^N(x,t) \ge 0$.

Similary, taking $z(x,t) = u_x^N(x,t)$, we obtain for the function z the problem

$$z_t = z_{xx} - 3x \left(u_x^N \right)^2 z_x - \left(u_x^N \right)^2 z,$$

$$z_x(0, t) = 0, \quad z(1, t) = u(1, t), \quad z(x, 0) = N.$$

The compatibility conditions again hold; therefore, $0 \leq z \leq N$. Thus, the estimate

$$\left| u^{N}(x, t) \right| + \left| u_x^{N}(x, t) \right| \leq 2N$$

holds true, whence the solvability of the problem follows.

Analogously we may establish the boundedness of

$$\left| u^{-N}(x, t) \right| + \left| u_x^{-N}(x, t) \right|.$$

Arbitrary smooth initial data $\varphi(x)$ $(\varphi(0) = 0)$ may be confined between two functions $\pm Nx$

$$-Nx \leq \varphi(x) \leq Nx.$$

Hence it follows that the solution with the initial function $\varphi(x)$ above satisfies the inequalities

$$u^{-N}(x, t) \leq u(x, t) \leq u^{N}(x, t).$$

Similarly, we may restrict $|u_x|$. Therefore, we see that there exists a solution uniformly bounded together with its derivatives, stabilized for $t \to \infty$ to the stationary solution.

The set of stationary solution of (4.5) such that $v(0, \alpha) = 0$ is the one-parameter set of the functions $v(\alpha) = \sin^{-1}(\alpha x)$. Using the second boundary condition we obtain

$$\theta(\alpha) = (v_x - v)\big|_{x=1} = \frac{\alpha}{(1 - \alpha^2)^{1/2}} - \sin^{-1}\alpha,$$

where

$$\theta(0) = 0, \quad \theta'(\alpha) = \frac{\alpha^2}{(1 - \alpha^2)^{3/2}}$$

As $\theta(\alpha)$ are monotone, the unique regular stationary solution of this boundary-value problem is the trivial solution $v(x,0) \equiv 0$. Thus, under our assumptions relative to initial data $\varphi(x)$, a solution to the parabolic problem exists for all positive values of t and uniformly tends to zero with its derivatives for $t \to \infty$, i.e., is stabilized.

Example 4.3 shows that if the condition (B) fails, the derivative of a bounded solution of the parabolic problem may infinitely increase. Unlike the other examples, this example a) is rather simple; b) the right hand of the equation may be represented as the Euler-Lagrange equation; and c) we have a demonstration of gradient blowing-up effect dependence on the boundary conditions.

Parabolic equations whose right hand side is the Euler-Lagrange equation for Lagrangians from examples like example 2.2 chapter 2 are not investigated.

In conclusion, we may recall example 3.1 chapter 2 and note that the derivative may infinitely grow in the interior point x (see example 4.5 given below).

For more detailes about blowing-up solutions we refer the reader to Samarskii *et al.* , (1995).

Remark. Note, that in this example the global attractor depends on the form of the boundary conditions. Note, also, that the dependence of the global attractor on boundary conditions was investigated by Fiedler (1996).

Example 4.4. (Fila and Lieberman, 1994). It was proved that the problem

$$u_t = u_{xx} + e^{u_x},$$

$$u(0,t) = 0, \qquad u(a,t) = 0,$$

$$u(x,0) = u_0(x)$$

sometimes has a bounded solution with unbounded derivative with respect to x. This gradient blowing-up effect depends on the length of the interval.

In particular, it was proved (Fila and Lieberman, 1994) that for the equation

$$u_t = u_{xx} + f(u_x),$$

where $f(s)$

1) is a smooth ($f \in C^2$), positive, and monotone increasing function,
2) possesses a bounded ratio $f'(s)/f(s)$ as $s \to \infty$,
3) possesses super quadratic growth at infinity

$$\int\limits_0^\infty \frac{s\,ds}{f(s)} < \infty;$$

there exists a number A, such that if $a > A$, for any u_0, the derivative u_x is unbounded

$$\sup_x |u_x(x,t)| \to \infty, \quad \text{as} \quad t \to T$$

for a proper $T \in (0, \infty)$.

Remark 4.2. Let us note that a family of singular solutions to the ordinary differential equation $y'' + f(y') = 0$ (which determines the steady states of the above parabolic problem) could be calculated directly.

In the given example the derivative $u_x(x,t)$ grows to infinity only in the edge point $x = 0$ (this may be proved for a number of such problems via the Liapunov functionals). Examples of interior gradient blowing-up are also available.

Example 4.5. (Angenent and Fila, to appear). A class of semilinear equations with bounded solutions whose derivative blows-up in finite time in the interior of the interval $(-1, 1)$ is as follows

$$u_t = u_{xx} + f(u)|u_x|^{m-1}u_x,$$

$$u(x,0) = u_0(x), \qquad u(\pm 1, t) = A_\pm$$

with $f \in C^1(R)$, $m > 2$. It was proved that the solution to the above problem may be singular only in points where $f(u)$ changes sign.

§3.5 Some qualitative properties of dissipative boundary-value problems for quasilinear parabolic equations with one spatial variable.

The so-called dissipative problems are often met both in theory and in practice (see, for example, Samarskii *et al.* , 1995).

Definition 5.1. We say that problem (1.1)-(1.3) is dissipative if such $R_0 > 0$ exists that for each $u_0 \in E$ (C^1-functions which satisfy the compatibility conditions) the solution $u(x, t; u_0)$ is determined for all $t \geq 0$ and the inequality

$$\overline{\lim_{t \to \infty}} \| u(x, t; u_0) \|_0 < R \tag{5.1}$$

holds.

Note, that estimate (5.1) and the condition (B) (see definition 1.2, §2.1) provide uniform boundedness of solution $u(x, t; u_0)$ in $C^1[0, 1]$. Then, by lemma 1.1,

$$\| u(x, t; u_0) \|_{3+\alpha} \leq R_3 \tag{5.2}$$

holds for $t \geq t_0 > 0$, where t_0 depends on $\| u_0 \|_1$.

Definition 5.2. A set $M(v) \subset E$ is said to be the attraction domain of a stationary solution $v(x)$, if for each $u_0 \in M(v)$ for $t \to \infty$ we have

$$\| u(x, t; u_0) - v(x) \|_1 \to 0. \tag{5.3}$$

Note some peculiarities of dissipative problems which simplify the analysis of the stationary solutions set and clarify its properties.

Theorem 5.1. (Vishnevskii, 1984, 1990). Let problem (1.1)-(1.3) be dissipative. Then it has the maximal stationary solution $v_{\max}(x)$ (this means that every other stationary solution $v(x)$ satisfies the inequality $v_{\max}(x) \geq v(x), \ x \in (0,1)$) stable from above (definition 4.1). Moreover,

$$M(v_{\max}) \supset \{u_0 \in E, \quad u_0(x) \geq v_{\max}(x), \quad x \in [0,1]\}.$$

Let $w(x) < v_{\max}$ be a stable from above stationary solution. Then there exists such an unstable from below stationary solution $v(x)$ that

$$M(w) \supset \{u_0 \in E, \quad v(x) \geq u_0(x) \geq w(x), \quad x \in [0,1]\}.$$

Moreover, there exists a unique monotone by t solution $\eta(x,t)$ determined for $t \in R$ which tends to $v(x)$ for $t \to -\infty$ and tends to $w(x)$ for $t \to +\infty$.

This theorem will be proved in a somewhat more general case in chapter 5. Therefore, we omit the proof. Note that in the one-dimensional case instead of liminations on the growth of nonlinear terms by the gradient of an unknown function which will be imposed in chapter 5 we may impose the conditions 1.2, §2.1 chapter 2.

The analogous statement holds for the minimal stationary solution of the problem.

Thus, if the dissipative problem has only a finite number of stationary solutions, then monotone connected orbits $\eta(x,t)$, which connect the unstable from below (from above) stationary solution with the unstable from above (from below) solution naturally give the partial order in the set of stationary solutions.

Note that a monotone connected orbit $\eta(x,t)$ for all t is (depending on its increasing or decreasing) the lower or upper function (solution) of the problem (Fiedler and Rocha, 1994).

These connected orbits (not necessarily monotone) are of great importance for investigating the behavior of parabolic equations for extended time, for example, when constructing global attractors. Investigation and construction of connected orbits one may find, for example, in Brunovsky and Fiedler (1988, 1989) and in Hartman (1970).

Remark 5.1. If (1.1)-(1.3) is dissipative and functions $a(x, \xi, \eta)$, $b(x, \xi, \eta)$ are analytic by ξ, η, then the problem has only a finite number of stationary solutions.

Really, considering for simplicity the Dirichlet problem, we obtain that $\varphi(0, 1, 0, \alpha)$ (see definition 1.2, §2.1) is analytic by α. By the dissipativity, $\varphi(0, 1, 0, \alpha) \neq 0$ and $\varphi(0, 1, 0, \alpha) \neq 0$ for $|\alpha| > R_3$. Therefore, the function $\varphi(0, 1, 0, \alpha)$, the zeros of which determine all stationary solutions, may become zero only in a finite number of points.

Consider now which properties of the structure of stationary solutions set provide the dissipativity of the problem. As before, to simplify the calculation, consider the first boundary-value problem.

We say that a stationary solution $\tilde{v}(x)$ is quasimaximal if each stationary solution either intersects its graph or is under it for $x \in (0, 1)$.

Theorem 5.2. Let problem (1.1)-(1.3) have no more than a countable set of stationary solutions and suppose that the condition (B) holds (definition 1.2, §2.1).

If a quasimaximal stationary solution $\tilde{v}(x)$ is unstable from above, then there exists such a monotone solution $\eta(x, t)$ determined for all $t \in R$ that converges to $\tilde{v}(x)$ for $t \to -\infty$ and

$$\limsup_{t \to \tau} \limits_{x} |\eta(x, t)| = \infty, \quad \tau \leq \infty.$$

In the case of example 4.2, the theorem states that if the quasimaximal solution $\theta_+(x, n)$ is unstable from above, then the problem could not be dissipative. The quasiminimal stationary solution may be determined analogously.

All this allows us to prove the following theorem.

Theorem 5.3. Let the condition (B) hold. In order that problem (1.1)-(1.3) be dissipative it is necessary and sufficient that all quasiminimal and quasimaximal stationary solutions be stable, correspondingly, from above and from below. In this case the quasimaximal (quasiminimal) solution is maximal (minimal). This means that each of the other stationary solutions is strictly less (greater) than it for $x \in (0, 1)$.

Proof. Necessity follows from theorem 5.2 and from its analog for quasiminimal solutions.

Sufficiency. (see Zelenyak, 1977). Denote by $v(x, \mu)$ the solution of the Cauchy problem

$$a(x, v, v')v'' + b(x, v, v') = 0, \quad v(0, \mu) = 0, \quad v'(0, \mu) = \mu$$

(recall that we consider the first boundary-value problem).

Show that for each initial data $u_0 \in E$ which is greater than $\tilde{v}(x)$ there exists such μ that

$$v(x, \mu) > u_0(x), \quad x \in (0, 1).$$

Then, by the comparison theorem,

$$v(x, \mu) > u(x, t; u_0), \quad t \geq 0.$$

By the condition (B) (theorem 7.2, Belonosov and Zelenyak, 1975) the solution $u(x, t; u_0)$ is bounded in $C^1[0, 1]$ and is stabilized (theorem 1.1).

By theorem 1.1, $u(x, t; u_0)$ is stabilized to a stationary solution which, by the quasimaximality, coincides with $\tilde{v}(x)$. By the arbitrariness of $u_0(x) \geq \tilde{v}(x)$, this proves the stability from above of $\tilde{v}(x)$.

Show that there exists such μ that

$$v(x, \mu) > u_0(x), \quad x \in (0, 1). \tag{5.4}$$

Without loss of generality we may consider that $\tilde{v}(x) \equiv 0$. Stability from above of the trivial solution is equivalent to (Belonosov and Zelenyak, 1975)

$$v_\mu(x,0) \geq 0, \quad x \in (0,1], \quad v_\mu \not\equiv 0.$$

As there are no positive stationary solutions, we have

$$v(1,\mu) > u_0(1), \quad \mu > 0.$$

Let for each μ inequality (5.4) be somewhere failed. Then the function $w(x,\mu) = v(x,\mu) - u_0(x)$ possesses the following properties

$$w(0,\mu) = 0, \quad w(1,\mu) > 0, \quad w(x^*,\mu) < 0 \qquad (5.5)$$

for all μ and for a certain $x^*(\mu) \in (0,1)$. Therefore, there exists such $x_1(\mu) \in (0,1)$ that

$$w_x(x_1,\mu) = 0, \qquad w(x_1,\mu) < 0,$$

i.e., for all μ

$$v_x(x_1,\mu) = u_0'(x_1), \qquad v(x_1,\mu) < u_0(x_1). \qquad (5.6)$$

The condition (B) and nonnegativeness of v yield

$$|v_x(x,\mu)| + v(x,\mu) \to +\infty, \quad \mu \to \infty \qquad (5.7)$$

uniformly by $x \in [0,1]$ (Belonosov and Zelenyak, 1975) which contradicts inequalities (5.6). This proves theorem 5.3.

In conclusion, consider the example (Zelenyak, 1977) which shows that monotone connected orbits $\eta(x,t)$ may establish only a partial order in the stationary solutions set (Vishnevskii, 1990), and the attraction domain described in theorem 3.4 (and the analogous attraction domains described further in chapter 5), generally speaking, are not uniquely defined.

Example 5.1.   For $(x,t) \in (-\pi, \pi) \times (0, \infty)$ consider the problem

$$u_t = u_{xx} + ue^{N\left(\sin^2 x - u^2\right)} , \tag{5.8}$$

$$u(\pi, t) = u(-\pi, t) = 0, \tag{5.9}$$

$$u(x, 0) = u_0(x). \tag{5.10}$$

The problem (5.8)-(5.10) is dissipative. Really, consider the new unknown function

$$u(x,t) = v(x,t)\cos(x/(2+p)), \quad p > 0,$$

i.e., in the interval $[-\pi, \pi]$ the multiplier is positive and is separated from zero. The function $v(x,t)$ solves the problem

$$v_t = v_{xx} - \frac{2v_x}{2+p}\tan\frac{x}{2+p} + ve^{N\left(\sin^2 x - v^2 \cos^2 \frac{x}{2+p}\right)} - \frac{v}{(2+p)^2},$$

$$v(\pi, t) = v(-\pi, t) = 0, \tag{5.11}$$

$$v(x, 0) = v_0(x) = u_0(x)\cos^{-1}\frac{x}{2+p}.$$

Evidently, the problems (5.8)-(5.10) and (5.11) are dissipative simultaneously. Dissipativity of (5.11) follows from the comparison theorem and from the dissipativity of the ordinary differential equation

$$\xi' = -\frac{\xi}{(2+p)^2} + \xi e^{N\left(1 - \cos^2 \frac{\pi}{2+p}\xi^2\right)}, \quad \xi(0) = \xi_0.$$

By the remark 5.1 the problem has only a finite number of stationary solutions. It is easy to see (theorem 5.3) that stable from above and from below respectively quasimaximal and quasiminimal solutions are $v_{max}(x)$ and $v_{min}(x)$, i.e., for each stationary solution $v(x)$

$$v_{max}(x) > v(x) > v_{min}(x), \quad x \in (0, 1).$$

By direct computation, we have that $v_{\max}(x)$ and $v_{\min}(x)$, just as steady states $v_{\pm}(x) = \pm \sin x$, are asymptotically stable.

In Zelenyak (1977), Belonosov and Zelenyak (1975) and Akramov and Zelenyak (1975) it was shown that each stationary solution $\pm \sin x$ is connected with a pair of such unstable stationary solutions $v_i^{\pm}(x)$ that if

$$v_1^+(x) < u_0(x) < v_2^+(x), \quad (v_1^-(x) < u_0(x) < v_2^-(x)), \quad x \in (-\pi, \pi),$$

then the solution $u(x, t; u_0)$ is stabilized to $\sin x$ (to $-\sin x$).

Thus, the monotone connected orbits $\eta(x, t)$ (theorem 5.1) select at least three subsets of stationary solutions

$$\{v_{\max}(x), 0, v_{\min}(x)\},$$

$$\{v_{\max}(x), v_2^+(x), \sin x, v_1^+(x), v_{\min}(x)\},$$

$$\{v_{\max}(x), v_2^-(x), -\sin x, v_1^-(x), v_{\min}(x)\}$$

ordered in the uniform norm.

Chapter 4

The stability criterion for the trivial solution to the mixed problem for the second order parabolic equation.

In the fourth chapter we have proved the criterion of exponential stability of the trivial solution of the mixed problem for a linear parabolic equation with coefficients depending on time.

The criterion of exponential stability of the trivial solution of the first boundary-value problem for parabolic equation with coefficients depending only on spatial variables was considered by Gaevoi (published in Zelenyak (1972)). The further generalization may be found in Zelenyak (1977); Belonosov and Zelenyak (1975). The closed problems were considered in Protter and Weinberger (1966, 1984) in the section weight maximum principle.

Generalization of the case when coefficients depend on time has been made by Vishnevskii (1984). Note, that as the Perron example shows (see, for example, Daletskii and Krein, 1970), instead of stability it should be considered as a case of exponential stability (the definition of exponential stability see in Demidovich (1977) in the case of ordinary differential equations). In Daletskii and Krein (1970) instead of the notion of exponential stability the equivalent notion is taken.

The second paragraph deals with the particular cases of the criterion of the periodic and almost periodic dependence on time. In this case the results of chapter 1 may be revised. Note, that using the results of Zelenyak (1966, 1967), in the case when coefficients of the problem depend only on spatial variables, and the results of Vishnevskii (1984) in the case of periodic dependence of coefficients on time, it is easy to obtain the criterion of stability of the trivial solution in another way. This method was considered by Hess and Kato (1980) and by Beltramo and Hess (1984).

In the third paragraph we have shown that if the trivial solution of the linear problem is exponentially stable, then the trivial solution of the disturbed nonlinear problem will also be exponentially stable. The essential difficulty (as the Perron example shows) is the fact that the linear problem coefficients depend not only on the spatial variables but on time also. For the more general case of parabolic systems this result was obtained by Vishnevskii (1993b).

In the fourth paragraph we consider the stability of periodic solutions of the Neumann problem in the case when the parabolic equation and the boundary conditions do not depend on spatial variables on the domain Ω, and where the sought-for solution is convex. The first results were obtained for stationary solutions of autonomous parabolic equations by Matano (1979) and by Costen and Holland (1978). It was proved that if the Neumann problem is spatially homogeneous, and the domain Ω is convex, then each stable solution is spatially homogeneous. Moreover, in Matano (1979) there is the example of a nonconvex domain, where we have a stable spatially inhomogeneous solution. Further investigations in this direction were made by Matano and Mimura (1983) and by Jimbo (1989) (see also the bibliography cited there).

The generalization on the case of periodic solutions and periodic dependence of the problem on time was considered by different methods in Vishnevskii (1984) and in Hess (1987).

The close results for certain classes of parabolic systems were obtained by Matano and Mimura (1983), Kishimoto and Weinberger (1985), and by Vishnevskii (1993b) (see also the bibliography cited in these papers).

§4.1 The stability criterion for the trivial solution to the linear problem.

As earlier, let Ω be a bounded domain in R^n with a boundary $\partial\Omega$ of the class $C^{2+\alpha}$. Consider the following linear equation in the cylinder $Q^\tau = \Omega \times (\tau, +\infty)$

$$u_t = \sum_{i,j=1}^{n} a_{ij}(x,t)\frac{\partial^2 u}{\partial x_i \partial x_j} + \sum_{i=1}^{n} a_i(x,t)\frac{\partial u}{\partial x_i} + a(x,t)u$$

$$= L(x,t)u, \qquad (x,t) \in Q^\tau, \quad \tau \in R. \qquad (1.1)$$

Assume that

$$\alpha_0 u + \alpha_1 \left(\sum_{i=1}^{n} \beta_i(x,t)u_{x_i} + \beta_0(x,t)u \right) = B(x,t)u = 0, \qquad (1.2)$$

where

$$(x,t) \in \Gamma^\tau = \partial\Omega \times (\tau, +\infty), \quad \tau \in R.$$

For $t = \tau$ we give the following initial data

$$u(x,\tau) = u_0(x), \quad x \in \Omega. \qquad (1.3)$$

We shall assume that

$$a_{ij}(x,t) \in C^3(\bar{Q}_0), \quad a_i(x,t) \in C^2(\bar{Q}_0), \quad a(x,t) \in C^1(\bar{Q}_0),$$

$$Q_0 = \Omega \times (-\infty, +\infty); \qquad i,j = 1, \ldots, n.$$

We impose, also, the uniform parabolicity condition on the coefficients $a_{ij}(x,t)$

$$\mu_1 |\xi|^2 \leq \sum_{i,j=1}^{n} a_{ij}(x,t)\xi_i\xi_j \leq \mu_2|\xi|^2, \quad (x,t) \in Q_0, \quad \mu_1, \mu_2 > 0.$$

Relative to the boundary data assume that the following smoothness conditions hold

$$\beta_i(x,t) \in H^{1+\alpha}(\Gamma_0), \qquad \Gamma_0 = \partial\Omega \times (-\infty, +\infty), \quad i = 0, \ldots, n.$$

Let, besides,

$$\sum_{i=1}^{n} \beta_i(x,t)\cos(n, x_i) \geq \mu_0 > 0,$$

where n is the outer normal to the boundary $\partial\Omega$ in the point $(x,t) \in \Gamma_0, \alpha_1^2 + \alpha_0^2 > 0, \quad \alpha_1 \geq 0$.

Gaevoi (see Belonosov and Zelenyak, 1975) (for the first boundary-value problem) and Zelenyak (1977) (for the third boundary-value problem) had considered the autonomous problem of the form (1.1)-(1.3), where the coefficients do not depend on time. They (Belonosov and Zelenyak, 1975; Zelenyak, 1977) had proved the theorem below.

Theorem 1.1. (Belonosov and Zelenyak, 1975; Zelenyak, 1977) The following three conditions are equivalent.

1). The trivial solution of autonomous problem (1.1)-(1.3) is asymptotically stable, i.e.,

$$\| u(x,t) \|_B \leq K \| u_0 \|_B \, e^{-\lambda(t-\tau)},$$

where λ and K are positive constants. The space B is one of the spaces $C^{k+\alpha}(\bar{\Omega})$, $W_2^k(\Omega)$, $k = 1, 2, 3$ with the essential restrictions. Here, by $W_2^k(\Omega)$ we have denoted the standard space of S.L. Sobolev consisting of functions which are square integrable with all derivatives up to the order k.

2). For autonomous problem (1.1)-(1.3) there exist the Liapunov functionals, i.e., there exist such positive functions $\sigma(x), r_i(x), i = 1, 2, 3$ that under the problem conditions, the following equality holds

$$\frac{\partial}{\partial t} \left(\int_{\Omega} r_1(x) u^2(x,t) dx \right) = - \int_{\partial\Omega} r_2(x) u^2(x,t) dS$$

$$- \sum_{i,j=1}^{n} \int_{\Omega} a_{ij}(x)\sigma(x)\,(r_3(x)u(x,t))_{x_i}\,(r_3(x)u(x,t))_{x_j}\,dx.$$

3). There exists a positive solution of the elliptic problem $L(x)\rho(x)=0$, $x \in \Omega$, which satisfies the boundary condition $B(x)\rho(x) = 1$ for $x \in \partial\Omega$.

Note, that the equivalence of the first and third conditions for the case of the first boundary-value problem was proved by Gaevoi (Belonosov and Zelenyak, 1975).

Results close to the point 3 were also achieved by Zelenyak (1977).

One of the most important applications of this theorem is the fact that the change $w(x,t)\rho(x) = u(x,t)$ allows us to apply the maximum principle to the function $w(x,t)$. This property has been given the special notation of the weight maximum principle in Protter and Weinberger (1966, 1984).

While considering the nonstationary problems of the form (1.1)-(1.3), it was important to obtain the analog of the above theorem. We shall give it below. Then, in this chapter, we shall consider some applications of the proved results. We shall restrict ourselves to the space $C^{2+\alpha}(\bar{\Omega})$ (as the space B), where the stability of the linear problem will be studied. Note, that the analogous theorem may be obtained in the case of spaces $C^{k+\alpha}(\bar{\Omega})$, $W_2^k(\Omega)$, $k = 1, 2, 3$.

Theorem 1.2. (Vishnevskii, 1984, 1996) The following three conditions are equivalent.

1). The trivial solution of problem (1.1)-(1.3) is exponentially stable, i.e., there exist such constants μ, K that

$$\| u(x,t) \|_{2+\alpha}^{\Omega} \leq K e^{-\mu(t-\tau)} \| u(x,\tau) \|_{2+\alpha}^{\Omega}, \quad t > \tau, \qquad (1.4)$$

where $u(x,t)$ solves (1.1)-(1.3).

2). There exist positive functions $\sigma(x,t), r_i(x,t)$, $i = 1, 2, 3$, $(x,t) \in Q_0$, such that for each solution $u(x,t)$ of (1.1)-(1.3) the equality

$$\frac{\partial}{\partial t}\left(\int_{\Omega} r_1(x,t)u^2(x,t)dx \right) = - \int_{\partial\Omega} r_2(x,t)u^2(x,t)dS$$

$$- \sum_{i,j=1}^{n} \int_{\Omega} a_{ij}(x,t)\sigma(x,t)\,(r_3(x,t)u(x,t))_{x_i}\,(r_3(x,t)u(x,t))_{x_j}\,dx.$$

$$(x,t) \in Q_0, \qquad \sigma(x,t) \geq \sigma_0 > 0$$

holds.

3). For each $\varepsilon > 0$ there exists bounded for all $t \in R$ positive solution $\rho(x,t)$ of the parabolic equation

$$\rho_t(x,t) = L(x,t)\rho(x,t) + \varepsilon\rho(x,t),$$

with the boundary conditions

$$B(x,t)\rho(x,t) = 1, \quad (x,t) \in \Gamma_0.$$

Proof of this theorem will be interspersed with certain lemmas, which are of independent interest.

First, consider the inhomogeneous boundary-value problem

$$u_t = L(x,t)u + f(x,t), \qquad (x,t) \in Q^{\tau};$$

$$B(x,t)u = \varphi(x,t), \qquad (x,t) \in \Gamma^{\tau}; \qquad u(x,\tau) = u_0(x). \tag{1.5}$$

The statement below is analogous to theorem 2.1 from chapter 1 in the case when the space E^+ is empty. The role of estimate (2.7) of chapter 1 is played by condition (1.4).

Lemma 1.3. Let (1.4) hold, $f(x,t) \in H^{\alpha}(Q^{\tau})$, and $\varphi(x,t) \in H^{1+\alpha}(\Gamma^{\tau})$. The initial data $u_0 \in H^{2+\alpha}(\Omega)$ satisfies the necessary compatibility conditions. Then the solution of (1.5) exists for all $t \geq \tau$, belongs to the class $H^{2+\alpha}(Q^{\tau})$, and satisfies the inequality

$$\| u(x,t) \|_{2+\alpha}^{Q^{\tau}} \leq K(\mu) \left(\| u_0 \|_{2+\alpha}^{\Omega} + \| f \|_{\alpha}^{Q^{\tau}} + \| \varphi \|_{1+\alpha}^{\Gamma^{\tau}} \right). \tag{1.6}$$

In the case of the first boundary-value problem the estimate (1.6) has the form

$$\| u(x,t) \|_{2+\alpha}^{Q^\tau} \leq K(\mu) \left(\| u_0 \|_{2+\alpha}^{\Omega} + \| f \|_{\alpha}^{Q^\tau} + \| \varphi \|_{2+\alpha}^{\Gamma^\tau} \right).$$

Proof. Without loss of generality we assume that $\tau = 0$. We must show that for each integer n, the function $u(x,n)$ is bounded in $C^{2+\alpha}(\bar{\Omega})$ by $K_1 M$, where

$$M = \| u_0 \|_{2+\alpha}^{\Omega} + \| f \|_{\alpha}^{Q^\tau} + \| \varphi \|_{1+\alpha}^{\Gamma^\tau}.$$

The constant K depends only on operators L and B. Apply the well-known Schauder estimates in the interval of a finite length (Ladyzhenskaya *et al*, 1967); the length is not greater than unity. We show that $u(x,n)$ are uniformly bounded.

Just as when proving theorem 2.1, chapter 1, represent $u(x,n)$ as follows

$$u(x,n) = \hat{u}(x,n) + \tilde{u}(x,n),$$

where $\tilde{u}(x,n) \in \tilde{E}(n)$. Here and further we denote by $\tilde{E}(\xi)$ the subspace from $C^{2+\alpha}(\bar{\Omega})$ consisting of the functions satisfying the compatibility conditions for $t = \xi$, where in the right side of (1.5) we have the function $f(x,t) = 0$, and in the boundary conditions, the function $\varphi(x,t) = 0$ (see chapter 1. §1.2). It was proved in the first chapter, §1.2 that

$$\| \hat{u}(x,t) \|_{2+\alpha}^{\Omega} \leq CM.$$

Denote by $V(t,\tau)$ the shift operator which transforms the function $\tilde{u}(x,\tau) \in \tilde{E}(\tau)$ into the function $\tilde{u}(x,t) \in \tilde{E}(t)$ (we assume that $t \geq \tau$). Denote by $w_k(x,n)$ the function

$$w_k(x,n) = u(x,n) - V(n,k)\tilde{u}(x,k), \quad n \geq k.$$

In particular,

$$w_k(x, k+1) = u(x, k+1) - V(k+1, k)\tilde{u}(x, k).$$

Then $w_k(x, t)$ solves (1.5) in the cylinder $\Omega \times (k, k+1)$ with the initial data $\hat{u}(x, k)$ for $t = k$. Then

$$\| w_k(x, k+1) \|_{2+\alpha}^{\Omega} \leq C_1 M.$$

The constant C_1 does not depend on k. Note, that the following relation

$$u(x, k+1) = V(k+1, k)\tilde{u}(x, k) + w_k(x, k+1)$$

holds.

Denote by $z_{k+1}(x)$ the following difference

$$z_{k+1}(x) = w_k(x, k+1) - \hat{u}(x, k+1).$$

Then, we may write out the recurrent relation

$$\tilde{u}(x, k+1) = V(k+1, k)\tilde{u}(x, k) + z_{k+1}(x).$$

This relation yields (as well as in the proof of theorem 2.1, chapter 1) the equality

$$\tilde{u}(x, m) = \sum_{k=1}^{m} V(m, m-k)z_k(x) + V(m, 0)\tilde{u}(x, 0). \qquad (1.7)$$

Taking into account condition (1.4), we obtain that series (1.7) converges in $C^{2+\alpha}(\bar{\Omega})$, and its sum for all m may be estimated as follows

$$\| \tilde{u}(x, m) \|_{2+\alpha}^{\Omega} \leq K_1 M.$$

Repeating the considerations from theorem 2.1, chapter 1 proof, where we applied the standard Schauder estimates in a finite interval, we obtain estimate (1.6). The lemma is proved.

Remark 1.1. A similar argument may be followed if, instead of regular Hölder classes, we consider weighted Hölder classes. In particular, instead of the space $H^{2+\alpha}(Q^\tau)$ we may consider the spaces $H_1^{2+\alpha}(Q^\tau)$ or $H_0^{2+\alpha}(Q^\tau)$ (since the principal part of L is self-adjoint). For example, for the space $H_1^{2+\alpha}(Q^\tau)$, estimate (1.6) will have the form

$$\| u(x,t) \|_{1,2+\alpha}^{Q^\tau} \le K(\mu) \left(\| u_0 \|_1^\Omega + \| f \|_{1,\alpha}^{Q^\tau} + \| \varphi \|_{1,1+\alpha}^{\Gamma^\tau} \right). \quad (1.6')$$

Corollary 1.4. Let ε be a positive number, $0 < \varepsilon < \mu$. There exists a constant $K_1(\mu, \varepsilon)$ such that, if the function

$$f(x,t)e^{(\mu-\varepsilon)(t-\tau)}$$

belongs to the space $H^\alpha(Q^\tau)$, and the function

$$\varphi(x,t)e^{(\mu-\varepsilon)(t-\tau)}$$

belongs to the space $H^{1+\alpha}(\Gamma^\tau)$, then solution $u(x,t)$ of problem (1.1)-(1.3) exists in Q^τ and satisfies the following inequality

$$\| u(x,t)e^{(\mu-\varepsilon)(t-\tau)} \|_{2+\alpha}^{Q^\tau} \le K_1(\mu, \varepsilon) \left(\| u_0 \|_{2+\alpha}^\Omega \right.$$

$$\left. + \| f(x,t)e^{(\mu-\varepsilon)(t-\tau)} \|_\alpha^{Q^\tau} + \| \varphi(x,t)e^{(\mu-\varepsilon)(t-\tau)} \|_{1+\alpha}^{\Gamma^\tau} \right). \quad (1.8)$$

Proof. Make the following change

$$u(x,t) = v(x,t)e^{-(\mu-\varepsilon)(t-\tau)}.$$

For the obtained problem, using lemma 1.3, we get the required estimate.

Lemma 1.5. Assume that for problem (1.1)-(1.3) condition (1.4) holds. Then for each positive ε, such that $0 < \varepsilon < \mu$, there exists a positive $q(\varepsilon)$ such that if

$$\| L_1(x,t)u(x,t) \|_\alpha^{Q_T^\tau} \leq q(\varepsilon) \| u(x,t) \|_{2+\alpha}^{Q_T^\tau}, \quad T > \tau,$$

then the solution $u(x,t)$ of the problem

$$u_t = L(x,t)u + L_1(x,t)u, \quad (x,t) \in Q_T^\tau;$$

$$B(x,t)u = 0, \quad (x,t) \in \Gamma_T^\tau; \quad u(x,\tau) = u_0(x)$$

exists for all $T > \tau$, belongs to the class $H^{2+\alpha}(Q_T^\tau)$, and satisfies the inequality

$$\| u(x,t) \|_{2+\alpha}^\Omega \leq K_1(\mu,\varepsilon)e^{-(\mu-\varepsilon)(t-\tau)} \| u(x,\tau) \|_{2+\alpha}^\Omega, \quad t \geq \tau. \quad (1.9)$$

Proof.   Make the change

$$u(x,t) = v(x,t)e^{-(\mu-\varepsilon)(t-\tau)}$$

Then $v(x,t)$ solves the boundary-value problem

$$v_t = L(x,t)v - (\mu - \varepsilon)v + L_1(x,t)v, \quad (x,t) \in Q_T^\tau;$$

$$B(x,t)v = 0, \quad (x,t) \in \Gamma_T^\tau; \quad v(x,\tau) = u_0(x). \quad (1.10)$$

It follows from (1.4) that the solution of the problem

$$w_t = L(x,t)w - (\mu - \varepsilon)w, \quad (x,t) \in Q_T^\tau;$$

$$B(x,t)w = 0, \quad (x,t) \in \Gamma_T^\tau; \quad w(x,\tau) = w_0(x).$$

satisfies the inequality

$$\| w(x,t) \|_{2+\alpha}^\Omega \leq Ke^{-(\mu-\varepsilon)(t-\tau)} \| w(x,\tau) \|_{2+\alpha}^\Omega .$$

Let $z(x,t) \in H^{2+\alpha}(Q_T^\tau)$ be an arbitrary function, and $v(x,t)$ solve the boundary-value problem

$$v_t = L(x,t)v - (\mu - \varepsilon)v + L_1(x,t)z, \quad (x,t) \in Q_T^\tau;$$

$$B(x,t)v = 0, \quad (x,t) \in \Gamma_T^\tau; \qquad v(x,\tau) = u_0(x). \qquad (1.11)$$

The solution of (1.11) $v(x,t)$ belongs to the class $H^{2+\alpha}(\bar{Q}_T^\tau)$, $v(x,t)$ satisfies (1.6); therefore,

$$\| v(x,t) \|_{2+\alpha}^{Q_T^\tau} \leq K_1(\mu,\varepsilon) \left(\| u_0(x) \|_{2+\alpha}^{\Omega} + q(\varepsilon) \| z(x,t) \|_{2+\alpha}^{Q_T^\tau} \right).$$

Consider the following mapping

$$G : z(x,t) \to v(x,t).$$

If $q(\varepsilon) < 1/2K_1$, then G transforms the ball of the radius $2K_1 \| u_0(x) \|_{2+\alpha}^{\Omega}$ from the space $H^{2+\alpha}(\bar{Q}_T^\tau)$ into itself and is contractive. Therefore, there exists a unique fixed point of G in this ball. This fixed point solves problem (1.10). Since this fixed point lies in the ball of the radius $2K_1 \| u_0(x) \|_{2+\alpha}^{\Omega}$, then

$$\| v(x,t) \|_{2+\alpha}^{\Omega} \leq 2K_1 \| u_0(x) \|_{2+\alpha}^{\Omega}.$$

Hence follows the inequality

$$\| u(x,t) \|_{2+\alpha}^{\Omega} \leq K_1(\mu,\varepsilon)e^{-(\mu-\varepsilon)(t-\tau)} \| u(x,\tau) \|_{2+\alpha}^{\Omega}, \quad t \geq \tau.$$

The lemma is proved.

Lemma 1.6. Assume that for problem (1.1)-(1.3) condition (1.4) holds and $f(x,t) \in H^{\alpha}(\bar{Q}_0), \varphi(x,t) \in H^{1+\alpha}(\Gamma_0)$. There exists a unique solution of (1.5) defined for all $t \in R$ from $H^{2+\alpha}(Q_0)$. Denote it by $\tilde{u}(x,t)$. Then

$$\| \tilde{u}(x,t) \|^{Q_0}_{2+\alpha} \le K \left(\| f(x,t) \|^{Q_0}_{\alpha} + \| \varphi(x,t) \|^{\Gamma_0}_{1+\alpha} \right). \qquad (1.12)$$

Note, that for the first boundary-value problem, inequality (1.12) will take the form

$$\| \tilde{u}(x,t) \|^{Q_0}_{2+\alpha} \le K \left(\| f(x,t) \|^{Q_0}_{\alpha} + \| \varphi(x,t) \|^{\Gamma_0}_{2+\alpha} \right).$$

If $f(x,t), \varphi(x,t)$ are positive, then, bounded in the whole axis solution $\tilde{u}(x,t)$ is also positive.

Proof. First we show that there exists a bounded in the whole axis solution of (1.5). For this purpose, denote by $u(x,t;\tau,u_0)$ the solution of (1.5) with the initial data

$$u|_{t=\tau} = u_0(x)$$

determined for $t \ge \tau$. Let R, ω be positive constants, which we shall determine further, m be an integer. Consider the sets

$$N(k,m) = \left\{ u(x, k\omega; (m-k)\omega, u_0), \| u_0 \|^{\Omega}_{2+\alpha} \le R \right\}.$$

Show that from the chapter 1 results it follows that the sets $N(k,m)$ are compact for $k \ge 1$ (see lemma 2.1 §1.2). Take positive numbers R, ω so that

$$N(k,m) \supseteq N(k+1,m), \quad k \ge 0.$$

To this end note that

$$N(k+1,m) = \left\{ u(x, (k+1)\omega; (m-k-1)\omega, u_0), \| u_0 \|^{\Omega}_{2+\alpha} \le R \right\}.$$

Denoting by

$$M = \| f(x,t) \|^{Q_0}_{\alpha} + \| \varphi(x,t) \|^{\Gamma_0}_{2+\alpha},$$

write the corollary of (1.6)

$$\| u(x, (m-k)\omega; (m-k-1)\omega, u_0) \|_{2+\alpha}^{\Omega} \le K R e^{-\mu\omega} + K M.$$

Set $R = 2KM$ and choose such positive ω that

$$2K e^{-\mu\omega} < 1.$$

Then, we obtain

$$\| u(x, (m-k)\omega; (m-k-1)\omega, u_0) \|_{2+\alpha}^{\Omega} \le R,$$

$$u(x, (m-k)\omega; (m-k-1)\omega, u_0) \in N(k, m);$$

therefore, $N(k, m) \supseteq N(k+1, m)$, $k \ge 0$.

Consider the intersection of the sets $\bigcap_{k \ge 1} N(k, m)$. This intersection, due to compactness, is not empty. Let $v(x, m)$ be a function from $\bigcap_{k \ge 1} N(k, m)$ for a certain fixed m. From (1.4) it follows that $v(x, m)$ is unique. Really, let $v_1(x, m)$ be another function from $\bigcap_{k \ge 1} N(k, m)$. For each integer k there exist the initial data $u_0, u_1 \in C^{2+\alpha}(\bar{\Omega})$ and bounded in $C^{2+\alpha}(\bar{\Omega})$ by the constant $2KM$ such that

$$u(x, k\omega; (m-k)\omega, u_0) = v(x, m),$$

$$u(x, k\omega; (m-k)\omega, u_1) = v_1(x, m).$$

Denote by $w(x, t)$ the difference

$$w(x, t) = u(x, k\omega; (m-k)\omega, u_0) - u(x, k\omega; (m-k)\omega, u_1)$$

$$= v(x, m) - v_1(x, m).$$

Then from (1.4) it follows that

$$\| \, w(x, k\omega) \, \|_{2+\alpha}^{\Omega} \leq 2KMe^{-k\mu\omega} \to 0, \quad k \to \infty.$$

Therefore, $v(x, m) = v_1(x, m)$.
Define the function $\tilde{u}(x, t)$ as follows

$$\tilde{u}(x, t) = u(x, t; m\omega, v(x, m))$$

for $m\omega \leq t \leq (m + 1)\omega$. The function $\tilde{u}(x, t)$ is defined for all $t \in R$. It is easy to see that

$$v(x, (m + 1)\omega) = u(x, t; m\omega, v(x, m));$$

therefore, $\tilde{u}(x, t) \in C^{2+\alpha}(Q_0)$.
    Note, that $\| \, \tilde{u}(x, t) \, \|_{2+\alpha}^{\Omega} \leq M$ yields inequality (1.12).
    Let, now, $f(x, t), \varphi(x, t)$ be positive. Choosing instead of $N(k, m)$ the sets

$$N_0(k, m) = \left\{ u(x, k\omega; (m - k)\omega, u_0), \| \, u_0 \, \|_{2+\alpha}^{\Omega} \leq R, \quad u_0 \geq 0 \right\}.$$

and repeating similar considerations, we obtain that the bounded in the whole axis solution $\tilde{u}(x, t)$ is positive. The lemma is proved.

Lemma 1.7. Assume that for (1.1)-(1.3) condition (1.4) holds; $v(x, t), w(x, t) \in H^{2+\alpha}(\bar{Q}_0)$ and

$$v_t \geq L(x, t)v, \quad w_t \leq L(x, t)w, \quad (x, t) \in Q_0;$$

$$B(x, t)(v - w) \leq 0, \quad (x, t) \in \Gamma_0.$$

Then $v(x, t) \geq w(x, t)$ for $(x, t) \in Q_0$.

Remark 1.2. Lemma 1.7 is different from the ordinary comparison theorem by the fact that the solutions are considered for all $t \in R$; therefore, the initial data are absent.

Proof. Let us introduce some notations

$$z(x,t) = v(x,t) - w(x,t), \quad \varphi(x,t) = B(x,t)(v - w),$$

$$f_1(x,t) = v_t - L(x,t)v, \quad f_2(x,t) = w_t - L(x,t)w.$$

The function $z(x,t)$ solves the following problem

$$z_t - L(x,t)z = f_1 - f_2, \quad (x,t) \in Q_0;$$

$$B(x,t)z = \varphi(x,t), \quad (x,t) \in \Gamma_0.$$

Consequently, from lemma 1.6 it follows that $z(x,t) \geq 0$. Lemma 1.7 is proved.

Proof of theorem 1.2. In lemma 1.6 it was proved that if (1.4) holds, then the problem

$$u_t = L(x,t)u + \varepsilon u, \quad (x,t) \in Q_0; \quad B(x,t)u = 1, \quad (x,t) \in \Gamma_0 \quad (1.13)$$

has a unique, bounded in the whole axis, solution $\rho(x,t)$. Show that there exists such $\delta > 0$ that

$$\rho(x,t) \geq \delta, \quad (x,t) \in Q_0. \quad (1.14)$$

Evidently, (1.14) holds in each finite cylinder. Assume that such a sequence (x_k, t_k), $t_k \to +\infty$ exists that $\rho(x_k, t_k) \to 0$, $t_k \to +\infty$. The case $t_k \to -\infty$ may be considered analogously. Let

$$\rho(x, -t + t_k + 1), \quad a_{ij}(x, -t + t_k + 1), \quad a_i(x, -t + t_k + 1),$$

$$a(x, -t + t_k + 1), \quad 0 \leq t \leq 1,$$

be functions which for each k belong to the space $H^{2+\beta}(\bar{Q}_1)$, $H^{2+\beta}(\bar{Q}_1)$,

$H^{1+\beta}(\bar{Q}_1), H^\beta(\bar{Q}_1),\ 0 < \alpha < \beta < 1$ respectively. Choosing from these sequences converging in $H^{2+\alpha}(\bar{Q}_1),\ H^{1+\alpha}(\bar{Q}_1),\ H^\alpha(\bar{Q}_1)$ subsequences, we obtain that the limit function $\rho_0(x,t)$ solves the problem

$$(\rho_0)_t = \tilde{L}(x,t)\rho_0 + \varepsilon\rho_0, \quad (x,t) \in Q_1,$$

$$\tilde{B}(x,t)\rho_0 = 1, \quad (x,t) \in \Gamma_1,$$

$$v(x,0) = \rho_0(x,0) \geq 0.$$

where $\tilde{L}$ and $\tilde{B}$ are obtained from L and B after passing to the limit as $t_k \to \infty$.

From the comparison theorems it follows that $\rho_0(x,1) > 0,\ x \in \Omega$. At the same time $\rho_0(x,1)$ must be zero at least in one point x_0 from Ω. We obtain the contradiction which shows that the initial assumption is false. Consequently, $\rho_0(x,t) \geq \delta$ for $(x,t) \in \bar{Q}_0$.

Now we shall prove the existence of the Liapunov functional for problem (1.1)-(1.3). For this purpose, as in lemma 1.5, consider the shift operator $V(t,\tau)$ which transforms a function $u(x,\tau) \in \tilde{E}(\tau)$ into $u(x,t) \in \tilde{E}(t)$ for $t > \tau$. The conjugated operator $V^*(\tau,t)$ transforms $v(x,t)$ into $v(x,t)$, where

$$v_t = -L^*(x,t)v, \quad (x,t) \in Q_t^\tau; \tag{1.15}$$

$$B^*(x,t)v = 0, \quad (x,t) \in \Gamma_t^\tau; \quad v(x,t) = v_0(x).$$

Operators $\left(L^*(x,t), B^*(x,t)\right)$ in (1.15) for each fixed t are formally conjugated with the operator $\left(L(x,t), B(x,t)\right)$ (see Miranda, 1956). This denotes that for each functions $\xi(x), \eta(x)$ from $C^2(\bar{\Omega})$ the following equality

$$\int_\Omega \left(\eta(x)L(x,t)\xi(x) - \xi(x)L^*(x,t)\eta(x)\right)dx$$

$$= \int_{\partial\Omega} \left(\eta(x)B(x,t)\xi(x) - \xi(x)B^*(x,t)\eta(x)\right)ds \tag{1.16}$$

holds for all $t \in R$.

From (1.4) it follows that

$$\| v(x,\tau) \|_{2+\alpha}^{\Omega} \leq K \| v(x,t) \|_{2+\alpha}^{\Omega} \, e^{\mu(t-\tau)}.$$

Make the change $u(x,t) = \rho(x,t)w(x,t)$, where $\rho(x,t)$ is a positive solution of (1.13) determined in the whole axis. In this case, $w(x,t)$ solves the problem

$$w_t = \sum_{i,j=1}^{n} \left(a_{ij}(x,t)w_{x_j} \right)_{x_i} + \sum_{i=1}^{n} \hat{a}_i(x,t)w_{x_i} \tag{1.17}$$

$$= M(x,t)w \qquad (x,t) \in Q^{\tau};$$

$$\sum_{i=1}^{n} \beta_i(x,t)\rho(x,t)w_{x_i} + w = B_1(x,t)w = 0,$$

$$(x,t) \in \Gamma^{\tau}; w(x,\tau) = u_0(x).$$

Here the coefficients $\hat{a}_i$ may be calculated by means of the function ρ and the coefficients of (1.13).

Write out the formally conjugated problem (for each fixed t the equality (1.16) holds, where (L,B) are replaced by (M,B_1)).

$$z_t = -M^*(x,t)z, \qquad (x,t) \in Q_-^{\tau} = \Omega \times (-\infty,\tau);$$

$$B_1^*(x,t)z = 0, \qquad (x,t) \in \Gamma_-^{\tau} = \Omega \times (-\infty.\tau); \tag{1.18}$$

$$z(x,t) = z_0(x).$$

For this problem the analog of inequality (1.4) holds with substitution of $(+)$ by $(-)$. Therefore, by lemma 1.6, the problem

$$z_t = -M^*(x,t)z - 1, \quad (x,t) \in Q_0; \quad B_1^*(x,t)z = 1, \quad (x,t) \in \Gamma_0$$

has a unique, determined in the whole axis positive solution from the space $H^{2+\alpha}(Q_0)$. Denote it by $\sigma(x,t)$.

By the definition of formally conjugated operators (1.16) (Miranda, 1956), for each of the two functions $\xi(x,t), \eta(x,t) \in C^2(\bar{\Omega})$, $t \in R$ the following equality

$$\int_\Omega (\eta(x,t)M(x,t)\xi(x,t) - \xi(x,t)M^*(x,t)\eta(x,t))\,dx$$

$$= \int_{\partial\Omega} (\eta(x,t)B_1(x,t)\xi(x,t) - \xi(x,t)B_1^*(x,t)\eta(x,t))\,dS$$

holds.

Substitute in this equality the functions

$$\eta(x,t) = \sigma(x,t), \qquad \xi(x,t) = w^2(x,t),$$

where $w(x,t)$ solves (1.14).

In this case note that

$$M(x,t)w^2(x,t) = 2\sum_{i,j=1}^n a_{ij}(x,t)w_{x_i}(x,t)w_{x_j}(x,t)$$

$$+2w(x,t)M(x,t)w(x,t),$$

$$B_1(x,t)w^2(x,t) = 2w(x,t)B_1(x,t)w(x,t) - w^2(x,t).$$

As $w(x,t)$ solves (1.14), then

$$w_t(x,t) = M(x,t)w(x,t); \quad B_1(x,t)w(x,t) = 0.$$

Finally, we obtain

$$\int_\Omega \left(\sigma(x,t)M(x,t)w^2(x,t) - w^2(x,t)M^*(x,t)\sigma(x,t)\right)dx$$

$$= \int_{\partial\Omega} \left(\sigma(x,t)B_1(x,t)w^2(x,t) - w^2(x,t)B_1^*(x,t)\sigma(x,t)\right)dS.$$

Hence follows

$$\int\limits_{\Omega} \left(\sigma(x,t)2w(x,t)w_t(x,t)+ \sum_{i,j=1}^{n} 2\sigma(x,t)a_{ij}(x,t)w_{x_i}(x,t)w_{x_j}(x,t) \right.$$

$$\left. +w^2(x,t)\sigma_t(x,t) \right) dx = \int\limits_{\partial\Omega} - \left(\sigma(x,t)+1\right) w^2(x,t)dS.$$

The latter equality write out in the form

$$\frac{\partial}{\partial t} \int\limits_{\Omega} \sigma(x,t)w^2(x,t)dx$$

$$= - \int\limits_{\Omega} \left(\sum_{i,j=1}^{n} 2\sigma(x,t)a_{ij}(x,t)w_{x_i}(x,t)w_{x_j}(x,t) \right) dx$$

$$- \int\limits_{\partial\Omega} \left(\sigma(x,t)+1\right) w^2(x,t)dS.$$

Denote by $r_1(x,t), r_2(x,t), r_3(x,t)$ the positive functions

$$\frac{\sigma(x,t)}{\rho^2(x,t)}, \quad \frac{\sigma(x,t)+1}{\rho^2(x,t)}, \quad \frac{\sqrt{2}}{\rho(x,t)}.$$

As a result we obtain

$$\frac{\partial}{\partial t} \int\limits_{\Omega} r_1(x,t)u^2(x,t)dx = - \int\limits_{\partial\Omega} r_2(x,t)u^2(x,t)dS$$

$$- \sum_{i,j=1}^{n} \int\limits_{\Omega} a_{ij}(x,t)\sigma(x,t) \left(r_3(x,t)u(x,t)\right)_{x_i} \left(r_3(x,t)u(x,t)\right)_{x_j} dx.$$

This establishes that problem (1.1)-(1.3) has a Liapunov functional.

To complete the proof, assume that in this case the trivial solution of (1.1)-(1.3) will be exponentially stable, i.e., the trivial solution will satisfy the condition (1.4) with certain positive constants K, μ.

There exist positive constants $C_1, C_2, \ 0 < \mu < \varepsilon$ such that

$$-C_1\rho(x,\tau) \le u_0(x) \le C_2\rho(x,t), \quad x \in \Omega,$$

where $\rho(x,t)$ is a positive bounded in the whole axis solution of the problem

$$\rho_t = L(x,t)\rho + \varepsilon\rho, \quad (x,t) \in Q_0;$$

$$B(x,t)\rho(x,t) = 1, \quad (x,t) \in \Gamma_0.$$

Note, that since $\rho(x,t)$ is positive, then

$$\rho_t \geq L(x,t)\rho + (\varepsilon - \mu)\rho, \quad (x,t) \in Q_0;$$

$$B(x,t)\rho(x,t) = 1, \quad (x,t) \in \Gamma_0.$$

Applying the comparison theorem, we obtain

$$-C_1 e^{-\mu(t-\tau)}\rho(x,t) \leq u(x,t;\tau,u_0) \leq C_2 e^{-\mu(t-\tau)}\rho(x,t).$$

Now, to prove inequality (1.4), we need to apply the standard Schauder estimates.

To finish the theorem proof, we show that from point 2, point 1 follows.

It is easy to show that such positive C_3 exists that

$$C_3 \int_\Omega r_1(x,t)u^2(x,t)dx \leq \int_{\partial\Omega} r_2(x,t)u^2(x,t)dS$$

$$+ \sum_{i,j=1}^n \int_\Omega a_{ij}(x,t)\sigma(x,t)\left(r_3(x,t)u(x,t)\right)_{x_i}\left(r_3(x,t)u(x,t)\right)_{x_j} dx.$$

Therefore, from the Gronuall inequality we obtain

$$\int_\Omega r_1(x,t)u^2(x,t)dx \leq e^{-C_3(t-\tau)} \int_\Omega r_1(x,\tau)u^2(x,\tau)dx.$$

Applying the standard Schauder estimates, from the latter inequality we obtain estimate (1.4). Theorem 1.2 is proved.

§4.2 The stability criterion of the trivial solution of the linear mixed problem for the second order parabolic equation with time coefficients that are periodic in time.

In this paragraph we shall consider problem (1.1)-(1.3) assuming that coefficients of the problem periodically depend on time. In this case theorem 1.2, proved in the first paragraph, may be utilized.

We shall prove that if coefficients of (1.1)-(1.3) ω-periodically depend on time, then

$$\rho(x,t), \sigma(x,t), r_i(x,t), \quad i = 1, 2, 3,$$

ω-periodically depend on time. We shall prove, also, the analogous result for almost periodical dependence on time.

One more point distinguishes theorem 1.2 in the case of periodic dependence of coefficients from the general case: in the point 3) of the theorem, we may set $\varepsilon = 0$.

The validity of the statements formulated above will become apparent as we prove some lemmas from which these statements will follow.

Lemma 2.1. For problem (1.1)-(1.3), let condition (1.4) hold, and suppose that the coefficients of the problem periodically depend on time. Assume, also, that the functions $f(x,t), \varphi(x,t)$ from problem (1.5) ω-periodically depend on time. In this case the unique bounded in the whole axis solution of problem (1.5) will be ω-periodic.

Proof. Parallel with problem (1.5), consider the problem

$$\tilde{u}_t = L(x, t + \omega)\tilde{u} + f(x, t + \omega), \quad (x,t) \in Q^\tau;$$

$$B(x, t + \omega)\tilde{u} = \varphi(x, t + \omega), \quad (x,t) \in \Gamma^\tau. \tag{2.1}$$

If $u(x,t)$ is a unique bounded in the whole axis solution of (1.5), then $u(x,t+\omega)$ is the unique bounded in the whole axis solution of problem (2.1). Due to the ω-periodicity, we obtain

$$L(x,t) = L(x,t+\omega), \qquad B(x,t) = B(x,t+\omega),$$

$$f(x,t) = f(x,t+\omega), \qquad \varphi(x,t) = \varphi(x,t+\omega).$$

Hence, by virtue of the uniqueness of bounded in the whole axis solution of problems (1.5), (2.1), we have $u(x,t) = \tilde{u}(x,t) = u(x,t+\omega)$. The lemma is proved.

We shall say that a function $f(x,t) \in H^{\alpha}(\bar{Q}_0)$ is almost periodical with respect to time in $H^{\alpha}(\bar{Q}_0)$, if for each positive ε there exists positive $l = l(\varepsilon)$ such that each interval $[\alpha, \alpha + l]$ contains at least one number ω, where

$$\| f(x,t+\omega) - f(x,t) \|_{\alpha}^{\Omega} < \varepsilon$$

for each $t \in R$.

Analogously may be determined the almost-periodicity of functions $\varphi(x,t) \in H^{1+\alpha}(\bar{Q}_0)$ and $u(x,t) \in H^{2+\alpha}(\bar{Q}_0)$.

Lemma 2.2. For the problem (1.1)-(1.3), let condition (1.4) hold, and let coefficients of problem (1.5) and functions $f(x,t), \varphi(x,t)$ be almost periodic in time. Then the unique bounded in the whole axis solution of (1.5) will almost periodically depend on time.

Proof. Set two positive numbers ε and ε_1. By the number ε_1, we find $l = l(\varepsilon_1)$ such that in each interval $[\alpha, \alpha+l]$ there exists at least one positive number ω which is the common almost-period for coefficients and right hand sides of problem (1.5). Consider problem (2.1) and denote the unique bounded in the whole axis solution of problem (2.1), similar to lemma 2.1, by $\tilde{u}(x,t)$. Since the bounded in the whole axis solution is unique, we obtain that $\tilde{u}(x,t) = u_\omega(x,t+\omega)$, where $u_\omega(x,t)$ solves the problem

$$(u_\omega)_t = L(x, t + \omega)u_\omega(x, t) + f(x, t + \omega), \quad (x, t) \in Q_0;$$

$$B(x, t + \omega)u_\omega = \varphi(x, t + \omega), \quad (x, t) \in \Gamma_0.$$

Denote $v(x, t) = u(x, t + \omega) - u(x, t)$, then

$$v_t = L(x, t)v + [L(x, t + \omega) - L(x, t)] u(x, t + \omega)$$

$$+f(x, t + \omega) - f(x, t), \qquad (x, t) \in Q_0;$$

$$B(x, t)v = - [B(x, t + \omega) - B(x, t)] u(x, t + \omega)$$

$$+\varphi(x, t + \omega) - \varphi(x, t), \qquad (x, t) \in \Gamma_0.$$

Using estimate (1.11), we obtain

$$\| v(x, t) \|_{2+\alpha}^{Q_0} < K_1\varepsilon_1.$$

Setting $K_1\varepsilon_1 = \varepsilon$, from this estimate, we obtain that $u(x, t)$ is almost periodical. The lemma is proved.

From lemmas 2.1, 2.2 it follows that the functions $\rho(x, t), \sigma(x, t)$ are periodical or almost periodical in time depending on whether the initial problem is periodical or almost periodical by time. Therefore, the functions $r_i(x, t), \ i = 1, 2, 3$, have the same property.

Lemma 2.3. Assume that all coefficients of problem (1.5) are periodic in time; the problem

$$u_t = L(x, t)u, \quad (x, t) \in Q_0; \quad B(x, t)u = 1, \quad (x, t) \in \Gamma_0 \qquad (2.2)$$

has a unique positive bounded in the whole axis periodical solution. Then there exists such positive ε, that problem (1.5) has bounded in the whole axis periodic solution.

Proof. Consider a completely continuous shift operator $\hat{V}(\tau + \omega, \tau)$ which maps $\hat{u}(x, \tau)$ into $\hat{u}(x, \tau + \omega)$, where

$$\hat{u}_t = L(x, t)\hat{u}, \quad (x, t) \in Q_0; \tag{2.3}$$

$$B(x, t)\hat{u} = 0, \quad (x, t) \in \Gamma_0; \quad \hat{u}(x, \tau) = \hat{u}_0(x),$$

for a certain fixed $\tau \in R$. As was shown in chapter 1, the spectrum of $\hat{V}(\tau + \omega, \tau)$ is discrete and has the unique limit point, zero. Note, that the spectrum of $\hat{V}(\tau + \omega, \tau)$ has no points lying outside the unit circle of the complex plane. Really, the operator $\hat{V}(\tau + \omega, \tau)$ transforms the cone of positive functions into itself and is completely continuous. Therefore, by the Krein-Rutman theorem (see, for example, Krein and Rutman, 1948) the maximal by modulus eigenvalue of the operator $\hat{V}(\tau + \omega, \tau)$ is real and simple. Denote by λ this maximal by the modulus eigenvalue. The eigenfunction which corresponds to this eigenvalue is positive and is an interior point of the cone of positive functions. Denote this function by $\hat{\varphi}'(x)$. Note, also, that the maximal by modulus eigenvalue of the operator $\hat{V}(\tau + \omega, \tau)$ is positive. Denote by $\hat{\varphi}(x, t)$ the solution of (2.3) with the initial data $\hat{\varphi}(x)$. Then the function

$$\hat{\psi}(x, t) = \hat{\varphi}(x, t)e^{\frac{\ln \lambda}{\omega}(t-\tau)}$$

is the ω-periodical solution of problem (2.3). Denote, as earlier, by $\hat{\rho}(x, t)$ the positive solution of problem (2.2). There exists such k that

$$0 \leq \hat{\varphi}(x, \tau) \leq k\hat{\rho}(x, \tau), \quad x \in \Omega.$$

Applying the analog of the comparison theorem - lemma 1.7, we obtain

$$0 \leq \hat{\varphi}(x, t)e^{\frac{\ln \lambda}{\omega}(t-\tau)} \leq k\hat{\rho}(x, t), \quad x \in \Omega, \ t \geq \tau.$$

The latter inequality holds for each $t > \tau$, in only $\ln \lambda \leq 0$.

We now show that $\lambda < 1$. Assume the contrary, namely that $\lambda = 1$. In this case the function $\hat{\varphi}(x,t)$ is the ω-periodical solution of (2.3). Consider the set of positive σ such that

$$0 < \sigma\hat{\varphi}(x,t) < \hat{\rho}(x,t), \quad x \in \Omega, \quad \sigma \in [0, \sigma_0).$$

Then, for $\sigma = \sigma_0$, we have

$$0 < \sigma_0\hat{\varphi}(x,t) \leq \hat{\rho}(x,t), \quad x \in \bar{\Omega},$$

and such a point $x_0 \in \bar{\Omega}$ exists that

$$\sigma_0\hat{\varphi}(x_0, \tau) = \hat{\rho}(x_0, \tau).$$

Denote by $\xi(x,t)$ the following function

$$\xi(x,t) = \sigma_0\hat{\varphi}(x_0, \tau) - \hat{\rho}(x_0, \tau).$$

The function $\xi(x,t)$ is the ω-periodical solution of the problem

$$\xi_t = L(x,t)\xi, \qquad (x,t) \in Q_0;$$

$$B(x,t)\xi = 1, \qquad (x,t) \in \Gamma_0;$$

$$\xi(x,0) \leq 0, \qquad x \in \Omega.$$

Applying to this problem the strict maximum principle, we obtain that

$$\xi(x,t) < 0, \quad x \in \Omega, \quad t > \tau;$$

but, due to periodicity, $\xi(x_0, \tau + \omega) = 0$. The obtained contradiction shows that the initial assumption $\lambda = 1$ was false. Therefore, $\lambda < 1$.

Denote by μ the following number

$$\mu = \frac{1}{2}(1 - \lambda).$$

Consider the problem

$$u_t = L(x,t)u + \varepsilon u, \quad (x,t) \in Q_0; \tag{2.4}$$

$$B(x,t)u = 0, \quad (x,t) \in \Gamma_0; \quad u(x,\tau) = u_0(x).$$

The spectrum of the shift operator for the problem (2.4) lies inside the unit circle of the complex plane, and, as was shown in the first chapter, for problem (1.1)-(1.3) condition (1.4) holds. The application of the lemmas from the first paragraph completes the proof of lemma 1.3.

§4.3   Justification of the linearization method for the bounded nonstationary solution of the parabolic equation.

Consider the boundary value problem for the nonlinear parabolic equation

$$u_t = \sum_{i,j=1}^n a_{ij}(x,t,u,\nabla u)\frac{\partial^2 u}{\partial x_i \partial x_j} + a(x,t,u,\nabla u), \quad (x,t) \in Q_0. \tag{3.1}$$

Here ∇u is the gradient of $u(x,t)$ by the spatial variables. Assume, also, that

$$\mu_1 |\xi|^2 \le \sum_{i,j=1}^n a_{ij}(x,t,u,p)\xi_i\xi_j \le \mu_2 |\xi|^2, \quad \text{for } (x,t) \in Q_0;$$

$$-\infty < u, p_i < +\infty, \quad i = 1, \ldots, n; \quad p = (p_1, \ldots, p_n),$$

where $a_{ij}(x,t,u,p), a(x,t,u,p)$ are three times differentiable by all their arguments; μ_1, μ_2 are positive numbers.

Assume, for simplicity, that the boundary conditions that the solution in Γ_0 satisfies, are, homogeneous

$$B(x,t)u = 0, \qquad (x,t) \in \Gamma_0, \qquad (3.2)$$

where the operator $B(x,t)$ is determined by (1.2).

The case when the boundary conditions are inhomogeneous is considered analogously.

Assume, also, that the initial data are given for $t = \tau$

$$u(x,\tau) = u_0(x). \qquad (3.3)$$

Let $\varphi(x,t) \in H^{2+\alpha}(\bar{Q}_0)$ be a bounded solution of (3.1)-(3.3). Linearize problem (3.1)-(3.3) on the nonstationary solution $\varphi(x,t)$. After this we obtain the linear problem of the form (1.1)-(1.3). Set

$$w(x,t) = u(x,t) - \varphi(x,t).$$

Then

$$w_t = L(x,t)w + F(w), \quad (x,t) \in Q^{\tau}; \qquad (3.4)$$

$$B(x,t)w = 0, \quad (x,t) \in \Gamma^{\tau}; \quad w(x,\tau) = w_0(x) = u_0(x) - \varphi(x,\tau).$$

Remark 3.1. Here

$$L(x,t)w = \sum_{i,j=1}^{n} a_{ij}(x,t)\frac{\partial^2 w}{\partial x_i \partial x_j} + \sum_{i=1}^{n} a_i(x,t)\frac{\partial w}{\partial x_i} + a(x,t)w,$$

$$B(x,t)w = \alpha_0 + \alpha_1 \left(\sum_{i=1}^{n} \beta_i(x,t)\frac{\partial w}{\partial x_i} + \beta_0(x,t)w \right),$$

where

$$a_{ij}(x,t) = a_{ij}\left(x,t,\varphi(x,t),\nabla\varphi(x,t)\right)$$

$$a_i(x,t) = \sum_{k,l=1}^{n} a_{klu_{x_i}}\left(x,t,\varphi(x,t),\nabla\varphi(x,t)\right)\frac{\partial^2\varphi}{\partial x_k \partial x_l}$$

$$+a_{u_{x_i}}\Big(x,t,\varphi(x,t),\nabla\varphi(x,t)\Big)$$

$$a(x,t)=\sum_{k,l=1}^{n} a_{klu}\Big(x,t,\varphi(x,t),\nabla\varphi(x,t)\Big)\frac{\partial^2\varphi}{\partial x_i\partial x_j}$$

$$+a_u\Big(x,t,\varphi(x,t),\nabla\varphi(x,t)\Big).$$

In the fifth chapter we shall denote the operators $L(x,t)$ and $B(x,t)$ by $L(x,t,\varphi)$ and $B(x,t,\varphi)$ respectively if $\varphi(x,t)$ is a periodic solution to the parabolic equation. In this case, $L(x,t,\varphi)$, $B(x,t,\varphi)$ periodically depend on time.

If $\varphi(x)$ is a stationary solution of an autonomous parabolic equation, then the linearized on $\varphi(x)$ equation and boundary conditions will be denoted by $L(x,\varphi)$ and $B(x,\varphi)$.

Applying to the difference $F(w_1)-F(w_2)$ the results of lemmas 1.1, 1.2, chapter 1, we obtain

$$\| F(w_1)-F(w_2)\|_{-1,\alpha}^{Q^\tau}\leq q(\xi)\| w_1-w_2\|_{1,2+\alpha}^{Q^\tau}, \qquad (3.5)$$

$$\xi=\max_{i=1,2}\left(\| w_i\|_{1,2+\alpha}^{Q^\tau}\right),$$

where $q(\xi)\to 0$ if $\xi\to 0$.

Denote by $u(x,t;u_0,\tau)$ a solution of (3.1)-(3.3). Assume that $u_0(x)\in C^1(\bar\Omega)$ and $Bu_0(x)=0$ for $x\in\partial\Omega$. Denote this set of initial data by E.

We shall say that the nonstationary solution $\varphi(x,t)$ is exponentially stable, if there exist such positive constants ε, K_1, μ_1 that for each $\tau\in R$ the solution $u(x,t;u_0,\tau)$ of problem (3.1)-(3.3) with the initial data $u_0\in E$, such that $\| u_0(x)-\varphi(x,\tau)\|_1^\Omega<\varepsilon$, exists for all $t>\tau$ and satisfies the inequality

$$\| u(x,t;u_0,\tau)-\varphi(x,t)\|_1^\Omega\leq K_1\| u_0(x)-\varphi(x,\tau)\|_1^\Omega e^{-\mu_1(t-\tau)}. \quad (3.6)$$

Note, that (3.6) and (1.6), chapter 1 yield the inequality

$$\| u(x,t;u_0,\tau)-\varphi(x,t)\|_{1,2+\alpha}^{Q^\tau}\leq K_1\| u_0(x)-\varphi(x,\tau)\|_1^\Omega. \quad (3.7)$$

Theorem 3.1. (Vishnevskii, 1993a,b) Let $\varphi(x,t)$ be a bounded solution of (3.1)-(3.3) from $H^{2+\alpha}(\bar{Q}_0)$, and let the linearized on $\varphi(x,t)$ problem (3.1)-(3.3) be exponentially stable; i.e., it satisfies all the conditions formulated in the point 1 of theorem 1.2. Then the nonstationary solution $\varphi(x,t)$ of nonlinear problem (3.1)-(3.3) is exponentially stable also.

Remark 3.2. Theorem 3.1 completes the sequence of some articles (see, for example, Belonosov and Vishnevskii (1977) and the bibliography there), where for different assumptions the linearization method is justified for nonlinear parabolic equations and systems.

Proof of the theorem. Let $0 < \mu_1 < \mu$, μ be defined in the first point of theorem 1.2, and $w(x,t)$ solve problem (3.4). Make the change of the function

$$v(x,t) = w(x,t)e^{\mu_1(t-\tau)}.$$

Then $v(x,t)$ solves the problem

$$v_t = L(x,t)v + \mu_1 v + F_1(v), \quad (x,t) \in Q^\tau,$$

$$B(x,t)w = 0, \quad (x,t) \in \Gamma^\tau; \quad v(x,0) = w_0(x), \qquad (3.8)$$

where

$$F_1(v) = e^{\mu_1(t-\tau)} \cdot F\left(e^{-\mu_1(t-\tau)} \cdot v\right).$$

Simple calculations show that $F_1(v)$ satisfies inequality (3.5), only, perhaps, with another multiplier $q_1(\xi)$. Moreover, the property $q_1(\xi) \to 0$ for $\xi \to 0$ holds for this multiplier $q_1(\xi)$ too.

Denote by M the subset of E with the following functions $\xi(x,t)$

$$\xi(x,\tau) = w_0(x); \quad \| \xi(x,t) \|_{1,2+\alpha}^{Q^\tau} \leq K_1 \| w_0 \|_1^\Omega.$$

The constant K_1 we shall define further. The set M is closed and bounded. Let the mapping G transform $\xi(x,t) \in M$ into $\eta(x,t)$ such that $\eta(x,t)$ solves the following boundary-value problem for the non-homogeneous parabolic equation

$$\eta_t = L(x,t)\eta + \mu_1\eta + F_1(\xi), \quad (x,t) \in Q^\tau;$$

$$B(x,t)\eta = 0, \quad (x,t) \in \Gamma^\tau, \tag{3.9}$$

$$\eta(x,0) = w_0(x).$$

We must show that we may choose the constants ε, K_1 such that G for these ε, K_1 transforms M into itself and is a contractive mapping in M.

Let $\| w_0 \|_1^\Omega < 1$; then

$$\| \xi(x,t) \|_{1,2+\alpha}^{Q^\tau} \leq K_1\varepsilon.$$

Relations (3.4), (3.5) yield $F_1(0) = 0$ and

$$\| F_1(\xi) \|_{-1,\alpha}^{Q^\tau} \leq q(K_1\varepsilon) \| \xi(x,t) \|_{1,2+\alpha}^{Q^\tau}.$$

Estimate (1.6) from chapter 1 yields

$$\| \eta(x,t) \|_{1,2+\alpha}^{Q^\tau} \leq C(\varepsilon + q_1(K_1\varepsilon)K_1\varepsilon).$$

Set $K_1 = 2C$ and choose $\varepsilon > 0$ so that

$$q_1(2C\varepsilon)2C < 1.$$

We may choose ε sufficiently small that the latter inequality holds, since $q_1(2C\varepsilon) \to 0$ for $\varepsilon \to 0$. Then

$$\| \eta(x,t) \|_{1,2+\alpha}^{Q^\tau} \leq 2C\varepsilon = K_1\varepsilon$$

and $\eta(x,t) \in M$.

We now show that for certain K, ε the mapping G is a contraction in M. To this end, consider $G(\xi_1) - G(\xi_2)$ and denote by η_1 the function $G(\xi_1)$ and by η_2, the function $G(\xi_2)$. Then

$$(\eta_1 - \eta_2)_t = L(x,t)(\eta_1 - \eta_2) + \mu_1(\eta_1 - \eta_2) + F(\xi_1) - F(\xi_2), \quad (x,t) \in Q^\tau;$$

$$B(x,t)(\eta_1 - \eta_2) = 0, \quad (x,t) \in \Gamma^\tau, (\eta_1 - \eta_2)(x,0) = 0. \qquad (3.10)$$

Taking into account estimate (1.6) from chapter 1, we have

$$\| \eta_1 - \eta_2 \|_{1,2+\alpha}^{Q^\tau} \leq C \| F(\xi_1) - F(\xi_2) \|_{-1,\alpha}^{Q^\tau}$$

$$\leq C q_1(2C\varepsilon) \| \xi_1 - \xi_2 \|_{1,2+\alpha}^{Q^\tau} \leq \frac{1}{2} \| \xi_1 - \xi_2 \|_{1,2+\alpha}^{Q^\tau}.$$

Hence follows that G is a contraction in M. Therefore, G has a unique fixed point in M. This fixed point solves problem (3.7). Passing from $v(x,t)$ to the function $w(x,t)$, we obtain the theorem statement.

§4.4   Stable solution of the Neumann problem.

We shall give here one application of the theorems from the two first paragraphs of this chapter. In order to obtain the basic result of this paragraph, consider the Neumann problem for the equation

$$u_t = \Delta u + f(u), \quad (x,t) \in Q_T = \Omega \times (0,T); \qquad (4.1)$$

$$\frac{\partial u}{\partial n} = 0, \quad (x,t) \in \Gamma = \partial\Omega \times (0,T); \quad u(x,0) = u_0(x).$$

Let $\varphi(x)$ be a stable stationary solution of problem (4.1). In Costen and Holland (1978) it was shown that if Ω is a convex domain, then each stable stationary solution is a constant. Moreover, in Matano (1979) the examples of nonconvex domains are given for which the stable stationary solution $\varphi(x)$ is not constant.

In Vishnevskii (1984a) was obtained the stability criterion stated in two first paragraphs of this chapter. It was shown there, that the investigation of exponentially stable solutions of the Neumann problem by this criterion allows us to generalize the results of Matano (1979) and of Costen and Holland (1978).

We can now set forth the results obtained. Consider the following problem

$$u_t = \sum_{i,j=1}^{n} a_{ij}(x,t,u,\nabla u)\frac{\partial^2 u}{\partial x_i \partial x_j} + a(x,t,u,\nabla u), \quad (x,t) \in Q_0, \quad (4.2)$$

where ∇u is the gradient of $u(x,t)$ with respect to the spatial variables.

We assume that

$$\mu_1|\xi|^2 \leq \sum_{i,j=1}^{n} a_{ij}(x,t,u,p)\xi_i\xi_j \leq \mu_2|\xi|^2, \quad \text{for} \quad (x,t) \in Q_0;$$

$$-\infty < u, p_i < +\infty, \quad i = 1,\ldots,n; \quad p = (p_1,\ldots,p_n),$$

for certain constants $\mu_i > 0$.

Let the following boundary Neumann conditions hold

$$\frac{\partial u}{\partial n} = 0, \quad (x,t) \in \Gamma_0. \tag{4.3}$$

Assume, also, that

$$u(x,\tau) = u_0, \quad t = \tau. \tag{4.4}$$

The functions $a_{ij}(x,t,p), a(x,t,p)$ are assumed to be three times differentiable by all arguments.

Let $\varphi(x,t)$ be bounded in the whole axis solution of (4.2)-(4.3); $\varphi(x,t) \in H^{2+\alpha}(\bar{Q}_0)$. From the assumptions on smoothness of nonlinear terms of the problem, it follows that

$$\varphi(x,t) \in H^{3+\alpha}(\bar{Q}_0).$$

The following theorem holds.

Theorem 4.1. (Vishnevskii 1984a, 1993b) Let $\varphi(x,t) \in H^{3+\alpha}(\bar{Q}_0)$ be an exponentially stable solution of (4.2)-(4.3), Ω be convex. Then $\varphi(x,t)$ does not depend on the spatial variables.

Proof. Theorem 4.1 will be established, if we prove that

$$\sum_{i=1}^{n} \varphi_{x_i}^2(x,t) \equiv 0 \quad \text{for all} \quad (x,t) \in Q_0.$$

Denote by $\rho(x,t)$ a positive bounded in the whole axis solution of the problem

$$u_t = L(x,t,\varphi)u + \varepsilon u, \quad (x,t) \in Q_0;$$

$$\frac{\partial u}{\partial n} = 1, \quad (x,t) \in \Gamma_0.$$

Here ε is a positive number from theorem 1.2. We denote by $L(x,t,\varphi)$ the linear operator which is obtained after linearization of (4.2)-(4.4) on the nonstationary solution $\varphi(x,t)$.

Recall, that this operator was defined in §1.3 of this chapter (see remark 3.1). There were written out the coefficients of the operator $L(x,t,\varphi)$. Note, that in this case $\varphi(x,t)$ is the periodic in time solution of the problem; therefore, the operator $L(x,t,\varphi)$ periodically depends on time.

It is well-known (see, for example, Payne, 1976) that if Ω is a convex domain, $w(x) \in C^3(\bar{\Omega})$, $\frac{\partial w}{\partial n} = 0$ for $x \in \partial\Omega$, then

$$\frac{\partial}{\partial n}\left(\sum_{i=1}^{n}\left(\frac{\partial w}{\partial x_i}\right)^2\right) \leq 0 \quad \text{for} \quad x \in \partial\Omega.$$

Apply this statement to the function

$$\psi(x,t) = \sum_{i=1}^{n} \varphi_{x_i}^2(x,t).$$

As the result we obtain that

$$\frac{\partial}{\partial n}\psi(x,t) \leq 0, \quad x \in \Omega, \quad t \in R.$$

Denote by $\eta_k(x,t)$ the function $\frac{\partial \varphi}{\partial x_k}(x,t)/\rho(x,t)$. After differentiating (4.2), we obtain

$$\left(\frac{\partial \varphi}{\partial x_k}\right)_t = L(x,t,\varphi)\left(\frac{\partial \varphi}{\partial x_k}\right), \quad (x,t) \in Q_0; \quad \left(\frac{\partial \varphi}{\partial x_k}\right) = \rho(x,t)\eta_k(x,t),$$

$$\rho_t\eta_k + \rho\eta_{kt} = \sum_{i,j=1}^{n} \tilde{a}_{ij}(x,t)\left(\rho_{x_i x_j}\eta_k + \rho_{x_j}\eta_{kx_i} + \rho_{x_i}\eta_{kx_j} + \rho\eta_{kx_i x_j}\right)$$

$$+ \sum_{i=1}^{n} \tilde{b}_i(x,t)\left(\rho_{x_i}\eta_k + \rho\eta_{kx_i}\right) + \tilde{c}(x,t)\rho\eta_k, \quad k = 1,\ldots,n.$$

Here, $\tilde{a}_{ij}(x,t), \tilde{b}_i(x,t), \tilde{c}(x,t)$ denote the coefficients of $L(x,t,\varphi)$. By the definition of $\rho(x,t)$, we obtain

$$\eta_{kt} = \sum_{i,j=1}^{n} \tilde{a}_{ij}(x,t)\eta_{kx_i x_j} + \sum_{i=1}^{n} \hat{b}_i(x,t)\eta_{kx_i} - \varepsilon\eta_k, \quad k = 1,\ldots,n.$$

Multiply this equality by $2\eta_k$. After simple calculations we obtain the following expression

$$(\eta_k^2)_t = \sum_{i,j=1}^{n} \tilde{a}_{ij}(x,t)(\eta_k^2)_{x_i x_j} + \sum_{i=1}^{n} \hat{b}_i(x,t)(\eta_k^2)_{x_i} - \varepsilon(\eta_k^2)$$

$$+ \sum_{i,j=1}^{n} \tilde{a}_{ij}(x,t)(\eta_k)_{x_i}(\eta_k)_{x_j}, \quad k = 1,\ldots,n.$$

Denote $\eta(x,t) = \sum_{k=1}^{2} \eta_k^2(x,t)$. Then, summing the above equalities, we obtain

$$\eta_t = \sum_{i,j=1}^{n} \tilde{a}_{ij}(x,t)\eta_{x_i x_j} + \sum_{i=1}^{n} \hat{b}_i(x,t)\eta_{x_i} - 2\varepsilon\eta$$

$$+ \sum_{i,j=1}^{n} \tilde{a}_{ij}(x,t)\eta_{x_i x_j}.$$

Note that $\rho^2\eta = \psi$. As $\frac{\partial\varphi}{\partial n} \leq 0$ for $(x,t) \in \Gamma_0$, then $\frac{\partial}{\partial n}(\rho^2\eta) \leq 0$ for $(x,t) \in \Gamma_0$. Or

$$\rho_2\left(\frac{\partial\eta}{\partial n}\right) \leq -2\rho\eta \quad \text{for} \quad (x,t) \in \Gamma_0.$$

The latter inequality rewrite in the form $\frac{\partial\eta}{\partial n} \leq -\left(\frac{2\eta}{\rho}\right)$. Applying the maximum principle and the Zhiro-Zaremba principle (Protter and Weinberger, 1966, 1984), we see that the maximum of a positive function $\eta(x,t)$ could not be either inside Q_0 or in the boundary Γ_0 of this cylinder. Denote

$$\zeta(t) = \max\{\eta(x,t); \ x \in \bar{\Omega}\}.$$

Prove that $\zeta(t)$ is absolutely continuous and $\zeta'(t) \in L_\infty(R)$, $\ t \in R$. Denote

$$M(t) = \{x \in \Omega : \eta(x,t) = \zeta(t)\}.$$

First, prove that $\zeta(t)$ is a Lipschitz function. Really, if $t_i \in R$, $\ x_i \in M(t_i)$, $\ i = 1,2$, $\ t_2 > t_1$, then

$$0 \leq \zeta(t_2) - \zeta(t_1) = \eta(x_2,t_2) - \eta(x_1,t_1)$$

$$\leq \eta(x_2,t_2) - \eta(x_2,t_1) = \eta(x_2,\check{t})(t_2 - t_1),$$

whence it follows that $\zeta(t)$ is a Lipschitz function. Hence follows that $\zeta(t)$ is absolutely continuous, $\zeta'(t) \in L_\infty(R)$. Now, we shall calculate the derivative of $\zeta(t)$. Let t_0 be a point where $\zeta(t)$ is differentiable, $t < t_0$, $\ x_0 \in M(t_0)$, $\ x \in M(t)$. Then

$$\frac{\zeta(t_0) - \zeta(t)}{t_0 - t} = \frac{u(x_0,t_0) - u(x_0,t)}{t_0 - t},$$

and hence we obtain that

$$\zeta'(t_0) \leq \eta_t(x_0, t_0) \ \text{ for } \ t \to t_0.$$

Analogously we may have

$$\zeta'(t_0) \geq \eta_t(x_0, t_0),$$

whence follows that $\zeta'(t_0) = \eta_t(x_0, t_0)$ for each $m \in M(t)$ and $t \in R$. For almost all t

$$\zeta'(t) \leq -2\delta\zeta(t) - \psi(t),$$

where $\psi(t)$ is a positive function. After the change $t = -\tau$, we see that

$$\zeta'(\tau) \leq -2\delta\zeta(\tau) - \psi(-\tau),$$

where $\zeta(\tau)$ is bounded for $\tau \to \infty$, which is possible only for $\eta(t) \equiv 0$. Therefore, $\sum\limits_{i=1}^{n} \varphi_{x_i}^2(x, t) \equiv 0$, which proves theorem 4.1.

Chapter 5

The attraction domains of stable stationary or stable periodic solutions.

The results of this chapter generalize on the case of many spatial variables the results on the attraction domains and other qualitative properties of the quasilinear parabolic equation with one spatial variable stated in chapter 3.

This generalization is not of a technical character, since, to obtain these results, we apply the methods which differ essentially from methods of chapter 3.

The point is that to obtain many of the deal of chapter 3 results we have used the following basic statements.

1) The theorem on unconditional stabilizing of each bounded (in corresponding norm) solution (this theorem fails for the parabolic equation with many spatial variables, as the example from Poláčik and Rubakovskii (1994) shows).

2) For the elliptic equation of second order with one spatial variable (ordinary differential equation) the well-posed Cauchy problem setting has no analogs for an elliptic equation with a spatial variable.

The results of this chapter is based mainly on the papers of Vishnevskii (1984a, b; 1990a, b; 1992). Note, that the existence of the greatest and the least stationary solutions under different assumptions was considered by Grandal. The case of periodic solutions was considered by Takač (1991, 1992b).

Vishnevskii (1984) established the existence of the attraction domain using the statement that between two stable stationary solutions there exists at least one unstable solution. An analogous statement is proved by Matano (1984).

The example of the existence of a periodic solution to the equation was constructed by Vishnevskii (1990b). Now we have many examples which essentially have clarified the qualitative behavior of solutions to nonlinear autonomous and periodical parabolic equations. These examples are in Hess and Danser (1991), Polačik (1992), Danser (1992), Takač (1992b), Polačik and Rubakovskii (1994).

§5.1   Some definitions and the preliminary results.

As proposed earlier, let Ω be a bounded domain in R^n with a boundary $\partial\Omega$ from the class $C^{2+\alpha}$. Consider in the cylinder $Q = \Omega \times (0, +\infty)$ the nonlinear equation

$$u_t = \sum_{i,j=1}^{n} a_{ij}(x,t,u,\nabla u)\frac{\partial^2 u}{\partial x_i \partial x_j} + a(x,t,u,\nabla u), \quad (x,t) \in Q_0, \quad (1.1)$$

where ∇u is the gradient of $u(x,t)$ by the spatial variables. The equation is assumed to be parabolic

$$\mu_1^{-1}|\xi|^2 \le \sum_{i,j=1}^{n} a_{ij}(x,t,u,p)\xi_i\xi_j \le \mu_1|\xi|^2,$$

for

$$(x,t) \in Q_0; \quad -\infty < u, p_i < +\infty, \quad i = 1,\ldots,n; \quad p = (p_1,\ldots,p_n).$$

The functions $a_{ij}(x,t,u,p), a(x,t,u,p)$ are three times differentiable by their arguments and are ω-periodic in time; μ_1 is certain positive number.

Suppose that the boundary conditions which the solution of the problem satisfies are as follows

$$B(x,t)u = g(x,t,u), \quad (x,t) \in \Gamma; \quad \Gamma = \partial\Omega \times (0,+\infty). \qquad (1.2)$$

The operator $B(x,t)$ is defined in the first paragraph of chapter 4. Additionally we assume that coefficients of $B(x,t)$ and $g(x,t,u)$ are ω-periodic in time.

Assume, also, that for $t = 0$ we are given the initial data

$$u(x,0) = u_0(x). \qquad (1.3)$$

Suppose that $u_0(x) \in C^1(\bar{\Omega})$ and $B(x,t)u_0 = g(x,t,u_0)$, or $u_0(x) = 0$ (in the case of the first boundary-value problem) for $x \in \partial\Omega$. Denote by E the set of initial data satisfying the above conditions.

The following theorem was proved in Vishnevskii (1987).

Theorem 1.1. For each $u_0 \in E$ there exists $T > 0$ such that the solution of (1.1)-(1.3) with the initial data $u_0 \in E$ exists for $t < T$ and belongs to $H_1^{2+\alpha}(Q_T)$.

Further, in this chapter, we shall suppose that this solution may be extended in the whole semi-infinite cylinder Q, and this solution $u(x,t)$ belongs to $H_1^{2+\alpha}(Q)$. We shall denote the solution of (1.1)-(1.3) with the initial data u_0 by $u(x,t;u_0)$.

From the estimates of the first chapter it follows that $u(x,t;u_0) \in H_1^{2+\alpha}(Q)$ if $\| u(x,t;u_0) \|_1^{\Omega}$ is bounded for all $t \geq 0$.

In this chapter we shall assume that for problem (1.1)-(1.3) the following restrictions hold (Ladyzhenskaya and Ural'tseva, 1986)

$$\left[\sum_{i,j=1}^{n} \left(\frac{\partial a_{ij}(x,t,u,p)}{\partial p_k} \right)^2 \right]^{\frac{1}{2}} \leq \frac{\mu_2(|u|)}{1+|p|}, \qquad (1.4)$$

$$\left[\sum_{i,j=1}^{n} \left(\frac{\partial a_{ij}(x,t,u,p)}{\partial u} \right)^2 \right]^{\frac{1}{2}} \leq \mu_3(|u|),$$

$$\left[\sum_{i,j=1}^{n} \left(\frac{\partial a_{ij}(x,t,u,p)}{\partial x_k} \right)^2 \right]^{\frac{1}{2}} \leq \mu_4(|u|)(1+|p|).$$

Here $p = (p_1, \ldots, p_n)$, $\mu_i(\cdot)$, $i = 1, 2, 4$, are monotonically increasing functions from R^+ into R^+, $\mu_i > 0$. It is easy to show that if all the above assumptions and (1.4) hold, then for each $u(x, t; u_0) \in H_1^{2+\alpha}(Q_T)$, the estimate

$$\| u(x, t; u_0) \|_0^{Q_T} < M_0$$

yields the estimate

$$\| u(x, t; u_0) \|_1^{Q_T} < M_1.$$

To establish this, it suffices to use the estimate in the weight Hölder classes which follows from the local existence theorem proved by Vishnevskii (1987) and then to use the results of Belonosov (1978, 1979) and of Solonnikov and Hachatryan (1980).

It is essential for the further consideration of this problem that if M_0 does not depend on T, then M_1 also does not depend on T (Akramov and Vishnevskii, 1992).

Let us give some definitions (see, also, ter 3).

Definition 1.1. Let $\varphi(x, t)$ be a periodic solution of (1.1)-(1.3). We say that $\varphi(x, t)$ is *stable from above*, if for each positive ε there exists such positive δ that from

$$u_0 \in E, \quad u_0 \geq \varphi(x, 0), \quad \| u_0 - \varphi(x, 0) \|_1^{\Omega} < \delta$$

it follows that $u(x, t; u_0)$ exists for all $t \geq 0$ and belongs to the class $H_1^{2+\alpha}(Q)$; moreover,

$$\| u(x, t; u_0) - \varphi(x, t) \|_{1, 2+\alpha}^{Q} < \varepsilon.$$

A stable from above periodic solution we shall call *asymptotically stable from above*, if , in addition

$$\lim_{t \to +\infty} \| u(x, t; u_0) - \varphi(x, t) \|_1^{\Omega} = 0.$$

Analogously we may define *the stability from below* and *the asymptotic stability from below*.

Definition 1.2. We say that a periodic solution is *stable (asymptotically stable)*, if it is stable (asymptotically stable) from above and from below.

Definition 1.3. A periodic solution $\varphi(x,t)$ of problem (1.1)-(1.3) we shall call *attracting from above* if such positive δ exists that from

$$u_0 \in E, \quad u_0 \geq \varphi(x,0), \quad \| u_0 - \varphi(x,0) \|_1^\Omega < \delta$$

it follows that the solution $u(x,t,u_0)$ exists for all $t \geq 0$, belongs to $H_1^{2+\alpha}(Q)$, and

$$\lim_{t \to +\infty} \| u(x,t;u_0) - \varphi(x,t) \|_1^\Omega = 0.$$

The periodic solution *attracting from below* may be defined analogously.

Definition 1.4. If a periodic solution is attracting from above and from below, then we say that it is *attracting*.

Note, that for dynamic systems in the plane the example may be constructed (see, for example, Nemytskii *et al.*, 1967), where the equilibrium state is attracting but not stable. In this chapter we shall show that for (1.1)-(1.3), if a periodic solution is attracting, then this solution is asymptotically stable.

The interesting case of (1.1)-(1.3) is the case when the coefficients of the problem do not depend on time explicitly, i.e., the problem is autonomous. In this case, instead of the periodic solution we need to consider the stationary solution. As well as for the periodic solution we may define the stability and the asymptotic stability from above and from below, the attracting from above and from below stationary solution.

Definition 1.5. A periodic solution $\varphi(x,t)$ of problem (1.1)-(1.3) we shall call *unstable*, if such positive ε exists, that for each positive δ there exist initial data $u_0 \in E$ and $t_1 \geq 0$, so that

$$\| u_0 - \varphi(x,0) \|_1^{\Omega} < \delta \quad \text{and} \quad \| u(x,t_1;u_0) - \varphi(x,t_1) \|_1^{\Omega} \geq \varepsilon.$$

The unstable stationary solution may be defined similarly.

Theorem 1.2. (Vishnevskii, 1993d) A periodic solution of autonomous problem (1.1)-(1.3) is unstable.

Proof. We must show that an ω-periodic solution $\varphi(x,t)$, $(\omega > 0)$ is unstable.

The linearized on $\varphi(x,t)$ problem (1.1)-(1.3) has the following form

$$v_t = L(x,t,\varphi)v, \qquad (x,t) \in Q;$$

$$B_1(x,t,\varphi)v = 0, \qquad (x,t) \in \Gamma;$$

$$v(x,0) = v_0(x).$$

The operator $L(x,t,\varphi)$ is writed out in detail in chapter 4 (Remark 3.1). Note, that in this case the operator $L(x,t,\varphi)$ is obtained as the result of linearizing problem (1.1)-(1.3) on the ω-periodic solution of the problem ω-periodic depending on time. Therefore, $L(x,t,\varphi)$ also ω-periodically depends on time. In the fourth chapter we have considered linear boundary conditions; consequently, we need not linearize the boundary operator. In this chapter the boundary conditions are nonlinear and

$$B_1(x,t,\varphi) = B(x,t) - g'_u(x,t,\varphi).$$

The operator $B_1(x,t,\varphi)$ also ω-periodically depends on time.

If λ_0 is sufficiently large and positive, then, since the spectrum of L is semibounded, the problem

$$v_t = L(x,t,\varphi)v - \lambda_0 v, \qquad (x,t) \in Q;$$

$$B_1(x,t,\varphi)v = 0, \qquad (x,t) \in \Gamma;$$

$$v(x,0) = v_0(x).$$

satisfies condition (1.4) of chapter 4. Therefore, there exists a positive ω-periodic solution $\rho(x,t)$ of the problem

$$\rho_t = L(x,t,\varphi)\rho - \lambda_0 \rho, \qquad (x,t) \in Q;$$

$$B_1(x,t,\varphi)\rho = 0, \qquad (x,t) \in \Gamma.$$

Let us make the following change

$$v(x,t) = \rho(x,t)w(x,t). \qquad (1.5)$$

For the function $w(x,t)$, we have the following boundary-value problem

$$w_t = L_1(x,t,\varphi)w, \qquad (x,t) \in Q;$$

$$\rho \sum_{i=1}^{n} b_i(x,t)\frac{\partial w}{\partial x_i} + w = B_2(x,t,\varphi)w = 0, \quad (x,t) \in \Gamma;$$

$$w(x,0) = w_0(x) = \frac{v_0(x)}{\rho(x,0)}. \qquad (1.6)$$

We denote here by $L_1(x,t,\varphi)$ the operator which is obtained from $L(x,t,\varphi)$ after the change (1.5). As earlier, denote by $V(\omega)$ the operator mapping the function $w(x,0) \in \tilde{E}$ (see chapter 1) into the function $w(x,\omega) \in \tilde{E}$. Here

$$\tilde{E} = \left\{ f(x) \in C^1(\bar{\Omega}), \quad B_2(x,t,\varphi)f(x) = 0 \right\}.$$

The operator $V(\omega)$ maps the cone of positive functions from the space $\tilde{E}$ into itself. Therefore, as shown in Vishnevskii (1992), the maximal by modulus eigenvalue σ_1 of the operator $V(\omega)$ is positive and simple. Corresponding to this eigenvalue, the eigenfunction is positive. Denote this eigenfunction by $\xi_1(x)$. Then

$$V(\omega)\xi_1(x) = \sigma_1\xi_1(x).$$

Denote by $\xi_1(x,t)$ the solution of problem (1.6) with the initial data $\xi_1(x)$. Let us show that $\sigma_1 \geq 1$. Note, that if $\varphi(x,t) \in H^{2+\alpha}(Q)$ then, using the conditions imposed on problem (1.1)-(1.3), we may prove that $\varphi(x,t) \in H^{4+\alpha}(Q)$. Consider in this theorem the case when (1.1)-(1.3) is autonomous, i.e., does not explicitly depend on time

$$u_t = \sum_{i,j=1}^{n} a_{ij}(x, u, \nabla u)\frac{\partial^2 u}{\partial x_i \partial x_j} + a(x, u, \nabla u), \quad (x, t) \in Q; \qquad (1.1')$$

$$B(x) \cdot u = g(x, u), \qquad (x, t) \in \Gamma, \qquad (1.2')$$

$$u(x, 0) = u_0(x), \qquad (1.3')$$

with the same assumptions on parabolicity and the complementary condition, which where imposed on operators (1.1)-(1.2) in the beginning of the chapter. Differentiate problem (1.1')-(1.3') by time setting $u = \varphi$

$$(\varphi_t)_t = L(x, t, \varphi)(\varphi_t), \qquad (x, t) \in Q;$$

$$B_1(x, t, \varphi)(\varphi_t) = 0, \qquad (x, t) \in \Gamma.$$

It denotes that $\varphi_t(x,t)$ is a nontrivial ω-periodic solution of linearized on $\varphi(x,t)$ problem (1.1)-(1.3), and the corresponding eigenvalue of the shift operator is equal to 1.

Therefore, $\varphi_t(x,0)/\rho(x,0)$ is the eigenfunction of $V(\omega)$ corresponding to the unit eigenvalue. Consequently, $\sigma_1 \geq 1$.

Assume that $\sigma_1 > 1$ and prove that in this case the periodic solution is unstable.

Let
$$u_0(x) = \varphi(x,0) + \varepsilon\rho(x,0)\xi_1(x,0).$$

Then the solution
$$u(x,t; \varphi(x,0) + \varepsilon\rho(x,0)\xi_1(x,0))$$

may be represented in the form
$$\varphi(x,t) + \varepsilon\rho(x,t)\xi_1(x,t) + \varepsilon^2\eta(x,t,\varepsilon),$$

where $\| \eta(x,t,\varepsilon) \|_{2+\alpha}^{Q_\omega} < C$ for sufficiently small ε.

Really, we shall look for the solution in the form
$$u\Big(x,t; \varphi(x,0) + \varepsilon\rho(x,0)\xi_1(x,0)\Big)$$

$$= \varphi(x,t) + \varepsilon\rho(x,t)\xi_1(x,t)e^{\lambda_1 t} + \varepsilon^2\eta(x,t,\varepsilon),$$

where $\lambda_1 = \ln\tau_1 > 0$.

Note that $u - \varphi = \xi(x,t)$ solves the problem

$$\xi_t = L(x,t,\varphi)\xi + F(\xi) \qquad (x,t) \in Q,$$

$$B(x,t,\varphi)\xi = \Phi(\xi) \qquad (x,t) \in \Gamma,$$

$$\xi(x,0) = \varepsilon\rho(x,t)\xi(x,t) + \varepsilon^2\eta(x,t,\varepsilon).$$

By the definition of $\xi_1(x,t)$, we obtain that if

$$v_1(x,t) = \rho(x,t)\xi_1(x,t)$$

$$v_{1t} = L(x,t,\varphi)v_1 \qquad (x,t) \in Q,$$

$$B_1(x,t,\varphi)v_1 = 0, \qquad (x,t) \in \Gamma;$$

therefore

$$\eta_t L(x,t,\varphi)\eta + \frac{1}{\varepsilon^2}F\left(\varepsilon\rho\xi_1 e^{\lambda_1 t} + \varepsilon^2\eta\right).$$

The mapping $F(\xi)$ contains only the terms that are higher than the second order of smallness with respect to φ; therefore,

$$\frac{1}{\varepsilon^2}F\left[\varepsilon\left(\rho\xi_1 e^{\lambda_1 t} + \varepsilon\eta\right)\right]$$

is bounded uniformly by ε and hence

$$\| \eta(x,t,\varepsilon) \|_{2+\alpha}^{Q_\omega} < C, \quad \text{for small} \quad \varepsilon.$$

Then

$$u(x,t;\varphi(x,0) + \varepsilon\rho(x,0)\xi_1(x,0)) - \left[\varphi(x,0) + \varepsilon\rho(x,0)\xi_1(x,0)\right]$$

$$= \varepsilon\rho(x,0)\left[(\sigma_1 - 1)\xi_1(x,0) - \varepsilon\eta(x,t,\varepsilon)\right]$$

and, if ε is sufficiently small, then the function

$$(\sigma_1 - 1)\xi_1(x,0) - \varepsilon\eta(x,t,\varepsilon)$$

is positive in Ω. In other words

$$u(x,\omega,u_0) - u_0 > 0, \quad x \in \Omega.$$

Let

$$z(x,t) = u(x,t+\omega;u_0) - u(x,t;u_0).$$

The function $z(x,t)$ may be considered as a solution of the homogeneous mixed problem for the linear parabolic equation with the initial data $u(x,\omega;u_0) - u_0$. Since $z(x,0) > 0, \quad x \in \Omega$, then, by the comparison theorem (see, for example, Protter and Weinberger, 1966, 1984), $z(x,t) > 0$ for $(x,t) \in Q$.

Note, that

$$z(x,0) = u(x,\omega;u_0) - u_0 > 0, \qquad x \in \Omega;$$

$$z(x,\omega) = u(x,2\omega;u_0) - u(x,\omega;u_0) > 0, \qquad x \in \Omega;$$

$$z(x,2\omega) = u(x,3\omega;u_0) - u(x,2\omega;u_0) > 0, \qquad x \in \Omega.$$

This means that the sequence $u(x, k\omega; u_0)$ monotonically increases in Ω; therefore, $\varphi(x,t)$ is unstable.

Consider the case $\sigma_1 = 1$. Then $\varphi_t(x,0)/\rho(x,0)$ is the eigenfunction of $V(\omega)$ corresponding to the unit eigenvalue. This eigenfunction must be positive; therefore, $\varphi_t(x,0) > 0$ for $x \in \Omega$. By the comparison theorem, $\varphi_t(x,t) > 0$ for $(x,t) \in Q$ which contradicts the periodicity of $\varphi(x,t)$. Theorem 1.2 is proved.

Consider now the case when (1.1)-(1.3) is ω-periodic in time. Then problem (1.1)-(1.3) may have two different types of periodic solutions.

1. The period of the periodic solution is commensurable with ω.
2. The period of the periodic solution is incommensurable with ω.

In the second case the following theorem holds.

Theorem 1.3. (Vishnevskii, 1993d) Let $\varphi(x,t)$ be a periodic solution of (1.1)-(1.3) with the period ω_1 incommensurable with ω. Then for each constant τ the function $\varphi(x,t+\tau)$ is also ω_1-periodic solution of (1.1)-(1.3).

Proof. Denote by $\tilde{L}(x,t,\varphi(x,\tau))$ the following expression

$$\tilde{L} = \sum_{i,j=1}^{n} a_{ij}(x,t,\varphi(x,\tau),\nabla\varphi(x,\tau))\frac{\partial^2\varphi(x,\tau)}{\partial x_i\partial x_j} + a(x,t,\varphi(x,\tau),\nabla\varphi(x,\tau)).$$

We shall prove that

$$\tilde{L}(x,t,\varphi(x,\tau)) = \tilde{L}(x,\tau,\varphi(x,\tau)), \quad t,\tau \in R. \tag{1.7}$$

Really, for each integer m and l, we have

$$\tilde{L}(x, \tau + m\omega_1 + l\omega, \varphi(x, \tau)) = \tilde{L}(x, \tau + m\omega_1, \varphi(x, \tau + m\omega_1))$$

$$= \tilde{L}(x, \tau, \varphi(x, \tau)).$$

As ω_1/ω is irrational, then the numbers $m\omega_1 + l\omega$, where m and l are arbitrary integers, form a dense set in R. Since $\tilde{L}(x, \tau, \varphi(x, \tau))$ is continuous by time, we obtain equality (1.7).

Denote

$$\tilde{B}(x, t, \varphi(x, \tau)) = B(x, t)\varphi(x, \tau) - g(x, t, \varphi(x, \tau)).$$

As before, we may show that

$$\tilde{B}(x, t, \varphi(x, \tau)) = \tilde{B}(x, \tau, \varphi(x, \tau)).$$

Note, that

$$\frac{\partial \varphi}{\partial t}(x, t + \tau) = \tilde{L}(x, t + \tau, \varphi(x, t + \tau)) = \tilde{L}(x, t, \varphi(x, t + \tau)), \quad t, \tau \in R;$$

$$\tilde{B}(x, t + \tau, \varphi(x, t + \tau)) = \tilde{B}(x, t, \varphi(x, t + \tau)).$$

Therefore,

$$\frac{\partial \varphi}{\partial t}(x, t + \tau) = \tilde{L}(x, t, \varphi(x, t + \tau)), \quad t, \tau \in R, \quad x \in \Omega;$$

$$\tilde{B}(x, t, \varphi(x, t + \tau)) = 0, \quad t, \tau \in R, \quad x \in \Omega.$$

Hence, $\varphi(x, t + \tau)$ is a periodic solution of (1.1)-(1.3). The theorem is proved.

Remark 1.1. Theorem 1.3 denotes that if autonomous problem (1.1)-(1.3) has at least one periodic solution and its period is incommensurable with ω, then this problem has infinitely many such solutions.

Remark 1.2. A periodic solution with a period ω_1 incommensurable with ω is not asymptotically stable. Moreover, acting similar to the proof of theorem 1.2, we may show that such solution is not stable.

Remark 1.3. In the fifth paragraph of this chapter we shall give an example of the autonomous problem which has periodic solutions. There is an example of a periodic problem which has solutions with the period incommensurable with the period of the problem.

Theorem 1.3 shows that periodic solutions with the period incommensurable with the period of the right sides are, in some sense, exclusive. Of greatest interest are those that have solutions in which the period is commensurable with ω. Further, in this chapter, we shall investigate these periodic solutions.

§5.2 The greatest and least periodic solutions of the mixed problem.

First, we shall give some definitions.

Definition 2.1. A function $u^+(x,t)$ will be called *the upper solution* of problem (1.1)-(1.3), if $u^+(x,t) \in H^{2+\alpha}(Q)$ and the inequlities

$$u_t^+ \geq \sum_{i,j=1}^{n} a_{ij}(x,t,u^+,\nabla u^+)\frac{\partial^2 u^+}{\partial x_i \partial x_j} + a(x,t,u^+,\nabla u^+); \quad (x,t) \in Q_0,$$

$$B(x,t)u^+ - g(x,t,u^+) \leq 0, \qquad (x,t) \in \Gamma$$

hold. In the case of the first boundary-value problem the second inequality is as follows

$$u^+(x,t) \geq 0, \qquad (x,t) \in \Gamma.$$

If the upper solution $u^+(x,t)$ is ω-periodic in time we shall say that $u^+(x,t)$ is an *upper ω-periodic solution* of problem (1.1)-(1.3). If the function $u^+(x,t)$ does not depend on time, we shall call it *the upper stationary solution* of problem (1.1)-(1.3).

We may similarly define *the lower solution, the lower ω-periodic solution,* and *the lower stationary solution.* This lower solution $u^-(x,t) \in H^{2+\alpha}(\bar{Q}_\omega)$ satisfies the following inequalities

$$u_t^- \leq \sum_{i,j=1}^n a_{ij}(x,t,u^-,\nabla u^-)\frac{\partial^2 u^-}{\partial x_i \partial x_j} + a(x,t,u^-,\nabla u^-), \quad (x,t) \in \bar{Q}_\omega;$$

$$B(x,t)u^- - g(x,t,u^-) \geq 0, \qquad (x,t) \in \Gamma_\omega,$$

$(u^-(x,t) \leq 0$ in the case of the first boundary-value problem). Denote the lower solution of problem (1.1)-(1.3) by $u^-(x,t)$.

Definition 2.2.   We shall say that problem (1.1)-(1.3) is *dissipative,* if there exists a positive number R_0 such that for each function $u_0 \in E$ the solution $u(x,t;u_0)$ of problem (1.1)-(1.3) is determined for all $t \geq 0$ and, besides,

$$\overline{\lim_{t \to +\infty}} \parallel u(x,t;u_0) \parallel_0^\Omega < R_0. \tag{2.1}$$

Note, that if condition (1.4) holds, then such constant R_1 exists that

$$\overline{\lim_{t \to +\infty}} \parallel u(x,t;u_0) \parallel_1^\Omega < R_1. \tag{2.2}$$

Definition 2.3.   Let $\varphi(x,t)$ be an attracting from above or attracting from below solution of problem (1.1)-(1.3). By *the domain of attraction* $A(\varphi)$ of solution $\varphi(x,t)$ is meant the set

$$A(\varphi) = \left\{ u_0 \in E, \ \lim_{t \to +\infty} \parallel u(x,t;u_0) - \varphi(x,t) \parallel_1^\Omega = 0 \right\}.$$

Theorem 2.1.   (Belonosov *et al.,* 1982; Vishnevskii, 1984a, b; 1990a, b) Let problem (1.1)-(1.3) have the upper and lower ω-periodic solutions $u^\pm(x,t)$ and

$$u^+(x,t) \geq u^-(x,t), \qquad (x,t) \in Q_\omega.$$

In this case problem (1.1)-(1.3) has ω-periodic solutions $M(x,t)$, $m(x,t) \in H^{2+\alpha}(Q_\omega)$ with the following properties
1. $M(x,t) \geq m(x,t)$, $(x,t) \in Q_\omega$.
2. Each $k\omega$-periodic (with natural k) solution $\varphi(x,t)$ of problem (1.1)-(1.3) satisfying the inequality

$$u^+(x,t) \geq \varphi(x,t) \geq u^-(x,t), \qquad (x,t) \in Q_{k\omega} \qquad (2.3)$$

satisfies, also, the inequality

$$M(x,t) \geq \varphi(x,t) \geq m(x,t), \qquad (x,t) \in Q_{k\omega}. \qquad (2.4)$$

3. If $M(x,t) \neq u^+(x,t)$, then $M(x,t)$ is an asymptotically stable from above periodic solution of problem (1.1)-(1.3), where the attraction domain $A(M)$ contains the set

$$\left\{ u_0 \in E, \quad u^+(x,0) \geq u_0(x) \geq M(x,0), \quad x \in \Omega \right\}.$$

Analogously, if $m(x,t) \neq u^-(x,t)$, then $m(x,t)$ is an asymptotically stable from below periodic solution and the set $A(m)$ contains the set

$$\left\{ u_0 \in E, \quad u^-(x,0) \leq u_0(x) \leq m(x,0), \quad x \in \Omega \right\}.$$

When problem (1.1)-(1.3) is autonomous, theorem 2.1 is as follows.

Definition 2.4 ω-periodic solutions $M(x,t)$ and $m(x,t)$ satisfying inequalities (2.3), (2.4) from the point 2 of theorem 2.1 we shall call $M(x,t)$-maximal ω-periodic solutions of (1.1)-(1.3) if they lie between $u^+(x,t)$ and $u^-(x,t)$ and $m(x,t)$-minimal ω-periodic solutions if they lie between $u^+(x,t)$ and $u^-(x,t)$.

Theorem 2.1'. Let problem (1.1)-(1.3) have the upper and lower stationary solutions $u^\pm(x)$, where $u^+(x) \geq u^-(x)$, $x \in \Omega$. Then problem (1.1)-(1.3) has the stationary solutions $M(x)$, $m(x) \in C^{2+\alpha}(\bar{\Omega})$ with the following properties

1. $M(x) \geq m(x), \quad x \in \Omega$.
2. Each either periodic or stationary solution $\varphi(x,t)$ of problem (1.1)-(1.3) satisfying the inequality

$$u^+(x) \geq \varphi(x,t) \geq u^-(x), \qquad (x,t) \in Q$$

satisfies, also, the inequality

$$M(x) \geq \varphi(x,t) \geq m(x), \qquad (x,t) \in Q.$$

3. If $M(x) \neq u^+(x)$, then $M(x)$ is an asymptotically stable from above stationary solution of problem (1.1)-(1.3), where $A(M)$ contains the set

$$\left\{ u_0 \in E, \ u^+(x) \geq u_0(x) \geq M(x), \ x \in \Omega \right\}.$$

Analogously, if $m(x) \neq u^-(x)$, then $m(x)$ is an asymptotically stable from below stationary solution and $A(m)$ contains the set

$$\left\{ u_0 \in E, \ u^-(x) \leq u_0(x) \leq m(x), \ x \in \Omega \right\}.$$

Proof. We must intersperse the proof of the theorem with some lemmas. First, let us modify the functions

$$a_{ij}(x,t,u,p), \quad a(x,t,u,p), \quad g(x,t,u),$$

$$p = (p_1, \ldots, p_n), \quad i,j = 1, \ldots, n.$$

Let $u(x,t)$ solve problem (1.1)-(1.3) and suppose that (2.3) holds. Then, by conditions (1.4), we have

$$\| u(x,t) \|_1^{\Omega} \leq C_1. \tag{2.5}$$

Extend the functions $a_{ij}(x,t,u,p), a(x,t,u,p), g(x,t,u)$, outside the ball with sufficiently large radius (see equalities below), saving the uniform parabolicity and so that the extended functions

$$\tilde{a}_{ij}(x,t,u,p), \quad \tilde{a}(x,t,u,p), \quad \tilde{g}(x,t,u),$$

are bounded. Besides, we require that for each function which satisfies inequality (2.5), the following identities hold:

$$a_{ij}(x,t,u,\nabla u) = \tilde{a}_{ij}(x,t,u,\nabla u);$$

$$a(x,t,u,\nabla u) = \tilde{a}(x,t,u,\nabla u);$$

$$g(x,t,u) = \tilde{g}(x,t,u).$$

Such an extension may be easily constructed (see, for example, Hartman, 1970).

Consider the following problem

$$u_t = \sum_{i,j=1}^{n} \tilde{a}_{ij}(x,t,u,\nabla u)\frac{\partial^2 u}{\partial x_i \partial x_j} + \tilde{a}(x,t,u,\nabla u), \quad (x,t) \in Q_0, \quad (2.6)$$

$$B(x,t)u = \tilde{g}(x,t,u), \quad (x,t) \in \Gamma, \quad u(x,0) = u_0(x).$$

If $u(x,t)$ solves problem (2.6) and (2.5) holds, then $u(x,t)$, simultaneously, solves problem (1.1)-(1.3).

Let

$$\mu > \max\left\{|\tilde{a}_u(x,t,u,p)|, \quad |g_u(x,t,u)|, \quad x \in \bar{\Omega},\right.$$

$$t \geq 0, \quad -\infty < u, p_i < +\infty, \quad i = 1,\ldots,n\Big\}.$$

Parallel with problem (2.6), consider the problem

$$u_t = \sum_{i,j=1}^{n} \tilde{a}_{ij}(x,t,u,\nabla u)\frac{\partial^2 u}{\partial x_i \partial x_j} + \tilde{a}(x,t,u,\nabla u)$$

$$-\mu u + \mu f(x,t), \quad (x,t) \in Q_0,$$

$$B(x,t)u = \tilde{g}(x,t,u) - \mu u + \mu g(x,t), \qquad (x,t) \in \Gamma, \qquad (2.7)$$

$$u(x,0) = u_0(x).$$

Lemma 2.2. Assume that $f(x,t) \in H_1^\alpha(Q)$, $g(x,t) \in H_0^{1+\alpha}(\Gamma)$; initial data $u_0 \in C^1(\bar{\Omega})$ satisfy the necessary compatibility conditions. Then problem (2.7) has a unique solution $u(x,t)$. It belongs to the class $H_1^{2+\alpha}(Q)$. If, in addition,

$$u^-(x,t) \le f(x,t) \le u^+(x,t) \qquad (x,t) \in Q_\omega,$$

$$u^-(x,t) \le g(x,t) \le u^+(x,t) \qquad (x,t) \in \Gamma_\omega, \qquad (2.8)$$

$$u^-(x,0) \le u_0(x) \le u^+(x,0) \qquad (x,0) \in \Omega,$$

then $u(x,t)$ satisfies inequality (2.3).

Proof. Fix $T > 0$. Applying the standard Leray-Shauder method, we may show that (2.7) has a unique solution $u(x,t)$ and it belongs to $H_1^{2+\alpha}(Q)$. Note, also, that by the choice of μ in (2.7) and by the maximum principle, we obtain

$$\| u(x,t) \|_{1,2+\alpha}^{Q_T} \le C_2,$$

where C_2 does not depend on T. Therefore, $u(x,t)$ may be extended in the whole semi-infinite cylinder Q, where

$$\| u(x,t) \|_{1,2+\alpha}^{Q} \le C_2.$$

Now show that if (2.8) holds, the solution $u(x,t)$ satisfies inequality (2.3).

Let $w(x,t)$ solve the following problem

$$w_t = \sum_{i,j=1}^n \tilde{a}_{ij}(x,t,v,\nabla v)\frac{\partial^2 w}{\partial x_i \partial x_j} + \tilde{a}(x,t,v,\nabla v)$$

$$-\mu w + \mu f(x,t), \qquad (x,t) \in Q_0,$$

$$B(x,t)w = \tilde{g}(x,t,v) - \mu v + \mu g(x,t), \qquad (x,t) \in \Gamma,$$

$$w(x,0) = u_0(x).$$

The mapping $G(v) = w(x,t)$, which maps $v(x,t)$ into $w(x,t)$, is completely continuous (see the results of chapter 1) and maps the set

$$M = \left\{ f : \| f(x,t) \|_{1,2+\alpha}^{Q} \le C_3; \ u^-(x,t) \right.$$

$$\left. \le f(x,t) \le u^+(x,t), \ (x,t) \in Q_\omega \right\}$$

into itself, if C_3 is taken sufficiently large. The fixed point of G solves problem (2.7) and belongs to the set M. Therefore, the solution $u(x,t)$ of problem (2.7) satisfies inequality (2.3). The lemma is proved.

Denote by $U(f,g,u_0)$ the solution of problem (2.7) constructed in lemma 2.2.

Lemma 2.3. Let functions f,g satisfy all conditions of lemma 2.2 and, in addition, are ω-periodic in time. Then there exists a unique ω-periodic solution of problem (2.7). (Besides, instead of initial data $u_0(x)$ for $t = 0$, we have to set the periodicity condition: $u(x,t+\omega) = u(x,t)$). If the functions f and g, in addition, satisfy inequality (2.8), then the ω-periodic solution satisfies inequality (2.3).

Assume that problem (1.1)-(1.3) is autonomous and the functions f and g do not depend on time. Then problem (2.7) has a unique stationary solution, and if f and g satisfy the inequalities

$$u^-(x) \le f(x) \le u^+(x), \ x \in \Omega;$$

$$u^-(x) \le g(x) \le u^+(x), \ x \in \partial\Omega;$$

then the stationary solution $u(x)$ satisfies the inequality

$$u^-(x) \leq u(x) \leq u^+(x), \quad x \in \Omega.$$

Proof. We shall prove the first part of the lemma. The second part may be proved analogously.

Consider the set M

$$M = \left\{ v(x,t) \in C^{2+\alpha}(Q_\omega), \ \| v(x,t) \|_{2+\alpha}^{Q_\omega} \leq C_3, \right.$$

$$\left. u^-(x,t) \leq v(x,t) \leq u^+(x,t), \quad (x,t) \in Q_\omega, \quad v(x,t+\omega) = v(x,t) \right\}.$$

Denote by $w(x,t)$ the unique ω-periodic solution of the problem

$$w_t = \sum_{i,j=1}^{n} \tilde{a}_{ij}(x,t,v,\nabla v) \frac{\partial^2 w}{\partial x_i \partial x_j} + \tilde{a}(x,t,v,\nabla v)$$

$$-\mu w + \mu f(x,t), \qquad (x,t) \in Q_0,$$

$$B(x,t)w = \tilde{g}(x,t,v) - \mu v + \mu g(x,t), \qquad (x,t) \in \Gamma,$$

$$w(x,0) = u_0(x),$$

where the funcions $\tilde{a}_{ij}, \tilde{a}$ and $\tilde{g}$ were defined in the beginning of the proof of theorem 2.1, changing C_1 to C_3.

The mapping $w(x,t) = G(v(x,t))$ for sufficiently large constant C_3 maps M into itself. As $\| w(x,t) \|_{2+\alpha}^{Q_\omega} \leq C_3$, then, G has a fixed point from M. This fixed point is an ω-periodic solution of problem (2.7). The uniqueness of this solution follows from the method of choosing μ and the maximum principle. The lemma is proved.

Denote the ω-periodic solution of problem (2.7), if f and g are ω-periodic, by $P(f,g)$ and the stationary solution of (2.7), correspondingly, by $S(f,g)$. In the second case, the functions f and g do not depend on time.

Lemma 2.4. Let the functions

$$f_i(x,t), \quad g_i(x,t), \quad u_{i0}(x), \qquad i = 1,2$$

satisfy the following inequalities

$$u^-(x,t) \leq f_1(x,t) \leq f_2(x,t) \leq u^+(x,t), \quad (x,t) \in Q,$$

$$u^-(x,t) \leq g_1(x,t) \leq g_2(x,t) \leq u^+(x,t), \quad (x,t) \in \Gamma, \qquad (2.9)$$

$$u^-(x,0) \leq u_{01}(x) \leq u_{02}(x) \leq u^+(x,0), \quad (x,0) \in \Omega;$$

then

$$u^-(x,t) \leq U(f_1,g_1,u_{01}) \leq U(f_2,g_2,u_{02}) \leq u^+(x,t), \quad (x,t) \in Q. \tag{2.10}$$

Here, we must recall, that we denote by $U(f,g,u_0)$ the solution to problem (2.7) which was constructed in lemma 2.2.

If, in addition, we assume that the functions $f_i(x,t), g_i(x,t), i = 1,2$, are ω-periodic in time, then

$$u^-(x,t) \leq P(f_1,g_1) \leq P(f_2,g_2) \leq u^+(x,t), \quad (x,t) \in Q. \tag{2.11}$$

If problem (1.1)-(1.3) is autonomous and $f_i(x,t), g_i(x,t), \; i = 1,2$, do not depend on time, then

$$u^-(x,t) \leq S(f_1,g_1) \leq S(f_2,g_2) \leq u^+(x,t), \quad (x,t) \in \Omega.$$

Proof. We need prove only one of the lemma inequalities: the others are proved analogously. Let us put

$$w(x,t) = P(f_1,g_1) - u^-(x,t).$$

We must show that the ω-periodic function $w(x,t)$ is nonnegative. Really,

$$w_t \le \sum_{i,j=1}^{n} \hat{a}_{ij}(x,t)\frac{\partial^2 w}{\partial x_i \partial x_j} + \sum_{i,j=1}^{n} \hat{a}_i(x,t)\frac{\partial w}{\partial x_i} + \hat{a}(x,t)w$$

$$-\mu w + \mu f_1(x,t) - \mu u^-, \quad (x,t) \in Q_0,$$

$$B(x,t)w \le \hat{g}(x,t) - \mu w + \mu g_1(x,t) + \mu u^-, \quad (x,t) \in \Gamma.$$

Here the function $w(x,t)$ is considered as the solution of the linear evolutionary equation. Therefore,

$$\hat{a}_{ij}(x,t) = a_{ij}(x,t,P(f_1,g_1) \bigtriangledown P(f_1,g_1))$$

The other coefficients may be determined similarly.

Using lemma 1.7, we obtain that $w(x,t) \ge 0$ for $(x,t) \in Q$. The lemma is proved.

Lemma 2.5. Let $f_2(x,t), g_2(x,t)$ be ω-periodic functions, and $f_1(x,t), g_1(x,t)$ be such functions that for a certain $\varepsilon > 0$

$$\Big(f_2(x,t) - f_1(x,t)\Big)e^{\varepsilon t} \in H^{\alpha}_{-1}(Q),$$

$$\Big(g_2(x,t) - g_1(x,t)\Big)e^{\varepsilon t} \in H^{1+\alpha}_0(\Gamma).$$

Then, the following estimate

$$\left\|\Big(U(f_1,g_1,u_0) - P(f_2,g_2)\Big)e^{\varepsilon t}\right\|_{1,2+\alpha}^{Q}$$

$$\le C\left(\left\|u_0 - P(f_2,g_2)(x,0)\right\|_1^{\Omega} + \left\|\Big(f_2(x,t) - f_1(x,t)\Big)e^{\varepsilon t}\right\|_{-1,\alpha}^{Q}\right.$$

$$\left. + \left\|\Big(g_2(x,t) - g_1(x,t)\Big)e^{\varepsilon t}\right\|_{0,1+\alpha}^{\Gamma}\right). \tag{2.12}$$

holds.

Proof. Denote by

$$w(x,t) = \Big(U(f_1, g_1, u_0) - P(f_2, g_2)\Big)e^{\varepsilon t}.$$

Then the function $w(x,t)$ may be considered as a solution of the linear problem

$$w_t = \sum_{i,j=1}^{n} \hat{a}_{ij}(x,t)\frac{\partial^2 w}{\partial x_i \partial x_j} + \sum_{i,j=1}^{n} \hat{a}_i(x,t)\frac{\partial w}{\partial x_i} + \hat{a}(x,t)w$$

$$-\mu w + \varepsilon w + \mu(f_1(x,t) - f_2(x,t))e^{\varepsilon t}, \quad (x,t) \in Q,$$

$$B(x,t)w = \hat{g}(x,t) - \mu w + \mu\Big(g_1(x,t) - g_2(x,t)\Big)e^{\varepsilon t}, \quad (x,t) \in \Gamma,$$

$$w(x,t) = u_0(x) - P(f_2, g_2)(x,0).$$

Choose a number $\varepsilon > 0$ so that

$$\hat{a}(x,t) + \varepsilon - \mu < 0, \qquad (x,t) \in Q.$$

Hence follows that the spectrum of $V(\omega)$, corresponding to (2.5) chapter 1 lies inside the unit circle. Applying the estimate (2.16) from theorem 2.1 (chapter 1), we obtain inequality (2.12). The lemma is proved.

In order to prove the first point of theorem 2.1, we shall use the iterative process, which will find the solution $M(x,t)$. Set

$$M_1(x,t) = P\Big(u^+(x,t), u^+(x,t)\Big),$$

$$M_n(x,t) = P\Big(M_{n-1}(x,t), M_{n-1}(x,t)\Big), \quad n = 2,3,\ldots.$$

Note, that by the maximum principle

$$u^+(x,t) \geq M_1(x,t) \qquad \text{for} \qquad (x,t) \in Q.$$

By lemma 2.4, the sequence $\{M_n(x,t)\}$ of ω-periodic functions monotonically decreases in $C(Q_\omega)$. On the other hand, the restrictions imposed on the smoothness of the nonlinear terms yield that this sequence is bounded in $H^{2+\alpha'}(Q_\omega)$, $0 < \alpha < \alpha' < 1$. Applying the interpolational inequality in the multiplicated form (Ladyzhenskaya *et al.*, 1967), we see that the sequence converges in $H^{2+\alpha}(Q_\omega)$ to the function $M(x,t)$. In addition

$$M(x,t) = P\big(M(x,t), M(x,t)\big),$$

and, therefore, the function $M(x,t)$ is an ω-periodic solution of problem (2.6). The function $M(x,t)$ satisfies, by lemma 2.4, inequality (2.3); therefore, $M(x,t)$ is simultaneously an ω-periodic solution of problem (1.1)-(1.3).

Let $\varphi(x,t)$ be a $k\omega$-periodic solution of (1.1)-(1.3) satisfying inequality (2.6). Let $k\omega = \hat{\omega}$. Then all the functions $\tilde{a}_{ij}, \tilde{a}_i, \tilde{g}, u^\pm, M, m$ are $\hat{\omega}$-periodic in time. As $\varphi(x,t)$ is an $\hat{\omega}$-periodic solution, then

$$P\big(\varphi(x,t), \varphi(x,t)\big) = \varphi(x,t).$$

Inequality (2.11) yields

$$M_1(x,t) = P\left(u^+(x,t), u^+(x,t)\right) \geq P\big(\varphi(x,t), \varphi(x,t)\big)$$

$$= \varphi(x,t) \geq P\left(u^-(x,t), u^-(x,t)\right) = m_1(x,t).$$

If we continue this process, we obtain

$$M_n(x,t) = P\big(M_{n-1}(x,t), M_{n-1}(x,t)\big) \geq P\big(\varphi(x,t), \varphi(x,t)\big)$$

$$= \varphi(x,t) \geq P\big(m_{n-1}(x,t), m_{n-1}(x,t)\big) = m_n(x,t).$$

Passing to this limit gives inequality (2.4). This proves the second point of the theorem.

To prove the third point, note that if $M(x,t) \neq u^+(x,t)$ then the upper ω-periodic solution of (2.6) is not a periodic solution of this problem and, therefore,

$$u^+(x,t) > M_1(x,t) > M_2(x,t) > \ldots \quad \text{for} \quad (x,t) \in Q_\omega.$$

Let
$$u^+(x,0) \geq u_0(x) \geq M(x,0), \quad x \in \Omega, \quad u_0(x) \in E.$$

We now show that

$$\lim_{t \to +\infty} \| u(x,t;u_0) - M(x,t) \|_1^\Omega = 0. \tag{2.13}$$

For this purpose consider the monotone sequence in $C(Q)$

$$u_1(x,t) = U\left(u^+(x,t), u^+(x,t), u_0(x)\right),$$

$$u_n(x,t) = U\left(u_{n-1}(x,t), u_{n-1}(x,t), u_0(x)\right), \quad n > 1.$$

Applying inequality (2.5) from chapter 1 to $u_1(x,t) - M_1(x,t)$, we obtain

$$\left\| \left(u_1(x,t) - M_1(x,t)\right)e^{\varepsilon t} \right\|_{1,2+\alpha}^Q \leq C \left\| u_0(x) - u^+(x,0) \right\|_1^\Omega .$$

This allows us to apply inequality (2.5) from chapter 1 to the difference $u_2(x,t) - M_2(x,t)$. As a result we obtain

$$\left\| \left(u_n(x,t) - M_n(x,t)\right)e^{\varepsilon t} \right\|_{1,2+\alpha}^Q \leq C(n).$$

On the other hand, the sequence $u_n(x,t)$ monotonically converges in $C(Q)$ to the solution $u(x,t;u_0)$ since

$$U\left(u(x,t;u_0), u(x,t;u_0), u_0\right) = u(x,t;u_0).$$

If $\delta > 0$ is arbitrary, then such t_0 exists that for $t \geq t_0$

$$\| u(x,t;u_0) - M(x,t) \|_0^\Omega < \delta.$$

Note, also, that $u(x,t;u_0) - M(x,t)$ is bounded in $C^{2+\alpha'}(\bar{\Omega})$. Therefore,

$$\| \, u(x,t;u_0) - M(x,t) \, \|_{2+\alpha}^{\Omega} < C\delta^{\alpha'-\alpha}.$$

As δ is arbitrary and positive, the latter inequality yields inequality (2.13). Theorem 2.1 is proved.

The conditions being formulated in theorem 2.1, that problem (1.1)-(1.3) has upper and lower ω-periodic solutions $u^{\pm}(x,t)$ such that

$$u^+(x,t) > u^-(x,t), \qquad (x,t) \in Q,$$

sometimes holds automatically. In the theorem below we point out the class of problems for which this condition always holds.

Theorem 2.6.   (Belonosov *et al.*, 1982; Vishnevskii, 1984a, b; 1990a, b).   Assume that problem (1.1)-(1.3) is dissipative (see definition 2.2). Then it has the maximal ω-periodic solution $M(x,t)$ and the minimal ω-periodic solution $m(x,t)$ such that

   1. $M(x,t) \geq m(x,t), \quad (x,t) \in Q_\omega.$

   2. Each periodic solution $\varphi(x,t)$ of problem (1.1)-(1.3) satisfies the inequality

$$M(x,t) \geq \varphi(x,t) \geq m(x,t), \quad (x,t) \in Q.$$

   3. The attraction domain $A(M)$ contains the set

$$\{u_0 \in E; \quad u_0(x) \geq M(x,0), \quad x \in \Omega\}.$$

Analogously, $A(m)$ contains the set

$$\{u_0 \in E; \quad u_0(x) \leq m(x,0), \quad x \in \Omega\}.$$

   4. If problem (1.1)-(1.3) is autonomous, then the stationary solutions exist and satisfy properties 1-3.

Proof. Consider, firstly, the first boundary-value problem. In the case of the other boundary conditions the proof is analogous.

Denote by $\xi_1(x)$ the first eigenfunction of the Laplace operator in the domain Ω. As it is known (Krasnoselskii, 1962), if

$$\| \psi(x) \|_1^\Omega \le 1, \quad \psi(x) \in E,$$

then such positive β_1 exists that for $\beta \ge \beta_1$

$$|\psi(x)| < \beta \xi_1(x), \qquad x \in \Omega. \tag{2.14}$$

Consider the solution of (1.1)-(1.3) with the initial data $\beta \xi_1(x)$, where $\beta \ge \beta_1 R_1$ (the constant R_1 is taken from the dissipativity condition of the problem, more precisely $\overline{\lim_{t \to +\infty}} \| u(x, t; u_0) \|_1^\Omega \le R_1$, if $u_0 \in E$). There exists such k that

$$\| u(x, kw; \beta \xi_1(x)) \|_1^\Omega < R_1.$$

Therefore,

$$|u(x, kw; \beta \xi_1(x))| < \beta_1 R_1 \xi_1(x) \le \beta \xi_1(x), \quad x \in \Omega.$$

Consider the function

$$w(x, t) = u(x, t + kw; \beta \xi_1(x)) - u(x, t; \beta \xi_1(x)).$$

The function $w(x, t)$ may be considered as a solution of the linear homogeneous problem with the initial data

$$w(x, 0) = u(x, kw; \beta \xi_1(x)) - \beta \xi_1(x) < 0, \quad x \in \Omega.$$

The comparison theorem yields

$$\beta \xi_1(x) > u(x, kw; \beta \xi_1(x)) > u(x, 2kw; \beta \xi_1(x)) > \dots$$

$$\dots > u(x, lkw; \beta \xi_1(x)), \quad x \in \Omega.$$

Therefore, the sequence $u(x, lk\omega; \beta\xi_1(x))$, $l = 1, 2, \ldots$ monotonically decreases in $C(\bar{\Omega})$. On the other hand, the dissipativity and the assumptions on the nonlinear terms' smoothness, yield that this sequence is bounded in $C^{2+\alpha'}(\bar{\Omega})$, $0 < \alpha' < \alpha$. Therefore, $u(x, lk\omega; \beta\xi_1(x))$, $l = 1, 2, \ldots$ converges in $C^{2+\alpha'}(\bar{\Omega})$ to the function $M(x, \beta)$. Denote by $M(x, t, \beta)$ the solution of (1.1)-(1.3) with the initial data $M(x, \beta)$. The sequence $u(x, lk\omega + t; \beta\xi_1(x))$, $l = 1, 2, \ldots$ converges in $H^{2+\alpha}(Q_{k\omega})$ to $M(x, t, \beta)$. Therefore, both $u(x, lk\omega; \beta\xi_1(x))$, $l = 1, 2, \ldots$ and $u(x, (l+1)k\omega; \beta\xi_1(x))$, $l = 1, 2, \ldots$, converge to $M(x, \beta)$. Hence follows that $M(x, t, \beta)$ is an ω-periodic solution of the problem in question.

Lemma 2.7. The solution $M(x, t, \beta)$ does not depend on β for $\beta > \beta_1 R_1$ and is ω-periodic by time.

Proof. Let $\beta, \beta' \geq \beta_1 R_1$. Then

$$\beta'\xi_1(x) > u(x, lk\omega; \beta\xi_1(x)), \qquad x \in \Omega$$

for a certain natural l. By the comparison theorem, we obtain

$$u(x, t; \beta'\xi_1(x)) > u(x, t + lk\omega; \beta\xi_1(x)), \quad (x, t) \in Q.$$

Passing to the limit, we obtain $M(x, t, \beta) \geq M(x, t, \beta')$. Analogously, $M(x, t, \beta') \geq M(x, t, \beta)$. Therefore,

$$M(x, t, \beta) = M(x, t, \beta') = M(x, t).$$

Now, prove that the period of $M(x, t)$ is equal to ω. For this purpose consider the sequence $u(x, lk\omega + \omega; \beta\xi_1(x))$, $l = 1, 2, \ldots$ which converges to the function $M(x, \omega)$. Choose β' and β so that

$$\beta'\xi_1(x) > u(x, \omega; \beta\xi_1(x)) > M(x, 0), \quad x \in \Omega.$$

Then, by the comparison theorem, for all $t > 0$

$$u(x,t; \beta' \xi_1(x)) > u(x, t + \omega; \beta \xi_1(x)) > M(x,t).$$

Going to the limit for $t = lk\omega$, $l \to \infty$, we obtain

$$M(x,0) \geq M(x,\omega) \geq M(x,0).$$

Hence follows that M is an ω-periodic solution.

The solution $m(x,t)$ is constructed analogously. The inequality

$$M(x,t) \geq m(x,t)$$

is evident.

We shall prove, firstly, the third point of the theorem, then the second. If $u_0 \in E$, $u_0 \geq M(x,0)$, $x \in \Omega$, then such positive β exists that

$$\beta \xi_1(x) > u(x) \geq M(x,0), \quad x \in \Omega.$$

By the comparison theorem, we obtain

$$u(x,t; \beta \xi_1(x)) > u(x,t; u_0) \geq M(x,t), \quad (x,t) \in Q.$$

Therefore, the solution $u(x,t; u_0)$ converges in $C(\bar{\Omega})$ to $M(x,t)$. From (1.4) and the smoothness assumptions it follows that the norm $\| u(x,t; u_0) \|_{2+\alpha}^{\Omega}$ is bounded. Therefore,

$$\lim_{t \to +\infty} \| u(x,t; u_0) \|_{2+\alpha}^{\Omega} = 0.$$

This proves the third point of the theorem.

To prove the second point of the theorem, note that if a periodic solution intersects the solution $M(x,t)$ then there exist arbitrary large values of t, such that for these values in a certain point $x \in \Omega$ the inequality

$$\varphi(x,t) - M(x,t) > \delta > 0$$

will hold. Here δ is a certain positive number which does not depend on t and x. Let a number β be such that

$$\beta\xi_1(x) > \varphi(x,0), \qquad x \in \Omega.$$

Then

$$u(x,t;\beta\xi_1(x)) > \varphi(x,t) \quad \text{for} \quad t \geq 0$$

and the solution $u(x,t;\beta\xi_1(x))$ could not converge to $M(x,t)$ for $t \to +\infty$. This contradicts the statement proved above. Analogously we may establish that $\varphi(x,t) \geq m(x,t)$.

If problem (1.1)-(1.3) is autonomous, then as ω we may take any positive number. Hence follows that $M(x,t)$ and $m(x,t)$ are smooth and periodic in time functions with any positive period. Therefore, $M(x), m(x)$ are the stationary solution of problem (1.1)-(1.3). The theorem is proved.

§5.3   The attraction domains of a stable periodic solution.

Here we shall assume that the problem in question has upper and lower ω-periodic solutions. Theorems 2.1 and 2.6 from the previous paragraph give the example of sufficient conditions for this assumption to hold.

Assume problem (1.1)-(1.3) to be dissipative (see definition 2.2) and $M(x,t)$ to be the maximal ω-periodic solution of this problem. The definition of maximal and minimal ω-periodic solutions was given in §5.2. In this case, since the problem is dissipative, it was shown that $u^+(x,t)$ and $u^-(x,t)$ may be chosen so that all stationary solutions are situated between $u^+(x,t)$ and $u^-(x,t)$. Therefore, the maximal solution of the problem lying between $u^+(x,t)$ and $u^-(x,t)$ is the maximal solution to the problem: inequality (2.4) from §5.2 holds for each periodic solution $\varphi(x,t)$. Linearize problem (1.1)-(1.3) on $M(x,t)$. Denote by $\xi(x,M)$ the first positive eigenfunction of the shift operator $V(\omega)$ of the linearized problem. It will be shown below that

$$M(x,t) + \beta\xi_1(x,t)$$

is the upper solution of the problem and $m(x,t) - \beta\xi_1(x,t)$ is the lower solution, if β is positive and sufficiently small. All the other periodic solutions lie between $M(x,t)$ and $m(x,t)$. Therefore, when the problem is dissipative, we may also suppose that upper and lower ω-periodic solutions of the problem are given.

Theorem 3.1. (Vishnevskii, 1984a, b; 1990a, b). Let $\varphi(x,t)$ be a $k\omega$-periodic (with natural k, see remark) solution of the problem (1.1)-(1.3) lying between $u^+(x,t), u^-(x,t)$ (see definition 2.1)

$$u^+(x,t) \le \varphi(x,t) \le u^-(x,t), \qquad (x,t) \in Q$$

and being attractive from above. Then the solution $\varphi(x,t)$ is asymptotically stable from above, and the following two variants are possible.
   1. $\varphi(x,t) = M(x,t)$.
   2. There exists a unstable from below $k\omega$-periodic solution $\psi(x,t)$ such that

$$M(x,t) \ge \psi(x,t) \ge \varphi(x,t), \qquad (x,t) \in Q, \qquad (3.1)$$

and the attraction domain of a stable from above $k\omega$-periodic solution $\varphi(x,t)$ contains the set

$$\{u_0 \in E, \quad \psi(x,0) \ge u_0(x) \ge \varphi(x,0), \quad u_0(x) \not\equiv \psi(x,0)\}.$$

If (1.1)-(1.3) is autonomous, then the analogous theorem holds, which may be formulated as follows.

Remark 3.1. In the considered case of ω-periodic with respect to time coefficients the solutions both with the period ω and with the period $k\omega$ with natural k may be stable and asymptotically stable.

Theorem 3.1′.   Let $\varphi(x)$ be a stationary solution of problem (1.1)-(1.3) attracting from above and

$$u^+(x,t) \le \varphi(x) \le u^-(x,t), \qquad (x,t) \in Q.$$

Then $\varphi(x)$ is stable from above, and the following two variants are possible
   1. $\varphi(x) = M(x)$.
   2. There exists an unstable from below stationary solution $\psi(x)$

$$M(x) \ge \psi(x) \ge \varphi(x), \qquad x \in Q,$$

such that the attraction domain $A(\varphi)$ contains the set

$$\{u_0 \in E, \ \psi(x) \ge u_0(x) \ge \varphi(x), \ u_0(x) \not\equiv \psi(x)\}.$$

First, we must prove some auxiliary statements. Let a $k\omega$-periodic solution $\chi(x,t)$ of the problem in question be such that

$$u^+(x,t) \ge \chi(x) \ge u^-(x,t), \qquad (x,t) \in Q.$$

We shall establish some properties of this solution.
   First, linearize problem (1.1)-(1.3) on the solution $\chi(x,t)$

$$v_t = L(x,t;\chi)v, \qquad (x,t) \in Q;$$

$$B_1(x,t;\chi)v = 0, \qquad (x,t) \in \Gamma;$$

$$v(x,0) = v_0(x).$$

Here by $L(x,t;\chi)$ we denote nonlinear operator (1.1) linearized on the solution $\chi(x,t)$. By $B_1(x,t;\chi)$ we denote boundary conditions (1.2) linearized on the solution $\chi(x,t)$. Denote by $\rho(x,t)$ the solution of the following problem (which exists according to the result of the fourth chapter for sufficiently large λ_0)

$$\rho_t = L(x, t; \chi)\rho - \lambda_0 \rho, \qquad (x, t) \in Q;$$

$$B_1(x, t; \chi)\rho = 1, \qquad (x, t) \in \Gamma.$$

Make the change $v(x, t) = \rho(x, t)w(x, t)$. Then

$$w_t = L_1(x, t; \chi)w, \qquad (x, t) \in Q;$$

$$\rho(x, t) \sum_{i=1}^{n} b_i(x, t) \frac{\partial w}{\partial x_i} + w = 0; \quad (x, t) \in \Gamma;$$

$$w(x, 0) = w_0(x) = \frac{v_0(x)}{\rho(x, 0)}. \tag{3.2}$$

Recall, that the operator $L_1(x, t, \chi)$ is obtained from operator $L(x, t, \varphi)(\rho w)$ after removing the parentheses and collecting similar terms. Evidently, we have to take into account that $\rho(x, t)$ solves the above parabolic equation.

The operator in the boundary conditions is written out analogously.

Let $V(k\omega)$ be the shift operator $w(x, 0) \to w(x, k\omega)$, where $w(x, t)$ solves problem (3.2). As was shown by Vishnevskii (1992), the maximal by modulus eigenvalue σ_1 of $V(k\omega)$ is positive and simple. Corresponding eigenfunction $\xi_1(x)$ is positive. If $\xi_1(x, t)$ solves (3.2) with initial data $\xi_1(x)$, then

$$\xi_1(x, k\omega) = \sigma_1 \xi_1(x, 0) = \sigma_1 \xi_1(x).$$

Denote by $\tilde{E}(t)$ the subspace of the space $C^1(\bar{\Omega})$ consisting of the functions $\bar{w}(x)$ satisfying the compatibility conditions

$$\rho(x, t) \sum_{i=1}^{n} b_i(x, t) \frac{\partial \tilde{w}}{\partial x_i} + \tilde{w} = 0; \quad x \in \partial\Omega.$$

Make the change

$$u(x, t) - \chi(x, t) = z(x, t)\rho(x, t)e^{t \ln \sigma_1 / k\omega}.$$

As a result we obtain

$$z_t = L_1(x,t;\chi)z - \frac{t\ln\sigma_1}{k\omega}z + F_1(x,t,z), \quad (x,t) \in Q;$$

$$B_2(x,t;\chi)z = G_1(x,t,z), \quad (x,t) \in \Gamma; \quad z(x,0) = z_0(x). \qquad (3.3)$$

Note, that in a certain ε_0 - neighborhood of zero in $H^{2+\alpha}(Q^\tau_{\tau+T})$ the mapping F_1, G_1 will satisfy the inequalities from lemmas 1.1 and 1.2 (chapter 1), where ε_0 does not depend on T, τ.

Decompose the space $\tilde{E}(t)$ into the direct sum

$$\tilde{E} = \tilde{E}_-(t) \oplus \tilde{E}_0(t).$$

The space $\tilde{E}_0(t)$ is one-dimensional and is as follows

$$\gamma\xi_1(x,t)e^{-\ln\sigma_1/k\omega}.$$

Using the results from chapter 1, we see that problem (3.3) for $l = q(\varepsilon_0)$, which always may be done since $\lim q(\varepsilon_0) = 0$, has the integral set

$$\gamma\xi_1(x,t) + \Psi_0(t, \gamma\xi_1(x,t)).$$

The constant γ must be sufficiently small, so that for $0 \le t \le k\omega$, we would have

$$\| \gamma\xi_1(x,t) + \Psi_0(t, \gamma\xi_1(x,t)) \|^\Omega_{2+\alpha} < \varepsilon_0.$$

From the inequality (3.12) of chapter 1

$$\| \Psi_0(t, \gamma_1\xi_1(x,t)) - \Psi_0(t, \gamma_2\xi_2(x,t)) \|^\Omega_{2+\alpha} \le cq(\varepsilon_0)|\gamma_1 - \gamma_2|. \qquad (3.4)$$

Returning to the function $u(x,t)$, we obtain that

$$\chi(x,t) + \left[\gamma\xi_1(x,t)\rho(x,t) + \Psi_0(t,\gamma\xi_1(x,t)\rho(x,t))\right]e^{t\ln\sigma_1/k\omega}$$

is the integral set $S(t,\gamma)$ of problem (1.1)-(1.3), if γ is sufficiently small. Let initial data be such that

$$u_0(x,\gamma) = \chi(x,0) + \left[\gamma\xi_1(x,0)\rho(x,0) + \Psi_0(0,\gamma\xi_1(x,0)\rho(x,0))\right].$$

Take γ_0 so that the solution $u(x,t;u_0(x,\gamma))$ lies in the integral set $S(t,\gamma)$ if $0 \leq t \leq k\omega$, $|\gamma| < \gamma_0$. This always may be done by diminishing γ_0 if it is necessary. The function $u(x,k\omega;u_0(x,\gamma))$ lies in the integral set; therefore, such γ_1, exists that

$$u(x,k\omega;u_0(x,\gamma)) = \chi(x,0) + \left[\gamma_1\xi_1(x,0)\rho(x,0) + \Psi_0(0,\gamma_1\xi_1(x,0)\rho(x,0))\right].$$

Determine for $|\gamma| < \gamma_0$ the mapping G which transforms γ into γ_1.

Lemma 3.2. The solution $u(x,t;u_0(x,\gamma))$ for $0 < \gamma < \gamma_0$ satisfies one of the following relations

1. $u(x,k\omega;u_0(x,\gamma)) = u(x,0;u_0(x,\gamma)).$

2. $u(x,k\omega;u_0(x,\gamma)) > u(x,0;u_0(x,\gamma)) \quad x \in \Omega.$ (3.5)

3. $u(x,k\omega;u_0(x,\gamma)) < u(x,0;u_0(x,\gamma)) \quad x \in \Omega.$

In the first case $u(x,t;u_0(x,\gamma))$ is a $k\omega$-periodic solution of (1.1)-(1.3). In the second and third cases the solution $u(x,t;u_0(x,\gamma))$ converges for $t \to +\infty$ to a $k\omega$-periodic solution. In the second case this solution is less than $u(x,t;u_0(x,\gamma))$, and in the third case it is greater than $u(x,t;u_0(x,\gamma))$.

Proof.   As we have noted

$$u(x, k\omega; u_0(x, \gamma)) = \chi(x, 0) + \Big[G(\gamma)\xi_1(x, 0)\rho(x, 0)$$

$$+ \Psi_0(0, G(\gamma)\xi_1(x, 0)\rho(x, 0)) \Big],$$

since $\gamma_1 = G(\gamma)$. Therefore,

$$u(x, k\omega; u_0(x, \gamma)) - u_0(x, \gamma) = \chi(x, 0)$$

$$+ \Big[(G(\gamma) - \gamma)\xi_1(x, 0)\rho(x, 0) + \Psi_0(0, (G(\gamma) - \gamma)\xi_1(x, 0)\rho(x, 0)) \Big]$$

$$= (G(\gamma) - \gamma)(\xi_1(x) + q(\varepsilon_0)\eta(x)).$$

If $G(\gamma) - \gamma \neq 0$, we may consider the function

$$\eta(x) = \frac{\Psi_0(0, G(\gamma)\xi_1(x)) - \Psi_0(x, 0)}{(G(\gamma) - \gamma)q(\varepsilon_0)}.$$

Note, that if $G(\gamma) - \gamma = 0$, then

$$u(x, k\omega; u_0(x, \gamma)) - u_0(x, \gamma) = 0;$$

therefore, $u(x, t; u_0(x, \gamma))$ is a $k\omega$-periodic solution, and the first point of lemma 3.2 holds.

Consider the case $G(\gamma) - \gamma \neq 0$. Then as follows from (3.4),

$$\| \eta(x) \|_{2+\alpha}^{\Omega} < C.$$

Therefore, for sufficiently small γ_0 the sign of the difference

$$u(x, k\omega; u_0(x, \gamma)) - u_0(x, \gamma)$$

for all $x \in \Omega$ is equal to the sign of $G(\gamma) - \gamma$. Really, by (3.4), for sufficiently small γ_0 the function $\xi_1(x) + q(\gamma_0)\eta(x)$ is positive for all $x \in \Omega$.

Let $G(\gamma) - \gamma$ be a positive number, i.e.

$$u(x, k\omega; u_0(x, \gamma)) > u(x, 0; u_0(x, \gamma)) \quad x \in \Omega.$$

Then $u(x, t; u_0(x, \gamma))$ converges for $t \to +\infty$ to a $k\omega$-periodic solution which is less than $u(x, t; u_0(x, \gamma))$. To prove this, consider the function

$$w(x, t) = u(x, t + k\omega; u_0(x, \gamma)) - u(x, t; u_0(x, \gamma)).$$

We may interpret this function as a solution of the linear homogeneous problem with the initial data

$$w_0(x) = u(x, k\omega; u_0(x, \gamma)) - u(x, 0; u_0(x, \gamma)) > 0 \quad \text{for } x \in \Omega.$$

By the comparison theorem, $w(x, t) > 0$ for $x \in \Omega$, $t > 0$. Therefore, the sequence $u(x, lk\omega; u_0(x, \gamma))$, $l = 1, 2, \ldots$ monotonically converges in the space $C(\bar{\Omega})$.

On the other hand, applying the comparison theorem, we see that

$$u^-(x, t) \leq u(x, t; u_0(x, \gamma)) \leq u^+(x, t), \quad x \in \Omega, \ t > 0.$$

For the problem in question, conditions (1.4) holds; therefore, taking into account the smoothness restrictions, we obtain

$$u(x, t; u_0(x, \gamma)) \in H^{2+\alpha'}(Q).$$

Hence follows that $u(x, lk\omega; u_0(x, \gamma))$, $l = 1, 2, \ldots$ converges for $l \to +\infty$ in $C^{2+\alpha}(\bar{\Omega})$ to a certain function $\zeta(x)$. Denote the solution of (1.1)-(1.3) with the initial data $\zeta(x)$ by $\zeta(x, t)$. Using the theorem on continuous dependence of a solution in the interval $[0, \omega]$, we see that $u(x, (l+1)k\omega; u_0(x, \gamma))$ converges for $l \to +\infty$ in $C^{2+\alpha}(\bar{\Omega})$ to $\zeta(x)$ and to $\zeta(x, k\omega)$. Therefore,

$$\zeta(x, 0) = \zeta(x, k\omega),$$

so, $\zeta(x, t)$ is a $k\omega$-periodic solution of (1.1)-(1.3).

The comparison theorem yields

$$M(x,t) \geq \zeta(x,t) > u(x, lk\omega; u_0(x,\gamma)) > \chi(x,0)$$

with l, an integer, $(x,t) \in Q_{k\omega}$. The third case is considered analogously. The lemma is proved.

Denote the solution constructed above by $\eta(x,t;\gamma,\chi)$. Note, that if one of relations (3.5) holds, then we may extend $\eta(x,t;\gamma,\chi)$ from the neighborhood of the $k\omega$-periodic solution $\chi(x,t)$ in the whole semi-infinite cylinder Q.

Further we shall investigate more thoroughly the properties of this solution. Now we have the following.

1) By $\eta(x,t,\gamma;\chi)$, $|\gamma| \leq \gamma_0$ we have denoted the solution

$$u(x,t;\chi(x,0)) + \gamma\Big[\xi_1(x,0)\rho(x,0) + \psi_0(0,\gamma\xi_1(x,0)\rho(x,0))\Big].$$

Here $\chi(x,t)$ is the $k\omega$-periodic solution of problem (1.1)-(1.3); the mapping ψ_0 is constructed in the first chapter.

2) In lemma 3.2 we have shown that one and only one of the following condition holds

$$\eta(x,k\omega,\gamma,\chi) = \eta(x,0,\gamma,\chi)$$

$$\eta(x,k\omega,\gamma,\chi) > \eta(x,0,\gamma,\chi)$$

$$\eta(x,k\omega,\gamma,\chi) < \eta(x,0,\gamma,\chi)$$

Let $\chi_1(x,t), \chi_2(x,t)$ be two different $k\omega$-periodic solutions of problem (1.1)-(1.3), where

$$\chi_1(x,t) < \chi_2(x,t), \qquad (x,t) \in Q_{k\omega}.$$

Lemma 3.3. Let the above assumptions of the lemma hold and such $\gamma > 0$ exist that

$$\eta(x, kw; -\gamma, \chi_2) > \eta(x, 0; -\gamma, \chi_2) > \chi_1(x, 0),$$

$$\eta(x, kw; -\gamma, \chi_1) < \eta(x, 0; -\gamma, \chi_1) < \chi_2(x, 0).$$

Then there exits a kw-periodic solution $\chi_3(x, t)$ of problem (1.1)-(1.3), where

$$\chi_1(x, t) < \chi_3(x, t) < \chi_2(x, t), \qquad (x, t) \in Q_{kw}. \qquad (3.6)$$

Proof. If one of the solutions $\eta(x, t; -\gamma, \chi_2)$ or $\eta(x, t; -\gamma, \chi_1)$ (γ is sufficiently small, positive) converges for $t \to +\infty$ to $\chi_1(x, t)$ or $\chi_2(x, t)$ correspondingly, then the lemma is trivial.

Consider the case when $\eta(x, t; -\gamma, \chi_2)$ converges for $t \to +\infty$ in $C^{2+\alpha}(\bar{\Omega})$ to the solution $\chi_2(x, t)$, and $\eta(x, t; \gamma, \chi_1)$ converges to $\chi_1(x, t)$. Define the following sets

$$Y = \{u_0 \in E; \chi_1(x, 0) \le u_0 \le \chi_2(x, 0)\},$$

$$Y_1 = \{u_0 \in E; \chi_1(x, 0) \le u_0 \le \eta(x, 0, \gamma, \chi_1)\},$$

$$V_1 = \{u_0 \in E; \chi_1(x, 0) < u_0 < \eta(x, 0, \gamma, \chi_1)\},$$

The sets Y_2 and V_2 may be defined analogously using the inequalities like

$$\eta(x, 0, \gamma, \chi_2) \le u_0 \le \chi_2(x, 0).$$

Consider the mapping

$$G : u_0(x) \to u(x, kw; u_0(x)).$$

By the comparison theorems, we obtain

$$G(Y) \subset Y, \quad G(Y_i) \subset Y_i, \quad G(V_i) \subset V_i, \quad i = 1, 2.$$

If the above assumptions hold, the mapping G has no fixed points in $\{Y_i \setminus V_i\}$. Therefore, applying theorem 14 from Amann (1976), we obtain that G has fixed points in the set $Y \setminus (Y_1 \cup Y_2)$. Denote one of these fixed points by $\chi_3(x)$. By the definition of the set $\{Y \setminus (Y_1 \cup Y_2)\}$, we obtain that

$$\chi_1(x,0) < \chi_3(x) < \chi_2(x,0).$$

Denote the solution of (1.1)-(1.3) with the initial data $\chi_3(x)$ by $\chi_3(x,t)$. As $\chi_3(x)$ is a fixed point of the mapping G, then $\chi_3(x,t)$ is a $k\omega$-periodic solution of the problem in question and

$$\chi_3(x,k\omega) = \chi_3(x,0) = \chi_3(x).$$

Inequality (3.6) follows from the comparison theorem. The lemma is proved.

We shall say that a periodic solution $\varphi(x,t)$ of problem (1.1)-(1.3) is isolated from above if there exists such $\varepsilon > 0$ that for each periodic solution $\psi(x,t)$ of (1.1)-(1.3), so that

$$\psi(x,t) \geq \varphi(x,t) \quad (x,t) \in Q$$

the inequality

$$\| \psi(x,t) - \varphi(x,t) \|_1^Q > \varepsilon$$

holds, only if $\psi(x,t)$ is not equal to $\varphi(x,t)$.

Analogously we may define an isolated from below periodic solution of (1.1)-(1.3).

Proof of theorem 3.1. First note, that since $\varphi(x,t)$ is an attracting from above periodic solution of problem (1.1)-(1.3), then $\varphi(x,t)$ is an isolated from above solution of (1.1)-(1.3). Consider the solution $\eta(x,t;\gamma,\varphi)$ which was determined earlier, after the proof of lemma 3.2. For sufficiently small positive γ, the inequality

$$\eta(x,0;\gamma,\varphi) \neq \eta(x,k\omega;\gamma,\varphi)$$

holds, since $\varphi(x,t)$ is isolated from above. By lemma 3.2, only two variants are possible: either

$$\eta(x, k\omega; \gamma, \varphi) > \eta(x, 0; \gamma, \varphi), \quad x \in \Omega;$$

or

$$\eta(x, k\omega; \gamma, \varphi) < \eta(x, 0; \gamma, \varphi), \quad x \in \Omega.$$

We must show that if $\varphi(x,t)$ is an attracting from above solution, then for sufficiently small γ

$$\eta(x, k\omega; \gamma, \varphi) < \eta(x, 0; \gamma, \varphi), \quad x \in \Omega.$$

Really, if we assume the contrary, then such a sequence $\gamma_n \to 0$, $\gamma_n > 0$ exists that

$$\eta(x, k\omega; \gamma_n, \varphi) > \eta(x, 0; \gamma_n, \varphi), \quad x \in \Omega.$$

By lemma 3.2, each $\eta(x, t; \gamma_n, \varphi)$ converges for $t \to +\infty$ to the greater than $\eta(x, t; \gamma_n, \varphi)$ $k\omega$-periodic solution.

Choose the initial data $u_0(x) > \varphi(x, 0)$, $x \in \Omega$ so that

$$\lim_{t \to +\infty} \| u(x, t; u_0) - \varphi(x, t) \|_1^{\Omega} = 0.$$

Let $\gamma_n > 0$ be such that

$$u_0(x) > \eta(x, 0; \gamma_n, \varphi) > \varphi(x, 0), \quad x \in \Omega.$$

From the comparison theorem it follows that

$$u(x, t; u_0(x)) > \eta(x, t; \gamma_n, \varphi) > \varphi(x, t), \quad (x, t) \in Q,$$

and the solution $\eta(x, t; \gamma_n, \varphi)$ converges for $t \to +\infty$ to a $k\omega$-periodic solution different from $\varphi(x, t)$. This contradicts the relation

$$\lim_{t \to +\infty} \| u(x, t; u_0) - \varphi(x, t) \|_1^{\Omega} = 0.$$

Consequently, for $0 < \gamma < \gamma_0$ the inequality

$$\eta(x, k\omega; \gamma_n, \varphi) < \eta(x, 0; \gamma_n, \varphi)$$

will hold true.

By the number γ_0, we find $\delta > 0$ such that

$$\| u_0(x) - \varphi(x, 0) \|_1^\Omega < \delta, \quad u_0(x) \in E, \quad u_0(x) \geq \varphi(x, 0), \quad x \in \Omega$$

yields

$$\eta(x, 0; \gamma_0, \varphi) > u_0(x) \geq \varphi(x, 0), \quad x \in \Omega.$$

Applying the comparison theorem, we see that $u(x, t, u_0)$ converges to $\varphi(x, t)$ in $C(\bar{\Omega})$. Then, taking into account (1.4) and the smoothness restrictions on the problem, we obtain that $\varphi(x, t)$ is stable from above. If $\varphi(x, t) = M(x, t)$, then the theorem is proved.

Let

$$\varphi(x, t) < M(x, t), \qquad (x, t) \in Q_{k\omega}.$$

Applying to $\eta(x, t; -\gamma, M)$ the same considerations as for $\eta(x, t; \gamma, \varphi)$, we see that for $0 < \gamma < \gamma_0$ the inequality

$$\eta(x, k\omega; -\gamma, M) > \eta(x, 0; -\gamma, M), \quad x \in \Omega$$

holds. Therefore, by lemma 3.3, between the periodic solutions M and φ is at least one $k\omega$-periodic solution. Consider the set of all $k\omega$-periodic solutions which are greater than the periodic solution $\varphi(x, t)$. Denote the least of them (such, evidently, exists) by $\psi(x, t)$. The statement that the solution $\psi(x, t)$ is the least solution greater that $\varphi(x, t)$ means the following: between the periodic solutions $\psi(x, t)$ and $\varphi(x, t)$ there are no $k\omega$-periodic solutions, and

$$\psi(x, t) > \varphi(x, t), \qquad (x, t) \in Q_{k\omega}.$$

We nou show that $\psi(x,t)$ satisfies the conditions formulated in theorem 3.1. We need to prove that $\psi(x,t)$ is unstable from below and the
set $A(\varphi)$ contains the set

$$\{u_0(x) \in E, \quad \psi(x,0) \geq u_0(x) \geq \varphi(x,0), \quad u_0(x) \not\equiv \psi(x,0)\}.$$

Consider the solution $\eta(x, k\omega; -\gamma, \psi)$, where $0 < \gamma < \gamma_0$, with γ_0
being sufficiently small and positive. There are three cases (3.5) of
lemma 3.2. The first and the second cases are impossible, since between
$\psi(x,t)$ and $\varphi(x,t)$ there would be a $k\omega$-periodic solution. In the third
case, $\eta(x, k\omega; -\gamma, \psi)$ may converge only to $\varphi(x,t)$. As γ is arbitrary,
the attraction domain of $\varphi(x,t)$ contains the set

$$\{u_0(x) \in E, \quad \psi(x,0) \geq u_0(x) \geq \varphi(x,0), \quad u_0(x) \not\equiv \psi(x,0)\},$$

and $\psi(x,t)$ is an unstable stationary solution. The theorem is proved.
Theorem 3.1' may be proved analogously to theorem 3.1.

Corollary 3.1.   If

$$u^+(x,t) \not\equiv M(x,t), \quad u^+(x,t) > M(x,t), \quad (x,t) \in Q_{k\omega},$$

then the solution $M(x,t)$ in theorem 2.1 is stable from above.
Similary, if

$$u^-(x,t) \not\equiv m(x,t), \quad u^-(x,t) < m(x,t), \quad (x,t) \in Q_{k\omega},$$

then the solution $m(x,t)$ is stable from below.

§5.4   The   classification   of   periodic solutions.

Consider the situation which arises for the problem (1.1)-(1.3) after
the establishment of theorem 3.1: $\varphi(x,t)$ is the stable from above $k\omega$-

periodic solution, and $\psi(x,t)$ is the unstable from below $k\omega$-periodic solution of this problem.

Let $\eta(x,t;-\gamma,\psi)$, $0 < \gamma < \gamma_0$ be a solution which converges for $t \to +\infty$ to $\varphi(x,t)$. It may, be extended for all negative $t < 0$. We can show, for example, how $\eta(x,t;-\gamma,\psi)$ may be extended into the interval $[-k\omega,0]$. Take, for this purpose, the mapping $G : [0,\gamma_0] \to [0,\gamma_0]$ from lemma 3.2. Let γ belong to the image of $[0,\gamma_0]$ under the mapping G. As it was shown in lemma 3.2, G is monotone; therefore, the image of the interval $[0,\gamma_0]$ is a certain interval $[0,\tilde{\gamma}_0]$. Consider the solution

$$\eta(x,t) = \eta(x,t+k\omega;G^{-1}(\gamma),\psi).$$

The function $\eta(x,t)$ is defined for all $t \in [-k\omega,0]$. Note, that

$$\eta(x) = \eta(x,k\omega;G^{-1}(\gamma),\psi) = \eta(x,0;\gamma,\psi).$$

Therefore, if we set

$$\eta(x,t;\gamma,\psi) = \eta(x,t), \quad t \in [-k\omega,0],$$

then the solution $\eta(x,t;\gamma,\psi)$ will be smoothly extended into the interval $[-k\omega,0]$. Analogously, $\eta(x,t;\gamma,\psi)$ may be extended into the interval $[-2k\omega,-k\omega]$ and so on. Note, that in this case

$$\| \psi(x,t) - \eta(x,t;\gamma,\psi) \|_{2+\alpha}^{\Omega} \to 0 \quad \text{for} \quad t \to \infty.$$

For the autonomous problem the sense of solutions $\eta(x,t;\gamma,\psi)$ is more clear. Let us consider this case more explicitly.

Let $\varphi(x)$ be a stable from above stationary solution of (1.1)-(1.3) and $\psi(x)$ be an unstable from below stationary solution of this problem. The following theorem holds.

Theorem 4.1. (Vishnevskii, 1990a, b; 1993b). For the above assumptions, the solution $\eta(x,t;\gamma,\psi)$ is a unique monotone solution of the autonomous problem which lies between stationary solutions $\varphi(x,t), \psi(x,t)$ and

$$\lim_{t \to -\infty} \| \eta(x,t;\gamma,\psi) - \psi(x) \|_{2+\alpha}^{\Omega} = 0,$$

$$\lim_{t \to +\infty} \| \eta(x,t;\gamma,\psi) - \varphi(x) \|_{2+\alpha}^{\Omega} = 0. \qquad (4.1)$$

Proof.　While constructing the solution $\eta(x,t;\gamma,\psi)$ (see lemma 3.2 and the definition after it) in the case of the ω-periodic in time problem (1.1)-(1.3), we have shown that

$$\eta(x, k\omega; \gamma, \psi) > \eta(x, 0; \gamma, \psi), \qquad x \in \Omega.$$

In the autonomous case, as ω, we may take any positive number; therefore, $\eta(x,t;\gamma,\psi)$ is monotonically increasing. The existence of the limits (4.1) for $\eta(x,t;\gamma,\psi)$, we have established earlier. Note, that for all $t \in R$, $\eta(x,t;\gamma,\psi)$ is less than the stationary solution $\psi(x)$ and converges to the stationary solution $\psi(x)$ for $t \to -\infty$. We must show that $\eta(x,t;\gamma,\psi)$ is a unique solution of (1.1)-(1.3) which is determined for all t and has the above properties.

Linearize problem (1.1)-(1.3) on the stationary solution $\psi(x)$. As the result we obtain

$$v_t = L(x,\psi)v, \qquad (x,t) \in Q;$$

$$B_1(x,\psi)v = 0, \qquad (x,t) \in \Gamma;$$

$$v(x,0) = v_0(x).$$

Let $\rho(x)$ be a positive solution of the problem

$$0 = L(x,\psi)\rho - \lambda\rho, \qquad x \in \Omega;$$

$$B_1(x,\psi)\rho = 1, \qquad x \in \partial\Omega.$$

Take λ sufficiently large, so that the spectrum of the auxiliary problem is lying in the left half-plane.

Make the change

$$w(x,t)\rho(x) = u(x,t) - \psi(x).$$

The new unknown function will solve the problem

$$w_t = L_1(x,\psi)w + F(x,w), \qquad (x,t) \in Q;$$

$$B_2(x,\psi)w = 0. \qquad (x,t) \in \Gamma; \qquad (4.2)$$

$$w(x,0) = w_0(x) = \big(u_0(x) - \psi(x)\big)/\rho(x).$$

The operators L_1 and B_2 are determined in §1.1 of this chapter.

There exists an ε_0 a neighborhood of zero in the space $H_1^{2+\alpha}(Q_{T+\tau}^\tau)$, such that the mappings F, G satisfy in this neighborhood all the restrictions which were imposed on the nonlinear terms in the first chapter, where ε_0 does not depend on T, τ. Let the spectral linearized problem

$$\lambda z = L_1(x,\psi)z, \qquad x \in \Omega;$$

$$B_2(x,\psi)z = 0, \qquad x \in \partial\Omega \qquad (4.3)$$

have the dimension of the eigenspace $E^0 \oplus E^+$ equal to m.

Choose in this space a foundation consisting of the eigenfunctions and adjoint eigenfunctions

$$\xi_1(x), \xi_2(x), \ldots, \xi_m(x),$$

where $\xi_1(x)$ is the eigenfunction corresponding to the eigenvalue with the greatest real part. Denote this eigenvalue, as earlier, by λ_1. By the Krein-Rutman theorem, $\xi_1(x) > 0$, $x \in \Omega$. All the other eigenfunctions and adjoint functions

$$\xi_2(x), \ldots, \xi_m(x), \xi_{m+1}(x), \ldots$$

change sign in Ω.

By theorem 3.3 from chapter 1, in order that $u(x, t; u_0)$ converges to a stationary solution $\psi(x)$ for $t \to -\infty$, it is necessary that

$$u_0(x) = \sum_{i=1}^{m} \theta_i \xi_i(x) + \Psi_0^+ \left(\sum_{i=1}^{m} \theta_i \xi_i(x) \right).$$

Moreover, for all $t \leq 0$ the solution $u(x, t; u_0)$ has to lie in the integral set

$$M_0^+ = \left\{ \sum_{i=1}^{m} \theta_i \xi_i(x) + \Psi_0^+ \left(\sum_{i=1}^{m} \theta_i \xi_i(x) \right); \quad \theta_i \in R \right\}.$$

Denote the projection of $u(x, t; u_0)$ into the subspace $E^0 \oplus E^+$ by $\sum_{i=1}^{m} \theta_i(t) \xi_i(x)$. Then

$$u(x, t; u_0) = \sum_{i=1}^{m} \theta_i(t) \xi_i(x) + \Psi_0^+ \left(\sum_{i=1}^{m} \theta_i(t) \xi_i(x) \right).$$

Substituting this relation into (1.1)-(1.3) and taking into account that the functions $\vec{\theta}(t)$ are continuously differentiable by time, we obtain for the system of ordinary differentiale equations of the order m

$$\theta_i'(t) = \lambda_i \theta_i + \Phi_i(\vec{\theta}(t)), \quad i = 1, \ldots, m. \tag{4.4}$$

In this case

$$\left\| \vec{\Phi}(\vec{\theta}_1) - \vec{\Phi}(\vec{\theta}_2) \right\| \leq Cl \left\| \vec{\theta}_1 - \vec{\theta}_2 \right\|,$$

where $\| \cdot \|$ is the norm in R^m.

To study this system (4.4) use the results of §8, chapter 10 from Hartman (1970). It was shown there that for a sufficiently small fixed l, we may select such positive number ε_1, that if the solution $\vec{\theta}(t)$ converges for $t \to -\infty$ to zero faster than $e^{(\lambda_1 - \varepsilon_1)t}$, then

$$\theta_1(t) = 0 \left(\sum_{i=2}^{m} \theta_i^2(t) \right)^{\frac{1}{2}}, \quad t \to -\infty.$$

Furthermore, it was shown there, that problem (4.4) has only one solution which converges to zero more slowly than $e^{(\lambda_1 - \varepsilon_1)t}$ for $t \to -\infty$.

Note, that the real number λ_1 is the eigenvalue with the maximal real part of the problem formally conjugated with problem (4.3). The eigenfunction denoted as $\xi_1^*(x)$ corresponds to this eigenvalue; $\sum\limits_{i=2}^{m} \theta_i(t)\xi_i(x)$ is always orthogonal to $\xi_1^*(x)$, therefore, is always alternating in Ω.

Inequality (1.35) yields that if $u(x, t; u_0)$ converges to the stationary solution $\psi(x)$ faster than $e^{(\lambda_1 - \varepsilon_1)t}$ for $t \to -\infty$, then this solution

$$u(x, t; u_0) = \psi(x) - \sum_{i=1}^{m} \theta_i(t)\xi_i(x) + \Psi_0^+\left(\sum_{i=1}^{m} \theta_i(t)\xi_i(x)\right)$$

has to be alternating in Ω. In other words, the difference $u(x, t; u_0) - \psi(x)$ for sufficiently small t becomes alternating, and $u(x, t; u_0)$ converges to the stationary solution $\psi(x)$. This means that $u(x, t; u_0)$ can not be monotonic at all moments. This also means that any other solution can not be monotonic at all moments. The theorem is proved.

Apply the obtained results for classification of periodic solutions which lie between the periodic solutions $M(x, t)$ and $m(x, t)$. To simplify considerations, assume, in addition, that the problem in question has only a finite number of periodic solutions; and the shift operator on the period of the linearized problem on the periodic solution has no eigenvalues out of the unit circle of the complex plane. In other words, if $\varphi(x, t)$ is a $k\omega$-periodic solution lying between $M(x, t)$ and $m(x, t)$, then the shift operator $V(k\omega)$ of the problem

$$v_t = L(x, t, \varphi)v, \qquad (x, t) \in Q;$$

$$B_1(x, t, \varphi)v = 0, \qquad (x, t) \in \Gamma;$$

$$v(x, 0) = v_0(x)$$

has no eigenvalues out of the unit circle.

If

$$M(x, t) \neq u^+(x, t), \qquad m(x, t) \neq u^-(x, t),$$

then, for the above assumptions, $M(x, t), m(x, t)$ are stable ω-periodic solutions of the problem.

Note, that since the monotone solution $\eta(x, t, \gamma, \psi)$ is unique, we obtain that for each sufficiently small γ_1 there exists such t_1 that $\eta(x, t + t_1, \gamma, \psi) \equiv \eta(x, t, \gamma_1, \psi)$.

Therefore, we may denote by $\eta(x, t; \varphi, \psi)$ the solution of (1.1)-(1.3), which we have considered in theorem 4.1 and in this paragraph, and which either converges to the periodic solution $\varphi(x, t)$ for $t \to -\infty$ and to the periodic solution $\psi(x, t)$ for $t \to +\infty$, or, conversely, solution $\eta(x, t, \varphi, \psi)$ converges to $\psi(x, t)$ for $t \to -\infty$ and to $\varphi(x, t)$ for $t \to +\infty$.

The solution which tends for $t \to -\infty$ to one stationary solution and for $t \to +\infty$ to another is called a connected orbit.

Connected orbits for evolutionary problems were investigated by many authors. The most complete study for the case of parabolic equation with one spatial variable may be found in Brunovskii and Fiedler (1988, 1989).

Denote by $\eta(x, t, \gamma, \psi)$ the monotone connected orbit which connects the stationary solutions φ and ψ. Theorem 4.1 yields that this monotone orbit is unique (to within the shift, since the solution $\eta(x, t + \gamma, \varphi, \psi)$, γ is a constant, is also a monotone connecting the stationary solutions φ and ψ, where the sets $\{\eta(x, t, \varphi, \psi), \quad t \in R\}$ and $\{\eta(x, t + \gamma, \varphi, \psi), \quad t \in R\}$ coincide).

There may be another connected orbits which connect the stationary solutions φ and ψ. But the monotone connected orbits are of great importance, since they give the partial order in the set of stationary solutions if the set of this solutions is finite.

Let us formulate more explicitly how this partial order is given. Consider the set of periodic solution of problem (1.1)-(1.3)

$$M(x, t) < \varphi_1(x, t) < \varphi_2(x, t) < \ldots < \varphi_l(x, t) < m(x, t), \quad (x, t) \in Q,$$

such that for each two neighboring solutions $\varphi_l(x, t), \varphi_{l+1}(x, t)$ there exists a solution $\eta(x, t; \varphi_l, \varphi_{l+1})$ which connects them. We denote such a set by a contour connecting the greatest and least periodic solutions.

The statements proved above allow us to formulate the following theorem.

Theorem 4.2. (Vishnevskii, 1990a, b; 1993b). If the above assumptions hold, then problem (1.1)-(1.3) has a finite number of con-

tours which connect the solutions $M(x,t)$ and $m(x,t)$. In each contour
the stable and unstable solutions alternate. Each periodic solution belongs to at least one contour.

§5.5 Solutions, periodic in time, of the mixed problems for autonomous parabolic equations.

In this paragraph we shall construct two examples important for under-
standing the results obtained in the previous chapters. First we shall
describe these examples. Let

$$u_t = \sum_{i,j=1}^n a_{ij}(x,u,\nabla u)\frac{\partial^2 u}{\partial x_i \partial x_j} + a(x,u,\nabla u), \quad (x,t) \in Q;$$

$$u(x,t) = 0, \quad (x,t) \in \Gamma; \qquad u(x,0) = u_0(x). \tag{5.1}$$

In the first paragraph we have shown that each periodic solution of
(5.1) is unstable. Here we shall construct an example of (5.1) which has
infinitely many periodic solutions. The set of these periodic solutions
together with those solutions which converge to them for $t \to +\infty$ is
the surface of the unit codimension. This surface divides the attraction
domains of two stationary solutions. It follows from theorem 3.1 that
(5.1) has a stationary solution which also divides the attraction domains
mentioned above. Thus, in this example, the main part of the boundary
of the attraction domain of two stable stationary solutions forms the
stable manifolds of periodic solutions.

By the stable manifold of a periodic solution here is meant, as usual,
the manifold of initial data $u_0(x)$ such that $u(x,t;u_0)$ converges to the
chosen periodic solution as $t \to +\infty$.

To simplify the presentation, we construct the example taking Ω as
the square $(0,\pi) \times (0,\pi)$. It will be clear, from the example, that an
analogous example may be constructed for each domain Ω such that the
second eigenvalue of the Laplace operator for the first boundary-value
problem has multiplicity two.

Consider the problem (Vishnevskii, 1990b).

$$u_t = \Delta u + \varepsilon \Big[b_1(x,y)u_x + b_2(x,y)u_y + c(\varepsilon)u \Big] + 5u,$$

$$(x,t) \in (0,\pi) \times (0,\pi) \times (0,+\infty); \tag{5.2}$$

$$u(x,0,t) = u(x,\pi,t) = u(0,y,t) = u(\pi,y,t) = 0;$$

$$u(x,y,0) = u_0(x,y)$$

and connected with it the spectral problem

$$\lambda v = \Delta v + \varepsilon \Big[b_1(x,y)v_x + b_2(x,y)v_y + c(\varepsilon)v \Big] + 5v,$$

$$(x,t) \in (0,\pi) \times (0,\pi) \times (0,+\infty); \tag{5.3}$$

$$v(x,0,t) = v(x,\pi,t) = v(0,y,t) = v(\pi,y,t) = 0.$$

For $\varepsilon = 0$, the problem (5.3) has the zero eigenvalue of multiplicity two. The corresponding eigenfunctions are $\sin 2x \sin y$, $\sin x \sin 2y$. Using the results of the first chapter, we shall show how to choose the functions $b_1(x,y), b_2(x,y)$ so that, for the small ε, problem (5.3) would have two complex eigenvalues. This will mean that problem (5.2) will have, for the small ε, a two-parameter family of periodic solutions, corresponding to these two complex numbers.

Parallel with problem (5.2) consider the nonlinear problem

$$u_t = \Delta u + \varepsilon \Big[b_1(x,y)u_x + b_2(x,y)u_y + c(\varepsilon)u \Big] + 5u + f(u),$$

$$(x,t) \in (0,\pi) \times (0,\pi) \times (0,+\infty); \tag{5.4}$$

$$u(x,0,t) = u(x,\pi,t) = u(0,y,t) = u(\pi,y,t) = 0;$$

$$u(x,y,0) = u_0(x,y),$$

with

$$f(u) = \begin{cases} 0, & \text{for } |u| < 1; \\ \\ -6u, & \text{for } |u| > 2, \end{cases}$$

where $f(u)$ is three times continuously differentiable. Such a function $f(u)$ may be easily constructed (see, for example, Hartman, 1970).

It is easy to show, taking as the upper and lower functions $\pm C$, $C > 2$, that problem (5.4) is dissipative. As we have shown before, each periodic solution of the problem is unstable; therefore, it divides the attraction domains of two stable stationary solutions. On the other hand, we have proved that there is a stationary solution which divides the attraction domains of the two mentioned stable stationary solutions. In the example in question, as it is easy to see, such stationary solution is the trivial solution.

Now we show how to choose $b_1(x,y), b_2(x,y)$ so that problem (5.2) would have a two-parameter family of periodic solutions.

For problem (5.2), theorem 3.5 chapter 1 holds. Eigenfunctions of problem (5.3) corresponding to the zero eigenvalue, for $\varepsilon = 0$, are equal to $\sin x \cdot \sin 2y$ and $\sin 2x \cdot \sin y$; therefore, the integral set M_0 constructed in theorem 3.5 is as follows

$$\theta_1 \sin x \cdot \sin 2y + \theta_2 \sin 2x \cdot \sin y + \Psi_0(\theta_1, \theta_2, \varepsilon).$$

As the operator Δ is self-ajoint, the following relations

$$\int_0^\pi \int_0^\pi \Psi_0(\theta_1, \theta_2, \varepsilon) \sin x \sin 2y \, dx \, dy = 0, \tag{5.5}$$

$$\int_0^\pi \int_0^\pi \Psi_0(\theta_1, \theta_2, \varepsilon) \sin 2x \sin y \, dx \, dy = 0$$

hold.

It is easy to show that

$$\left\| \Psi_0(\theta_1, \theta_2, \varepsilon) - \Psi_0(\tilde{\theta}_1, \tilde{\theta}_2, \varepsilon) \right\|_{2+\alpha}^\Omega \leq C\varepsilon \left(\left| \theta_1 - \tilde{\theta}_1 \right| + \left| \theta_2 - \tilde{\theta}_2 \right| \right). \tag{5.6}$$

Lemma 5.1. The mapping $\Psi_0(\theta_1, \theta_2, \varepsilon)$ constructed for problem (5.2) linearly depends on θ_1, θ_2.

Proof. All solutions which lie in M_0 for each $\gamma > 0$ satisfy the inequality (3.13) from theorem 3.5 chapter 1

$$\left\| u(x, t; u_0) e^{-\gamma |t|} \right\|_{1;2+\alpha}^{\Omega} \leq C(\gamma) \left\| u_0 \right\|_1^{\Omega}. \tag{5.7}$$

The inverse statement also holds: if solution $u(x, t; u_0)$ satisfies (5.7) then $u_0(x) \in M_0$.

Let $u_{10}(x), u_{20}(x) \in M_0$, then the solutions $u(x, t; u_{i0}(x))$, $i = 1, 2$, satisfy inequality (5.7); therefore, their linear combination solves the problem and satisfies inequality (5.7). Note, that since the problem is linear for $|u| < 1$

$$\theta_1(x, t; u_{10}(x)) + \theta_2(x, t; u_{20}(x)) = u(x, t; \theta_1 u_{10}(x) + \theta_2 u_{20}(x)),$$

and, since this solution satisfies inequality (5.7), then

$$\theta_1 u_{10}(x) + \theta_2 u_{20}(x) \in M_0.$$

Set

$$u_{10}(x) = \sin x \sin 2y + \Psi_0(1, 0, \varepsilon),$$

$$u_{20}(x) = \sin 2x \sin y + \Psi_0(0, 1, \varepsilon).$$

As $\theta_1 u_{10}(x) + \theta_2 u_{20}(x) \in M_0$, then

$$\theta_1 \sin x \sin 2y + \theta_1 \Psi_0(1, 0, \varepsilon) + \theta_2 \sin 2x \sin y + \theta_2 \Psi_0(0, 1, \varepsilon)$$

$$= \theta_1 \sin x \cdot \sin 2y + \theta_2 \sin 2x \cdot \sin y + \Psi_0(\theta_1, \theta_2, \varepsilon).$$

This relation proves the lemma.

Denote the bounded mapping $\Psi_0(1, 0, \varepsilon)/\varepsilon$ acting from the neighborhood of zero into the space $C^{2+\alpha}(\bar{\Omega})$ by $\Psi_1(\varepsilon)$ and the mapping

$\Psi_0(0, 1, \varepsilon)/\varepsilon$ from the neighborhood of zero into the space $C^{2+\alpha}(\bar{\Omega})$ by $\Psi_2(\varepsilon)$. Then

$$\Psi_0(\theta_1, \theta_2, \varepsilon) = \varepsilon\Psi_1(\varepsilon)\theta_1 + \varepsilon\Psi_2(\varepsilon)\theta_2.$$

If $u_0 \in M_0$, then the solution $u(x, t; u_0)$ lies in the integral set M_0 for all t from the time interval, where the solution exists. Therefore,

$$\theta_1(t) = \int_0^\pi \int_0^\pi u(x, t; u_0) \sin x \sin 2y \, dx dy,$$

$$\theta_2(t) = \int_0^\pi \int_0^\pi u(x, t; u_0) \sin 2x \sin y \, dx dy.$$

This means that the solution may be represented as

$$u(x, t; u_0) = \theta_1(t) \sin x \sin 2y + \theta_2(t) \sin 2x \sin y$$

$$+ \varepsilon\theta_1(t)\Psi_1(\varepsilon) + \varepsilon\theta_2(t)\Psi_2(\varepsilon). \tag{5.8}$$

Substituting (5.8) into (5.2), we obtain

$$\theta_1'(t) = \varepsilon \int_0^\pi \int_0^\pi \Big\{ b_1(x, y) \Big[\theta_1(t) \cos x \sin 2y + \theta_2(t) 2 \cos 2x \sin y$$

$$+ \varepsilon\theta_1 (\Psi_1(\varepsilon))_x + \varepsilon\theta_2 (\Psi_2(\varepsilon))_x \Big] + b_2(x, y) \Big[\theta_1(t) \sin x 2 \cos 2y$$

$$+ \theta_2(t) \sin 2x \cos y + \varepsilon\theta_1 (\Psi_1(\varepsilon))_y + \varepsilon\theta_2 (\Psi_2(\varepsilon))_y \Big] \Big\}$$

$$\times \sin x \sin 2y \, dx dy + \varepsilon c(\varepsilon)\theta_1(t). \tag{5.9}$$

The analogous differential equation may be written for $\theta_2'(t)$. The explicit formulas for $b_1(x, y), b_2(x, y)$ will be as follows

$$b_1(x, y) = -\frac{40\omega}{27\pi} \sin x \sin 2y + \frac{40\gamma}{9\pi} \sin 2x \sin y,$$

$$b_2(x, y) = -\frac{40\omega}{9\pi} \sin x \sin 2y + \frac{40\gamma}{27\pi} \sin 2x \sin y. \qquad (5.10)$$

Substitute formulas (5.10) into (5.9). As the result, we obtain

$$\theta_1'(t) = \varepsilon \Big[(\gamma + c(\varepsilon))\theta_1 + \omega\theta_2 + \varepsilon k_{11}(\varepsilon)\theta_1 + k_{12}\varepsilon\theta_2 \Big],$$

$$\theta_2'(t) = \varepsilon \Big[(\gamma + c(\varepsilon))\theta_2 - \omega\theta_1 + \varepsilon k_{21}(\varepsilon)\theta_1 + k_{22}\varepsilon\theta_2 \Big]. \qquad (5.11)$$

Note, that in the general case, if the problem has for $\varepsilon = 0$ the zero as the eigenvalue of multiplicity two, and $\varphi_1(x, y), \varphi_2(x, y)$ are the eigenfunctions corresponding to this zero eigenvalue, then $b_1(x, y), b_2(x, y)$ also may be chosen in the form

$$b_1(x, y) = m_{11}\varphi_1(x, y) + m_{12}\varphi_2(x, y),$$

$$b_2(x, y) = m_{21}\varphi_1(x, y) + m_{22}\varphi_2(x, y),$$

where m_{ij}; $i, j = 1, 2$, are certain indeterminate coefficients. These coefficients for this problem may be obtained explicitly.

For the small $\varepsilon \neq 0$, the problem (5.11), as follows from the theory of ordinary differential equations (Hail, 1962; Hartman, 1970), has a two-parameter family of periodic solution, if $c(\varepsilon)$ is chosen so that problem (5.11) for small ε has two complex eigenvalues, the real parts of which are equal to zero.

Note, that if $\big(\theta_1(t), \theta_2(t)\big)$ is the periodic solution of problem (5.11), then

$$\theta_1(t) \sin x \sin 2y + \theta_2(t) \sin 2x \sin y + \varepsilon\theta_1(t)\Psi_1(\varepsilon) + \varepsilon\theta_2(t)\Psi_2(\varepsilon)$$

is the periodic solution of problem (5.2). So, the desired example is constructed.

Let $\omega_1(\varepsilon)$ be the period of a periodic solution of (5.2) for a certain fixed ε. Consider instead of (5.4) the problem

$$u_t = \Delta u + \varepsilon b_1(x,y)u_x + \varepsilon b_2(x,y)u_y + (c(\varepsilon) - 5)u + \varepsilon f(u)\sin(\lambda t),$$

$$(x,t) \in (0,\pi) \times (0,\pi) \times (0,+\infty); \quad u(x,y,0) = u_0(x,y);$$

$$u(x,0,t) = u(x,\pi,t) = u(0,y,t) = u(\pi,y,t) = 0. \qquad (5.12)$$

The coefficients $b_1(x,y), b_2(x,y)$ are determined by (5.10), the constant $c(\varepsilon)$ is determined so that problem (5.12) has two pure imaginary eigenvalues equal to $i\omega$. If the number $2\pi/\lambda$ is incommensurable with ω then, as may be shown using the disturbances method, problem (5.12) has a two-parameter family of periodic solutions with the period incommensurable with the period of the right hand sides of the equations.

Note, in conclusion, that some analogous and even more complicated examples of parabolic equations have been considered in recently published articles by Danser (1991), Polačik (1993) and by Rubakovskii (1994).

Chapter 6

On stabilization of mixed problem solutions for autonomous quasilinear parabolic equations.

In this chapter we shall prove that if the problem is dissipative, then for an open and dense set of initial data M, the solution $u(x, t; u_0)$, if $u_0 \in M$, is stabilized.

Further research, in particular, the examples of Vishnevskii (1990a), Polačik (1992), and Polačik and Rubakovskii (1994), have shown that this result could not be improved for the parabolic equation with two or more spatial variables.

Lions (1984) suggested stabilization of not every bounded solution but of those bounded solutions for which the initial data belong to an open set: Lions (1984) realized this idea for the semi-linear parabolic equation. For the general quasilinear parabolic equation this result was proved by Vishnevskii (1987, 1988); and for weakly connected cooperative parabolic systems, by Vishnevskii (1992).

Another approach was considered by Hirsch (1988). There was selected a class of evolutionary problems for which the ω-limit set contains only stationary solutions. Under certain rather essential assumptions this class contains quasilinear autonomous parabolic equations. For these equations Polačik (1989a) proved that if u_0 belongs to an open dense set M, then the solution constructed by these initial data is stabilized. For this case of nonperiodic solutions the results on stabilization for almost all initial data has been considered by Poláčik and Tereščák (1991, 1993) and others.

In this chapter we shall consider the solutions which, beginning at a certain moment, become strictly monotone. The importance of such investigation has been pointed out by Hirsch (1988). Further results were obtained by Poláčik (1989a, b), Vishnevskii (1993b, 1994) and Mierczinskii (1991, 1994). In particular, here we present the results of Vishnevskii (1993c). If the problem analytically depends on an unknown function and on its gradient, and is dissipative, then for an open dense set of initial data M_1, the solution constructed by these initial data, from a certain moment, becomes strictly monotone.

Theorem 4.3 was proved by Vishnevskii (1987). Its generalization was expounded in Vishnevskii (1992, 1993a). A similar result was obtained in another way by Smith (1991).

§6.1   Setting   of   the   problem   and preliminary   results.

In this chapter we shall explore the autonomous problem

$$u_t = \sum_{i,j=1}^{n} a_{ij}(x,u,\nabla u)\frac{\partial^2 u}{\partial x_i \partial x_j} + a(x,u,\nabla u), \quad (x,t)\in Q, \qquad (1.1)$$

$$Bu = \alpha_1\left(\sum_{i=1}^{n} b_i(x)\frac{\partial u}{\partial x_i} + g(x,u)\right) + \alpha_0 u, \quad (x,t)\in \Gamma, \qquad (1.2)$$

$$u(x,0) = 0. \qquad (1.3)$$

Assume that for problem (1.1)-(1.3) all the restrictions, which were imposed in the first paragraph of chapter 5, hold. All the notations of chapter 5 are preserved.

Let problem (1.1)-(1.3) be dissipative, (the definition of dissipativity, see in previous chapters). Then, it is easy to show that each solution $u(x, t; u_0)$ of (1.1)-(1.3) with the initial data $u_0 \in E$ belongs to $H_1^{2+\alpha'}(Q)$, $0 < \alpha < \alpha' < 1$.

A function $\varphi(x)$ belongs to the ω-limit set $\omega(u_0)$ of solution $u(x, t; u_0)$ of problem (1.1)-(1.3) with the initial data $u_0 \in E$ if such a sequence $t_i \to +\infty$ exists that

$$\lim_{i \to \infty} \| u(x, t_i; u_0) - \varphi(x) \|_{2+\alpha}^{\Omega} = 0.$$

Note, that taking into account the restrictions imposed on the problem, if $\varphi(x) \in \omega(u_0)$, then $\varphi \in C^{4+\alpha}(\bar{\Omega})$.

If $u(x, t; u_0) \in H^{2+\alpha}(Q)$, then $\omega(u_0)$ is not empty.

Problem (1.1)-(1.3) with one spatial variable was considered considered, for example, Belonosov and Zelenyak (1975) and Fiedler (1989). There was proved (in different ways) the theorem of stabilization which states that if $u(x, t; u_0)$ is bounded in $C^2[0, 1]$ by a constant not depending on time, then $\omega(u_0)$ consists of one stationary solution. Further, the authors tried to transfer this property onto problem (1.1)-(1.3), i.e., to prove that each bounded solution of (1.1)-(1.3) is stabilized. These attempts failed, since the example constructed in the fifth paragraph of chapter 5 shows that without additional restrictions we have no stabilization of each bounded solution for $n = 2$ even for problem

$$u_t = \Delta u + \varepsilon b_1(x, y)u_x + \varepsilon b_2(x, y)u_y + (\varepsilon c(\varepsilon) - 5)u + f(u),$$

$$(x, t) \in Q; \quad u(x, t) = 0, \quad (x, t) \in \Gamma; \quad u(x, 0) = u_0(x). \qquad (1.4)$$

In this simplest case the nonlinearity of (1.4) depends only on the function; nevertheless, problem (1.4) has only periodic solutions. As we have proved in chapter 4, these periodic solutions have to be unstable.

It is not clear, whether each bounded solution of (1.4) would be stabilized if $b_1(x, t) = b_2(x, t) = 0$. In this case, the ω-limit set of each bounded solution consists only of stationary solutions of (1.4). But it is not clear whether $\omega(u_0)$ consists of the set of stationary solutions.

The analogous problem of stabilizing each bounded solution may be considered for the parabolic equations system. Consider the system of two parabolic equations with one spatial variable

$$u_t = u_{xx} + f_u(u, v); \quad v_t = v_{xx} + f_v(u, v) \qquad (1.5)$$

$$(x, t) \in Q = (0, 1) \times (0, +\infty);$$

$$u_x(0, t) = u_x(1, t) = v_x(0, t) = v_x(1, t);$$

$$u(x, 0) = u_0(x), \quad v(x, 0) = v_0(x).$$

All solutions $(\xi(t), \eta(t))$ of the system of ordinary differential equations

$$\xi(t) = f_u(\xi, \eta), \quad \eta(t) = f_v(\xi, \eta), \qquad (1.6)$$

are, simultaneously, solutions of (1.5) with the proper $u_0(x), v_0(x)$.

In Fokin (1981), parallel with the other results, an example of system (1.6) was constructed, where each nonstationary solution is not stabilized. At the same time, all the solutions are bounded and their ω-limit sets consist of the set of stationary solutions.

A somewhat different approach to the problem of stabilization of bounded solutions of parabolic equations was applied in Lions (1984). There was considered the following problem

$$u_t = \Delta u + f(x, u), \quad (x, t) \in Q;$$

$$u(x, t) = 0, \quad (x, t) \in \Gamma; \qquad (1.7)$$

$$u(x, 0) = u_0(x).$$

The following condition

$$\overline{\lim_{|u| \to \infty}} \frac{f(x, u)}{u} < \lambda_1(\Omega)$$

was assumed to hold. Here, $\lambda_1(\Omega)$ is the first eigenvalue of the Laplace operator in Ω. This condition allows us to prove that for (1.7) the ω-limit set of each bounded solution consists only of stationary solutions. Although we could not prove that each bounded solution is stabilized, Lions (1984) proved the following important property. The set of initial data $u_0(x)$ such that the constructed solution $u(x,t;u_0)$ is stabilized, i.e., the set $\omega(u_0)$ consists of a unique stationary solution, and contains an open and dense in E subset (in the norm of $C^1(\bar\Omega)$).

If we pass from (1.7) to problem (1.1)-(1.3), then the task will be more complicated. Really, the example constructed in §5 chapter 5, shows that (1.1)-(1.3) may have a periodic solution for which the ω-limit set contains this periodic solution and, therefore, does not contain only stationary solutions. It is clear that the periodic solution of the problem could not be stabilized.

The main result of this chapter (Vishnevskii, 1987, 1988) is that for (1.1)-(1.3) the analogous result, as was established for (1.7), holds true. In other words, the set of initial data $u_0(x) \in E$ of (1.1)-(1.3), such that $u(x,t;u_0)$ is stabilized, contains an open and dense in E subset.

Note, that even for parabolic systems of the form (1.5) without additional restrictions, the statement that the set of initial data, which determines the stabilizing solutions, contains on open and dense subset, is false.

Let us examine an example of the system, where almost all solutions are not stabilized but are bounded; and the ω-limit set of each solution contains only stationary solutions.

Following Vishnevskii (1987), construct $f(u,v)$ so that all solutions of (1.6) are bounded, but each nonconstant solution is not stabilized. Consider the parabolic system

$$u_t = u_{xx} + \varepsilon f_u(u,v); \quad v_t = v_{xx} + \varepsilon f_v(u,v) \qquad (1.8)$$

$$(x,t) \in Q = (0,1) \times (0,+\infty);$$

$$u_x(0,t) = u_x(1,t) = v_x(0,t) = v_x(1,t);$$

$$u(x,0) = u_0(x), \quad v(x,0) = v_0(x).$$

This system satisfies all the assumptions of theorem 1.6 chapter 1. Therefore, if $|\varepsilon| < \varepsilon_0$, ε_0 is a sufficiently small positive number, then each solution of (1.8) tends for $t \to +\infty$ to $(\xi(t/\varepsilon), \eta(t/\varepsilon))$. Here by $(\xi(t), \eta(t))$ we denote a solution of (1.6). Therefore, for an open dense set of initial data the solution of the problem (1.8) $(u(x, t; u_0, v_0), v(x, t; u_0, v_0))$ is bounded in $C^{2+\alpha}[0, 1] \times C^{2+\alpha}[0, 1]$ but is not stabilized.

For a long time it was not clear whether the stabilization theorem holds for the following parabolic equation

$$u_t - \Delta u = g(x, u), \qquad x \in \Omega, \quad t \geq 0,$$

$$u(x, t) = 0, \qquad (x, t) \in \Gamma,$$

$$u(x, 0) = u_0(x).$$

We may construct for this parabolic equation the Liapunov functional

$$\int_\Omega u_t^2 = \frac{d}{dt} \int_\Omega \left(\frac{|\nabla u|^2}{2} - G(x, u) \right) dx,$$

where

$$G'_u(x, u) = g(x, u).$$

Therefore, the ω-limit set for a bounded solution, as in the previous case, consists only of stationary solutions. There was the hypothesis that each bounded solution is stabilized. Nevertheless, Poláčik and Rubakovskii (1994) constructed an example of a bounded and not stabilized solution of the parabolic equation mentioned above.

However, if we consider weakly connected cooperative parabolic systems, then the result on stabilization of bounded solutions for almost all initial data will be hold true (see, for example, Vishnevskii, 1992b).

Another way of considering the problem of stabilizing the bounded solutions of evolutionary equations was suggested by Hirsch (1988) and Poláčik (1989a). The most important result was obtained by Hirsch (1988). There the nonlinear semigroup in the ordered Banach space was considered. It was proved that for some additional restrictions, for "almost all" initial data, the ω-limit set consists only of stationary solutions. In other words, almost all solutions of the evolutionary problem are quasistabilized (see definition in chapter 3). A further investigation was reported in Poláčik (1989b) (see also Poláčik and Tereščak, 1991, 1993) and the bibliography therein.

Let us investigate in greater detail the structure of $w(u_0)$ for problem (1.1)-(1.3).

Lemma 1.1.   Let $u(x, t; u_0) \in H_1^{2+\alpha}(Q)$ solve problem (1.1)-(1.3), $u_0(x) \in E$. Assume that $\varphi(x) \in w(u_0)$. Then the solution $u(x, t; \varphi(x))$ of (1.1)-(1.3) may be extended in the whole real axis, where

$$u(x, t; \varphi(x)) \in w(u_0) \quad \text{for} \quad -\infty < t < +\infty.$$

The set $w(u_0)$ is compact and connected.

Proof.   Let $\varphi(x) \in w(u_0)$. Then such a sequence $t_i \to +\infty$ exists that

$$\lim_{i \to \infty} \| u(x, t_i; u_0) - \varphi(x) \|_{2+\alpha}^{\Omega} = 0.$$

Assume that t is an arbitrary positive number, then

$$\| u(x, t + t_i; u_0) - u(x, t; \varphi(x)) \|_{2+\alpha}^{\Omega}$$

$$\leq C(t) \| u(x, t_i; u_0) - \varphi(x) \|_{2+\alpha}^{\Omega} \,.$$

Therefore, the sequence $u(x, t + t_i; u_0)$ converges to $u(x, t; \varphi(x))$ in the space $C^{2+\alpha}(\bar{\Omega})$.

Extend the solution $u(x, t; \varphi(x))$ for negative time. Firstly, extend the solution $u(x, t; \varphi(x))$ in the cylinder $\Omega \times (-1, 0)$. For this purpose consider the bounded in $C^{2+\alpha'}(\bar{\Omega})$ $(\alpha < \alpha' < 1)$ sequence $u(x, t_i - 1; u_0)$. Choose from this sequence the converging in $C^{2+\alpha}(\bar{\Omega})$ subsequence: its limit denote by $\varphi_1(x)$. Consider the solution $u(x, t; \varphi_1(x))$. The sequence $u(x, t_i; u_0)$ converges to $u(x, 1; \varphi_1(x))$ and to $\varphi(x)$; therefore,

$$u(x, 1; \varphi_1(x)) = \varphi(x).$$

Set

$$u(x, t; \varphi(x)) = u(x, t + 1; \varphi_1(x)) \quad \text{for} \quad 1 \le t \le 0.$$

Thus, we have extended the solution $u(x, t; \varphi(x))$ in the cylinder $\Omega \times (-1, 0)$. Analogously we may extend the solution into the cylinder $\Omega \times (-2, -1)$.

The sequence $u(x, t + t_i; u_0)$ converges to $u(x, t; \varphi(x))$ for each $t \in R$. Therefore,

$$u(x, t; \varphi(x)) \in \omega(u_0) \quad \text{for} \quad -\infty < t < +\infty.$$

The compactness of $\omega(u_0)$ follows from the boundedness of $\omega(u_0)$ in $C^{2+\alpha'}(\bar{\Omega})$ and from the connectedness of $\omega(u_0)$.

We now establish the connectedness of $\omega(u_0)$. Assume the contrary. Then

$$\omega(u_0) = \omega_1 \bigcup \omega_2,$$

where ω_1, ω_2 are two compact disjoint sets. Let $d > 0$ be the distance between them. There exist sequences, t_i^1, t_i^2 tending to $+\infty$ such that $t_i^1 < t_i^2 < t_{i+1}^1$, the distance between the function $u(x, t_i^1; u_0)$ and the set ω_1 is less than $d/3$, and the distance between $u(x, t_i^2; u_0)$ and ω_2 is less than $d/3$ for large enough i. By the solution continuity, there exists such t_i^3 than $t_i^1 < t_i^3 < t_i^2$, and the distance between $u(x, t_i^3; u_0)$ and the sets ω_1, ω_2 is greater than $d/3$. Choose, if necessary, from $u(x, t_i^3; u_0)$ the converging subsequence and denote its limit by $\varphi(x)$. By the definition of $\omega(u_0)$, we have $\varphi(x) \in \omega(u_0)$. On the other hand, the distance between the function $\varphi(x)$ and the sets ω_1, ω_2 is greater than $d/3$. Therefore, $\varphi(x)$ does not belong to ω_1 and ω_2. This contradicts the initial assumption that

$$\omega_1 \bigcup \omega_2 = \omega(u_0).$$

Therefore, the set $\omega(u_0)$ is connected. The lemma is proved.

§6.2 Stable ω-limit sets of solutions of the autonomous quasilinear parabolic equation.

In this section we shall investigate stable ω-limit sets of problem (1.1)-(1.3). The basisc result of this paragraph may be formulated as follows. We shall call a set $\omega(u_0)$ a stable ω-limit set, if there exists such $\delta > 0$, that from

$$\| u_0 - v_0 \|^{\Omega} < \delta$$

it follows that $\omega(u_0) = \omega(v_0)$. The stable ω-limit set of problem (1.1)-(1.3) may be only a stationary solution of this problem. The proof of this fact is based on certain theorems which are interesting in themselves.

Theorem 2.1. Let $u(x, t; u_0)$ solve (1.1)-(1.3) and belong to $H_1^{2+\alpha}(Q)$. Assume that such $t_1 \geq 0, \tau > 0$ exist that either

$$u(x, t_1 + \tau; u_0) \geq u(x, t_1; u_0), \quad x \in \Omega,$$

or

$$u(x, t_1 + \tau; u_0) \leq u(x, t_1; u_0), \quad x \in \Omega$$

hold.

Then, if
$$u(x, t_1 + \tau; u_0) \not\equiv u(x, t_1; u_0),$$

then the solution $u(x, t; u_0)$ of problem (1.1)-(1.3) is stabilized to the unique stationary solution $\varphi(x)$. In the first case, $\varphi(x)$ is asymptotically stable from below; in the second case, $\varphi(x)$ is asymptotically stable from above. Remind that we say that a stationary solution $\varphi(x)$ is stable from above, if for each $\varepsilon > 0$ there exists such $\delta > 0$ that from

$$\| \varphi - u_0 \|_1^{\Omega} < \delta, \qquad u_0 \geq \varphi(x)$$

it follows that the solution $u(x, t; u_0)$ is determined for all $t \geq 0$, where

$$\| u(x,t;u_0) - \varphi(x) \|_1^Q < \varepsilon.$$

If, in addition,

$$\lim_{t\to+\infty} \| u(x,t;u_0) - \varphi(x) \|_1^\Omega \to 0,$$

then we shall say that $\varphi(x)$ is asymptotically stable from above.

The stability and asymptotic stability from below may be defined analogously.

It is easy to show for problem (1.1)-(1.3) that from the stability from above and from below follows the stability of stationary solution $\varphi(x)$; and from the asymptotic stability from above and from below follows the asymptotic stability of $\varphi(x)$.

Proof.   Denote

$$w(x,t) = u(x,t_1 + \tau;u_0) - u(x,t_1;u_0).$$

This difference may be interpreted as the solution to the following linear problem

$$w_t = \sum_{i,j=1}^n a_{ij}^u(x,t)\frac{\partial^2 w}{\partial x_i \partial x_j} + \sum_{i=1}^n a_i^u(x,t) + a^u(x,t)\frac{\partial w}{\partial x_i}w, \quad (x,t) \in Q;$$

$$B^u(x,t)w = \alpha_1 \left(\sum_{i=1}^n b_i(x,t)\frac{\partial w}{\partial x_i} + b_0^u(x,t)w \right) + \alpha_0 w, \quad (x,t) \in \Gamma;$$

$$w(x,0) = u(x,t_1 + \tau;u_0) - u(x,t_1;u_0) \geq 0 \quad x \in \Omega.$$

Here by $a_{ij}^u(x,t), a_i^u(x,t), a^u(x,t), b_0^u(x,t)$ we have denoted the coefficients of the linear problem which solves $w(x,t)$.

Applying to this problem the maximum principle, we obtain

$$w(x,t) > 0; \quad x \in \Omega; \quad t > 0.$$

Therefore, in particular, we have

$$w(x, \tau) = u(x, t_1 + 2\tau; u_0) - u(x, t_1 + \tau; u_0) > 0, \quad x \in \Omega.$$

If we continue this process, we obtain that the sequence $u(x, t_1 + k\tau; u_0)$ monotonically increases and is bounded in $C(\bar{\Omega})$. Using the restrictions imposed on the smoothness of nonlinear terms, we see that the sequence $u(x, t_1 + k\tau; u_0)$ is bounded in $C^{2+\alpha'}(\bar{\Omega})$. Consequently, $u(x, t_1 + k\tau; u_0)$ converges in $C^{2+\alpha}(\bar{\Omega})$ to the function $\varphi(x) \in \omega(u_0)$.

The sequence $u(x, t_1 + (k+1)\tau; u_0)$ also converges to $\varphi(x)$ and

$$u(x, \tau; \varphi(x)) = \varphi(x).$$

The latter equality denotes that $u(x, t; \varphi(x))$ is a τ-periodic solution of (1.1)-(1.3).

We must now show that $u(x, t; \varphi(x))$ is a stationary solution of problem (1.1)-(1.3). Assume the contrary, i.e.,

$$u_t(x, t; \varphi(x)) \neq 0.$$

Consider the following linear problem

$$v_t = L\Big(x, t; u(x, t; \varphi(x))\Big)v, \qquad (x, t) \in Q; \qquad (2.1)$$

$$B\Big(x, t; u(x, t; \varphi(x))\Big)v = 0, \quad (x, t) \in \Gamma; \quad v(x, 0) = v_0(x).$$

Here we denote by $L\Big(x, t; u(x, t; \varphi(x))\Big)v$ the linear operator obtained after linearization of the nonlinear operator from the right side of (1.1) on the periodic solution. Analogously, by $B\Big(x, t; u(x, t; \varphi(x))\Big)v$ we denote linearized on $u(x, t; \varphi(x))$ boundary condition (1.2).

Taking into account the restrictions on smoothness of (1.1)-(1.3), we see that

$$u_t(x, t; \varphi(x)) \in H^{2+\alpha}(Q),$$

and that $u_t(x, t; \varphi(x))$ is a τ-periodic solution of linear problem (2.1).

Denote, as earlier (see chapter 1), by $V(\tau)$ the shift operator on the period of solutions of linear problem (2.1), i.e, the operator which maps $v(x, 0)$ into $v(x, \tau)$, where $v(x, t)$ solves (2.1).

Note, that the function $u_t(x, 0; \varphi(x))$ is the eigenfunction of $V(\tau)$ which corresponds to the unit eigenvalue. Denote by σ_1 the greatest by modulus eigenvalue of $V(\tau)$. As we have said above, this eigenvalue is real and simple. Corresponding to this value, eigenfunction $\xi_1(x)$ is positive in Ω.

It is clear that $\sigma_1 \geq 1$. Now we show that $\sigma_1 > 1$. Really, if we suppose that $\sigma_1 = 1$, then

$$\xi_1(x) = u_t(x, 0; \varphi(x)),$$

therefore, $u_t(x, 0; \varphi(x)) > 0, \quad x \in \Omega$. But

$$u(x, \tau; \varphi(x)) = \varphi(x) = \int_0^\tau u_t(x, \eta; \varphi(x))d\eta + \varphi(x),$$

which is impossible, since

$$\int_0^\tau u_t(x, \eta; \varphi(x))d\eta > 0, \quad x \in \Omega.$$

Hence, it follows that $\sigma_1 > 1$. Consider as in chapter 5 the one-parameter set of solutions of (1.1)-(1.3)

$$u(x, t; \varphi(x)) + \varepsilon\xi_1(x, t) + \varepsilon^2\eta(x, t, \varepsilon). \tag{2.2}$$

Substitute the expression into (1.1)-(1.3) and, using the standard disturbance method and the Shauder estimates in the finite interval, we see that there always exists a unique function $\eta(x, t, \varepsilon)$ which belongs to $H^{2+\alpha}(Q_\tau)$ such that (2.2) solves (1.1)-(1.3). Denote this solution by $z(x, t, \varepsilon)$. Then

$$z(x, \tau, \varepsilon) - z(x, 0, \varepsilon) = \varepsilon\Big((\sigma_1 - 1)\xi_1(x) - \varepsilon\eta(x, \tau, \varepsilon)\Big).$$

If it is necessary, diminishing ε, we may always take such negative ε that $z(x, \tau, \varepsilon) - z(x, 0, \varepsilon)$ is negative for $x \in \Omega$. Let

$$\zeta(x, t, \varepsilon) = z(x, t + \tau, \varepsilon) - z(x, t, \varepsilon).$$

The function $\zeta(x, t, \varepsilon)$ may be considered as a solution of a linear homogeneous parabolic problem with negative initial data. Therefore, the comparison theorem yields that $\zeta(x, t, \varepsilon)$ is negative. Then the sequence $z(x, k\tau, \varepsilon)$ monotonically decreases in $C(\bar{\Omega})$. The sequence $u(x, t_1 + k\tau; u_0)$ conversely, monotonically increasing, converges to the stationary solution $\varphi(x)$. Choosing ε sufficiently small and negative, we may always obtain

$$\varphi(x) > z(x, 0, \varepsilon) > u(x, t_1; u_0), \quad x \in \Omega.$$

By the comparison theorem

$$\varphi(x) > z(x, t, \varepsilon) > u(x, t_1 + t; u_0), \quad x \in \Omega.$$

In particular, for $t = k\tau$, we have

$$\varphi(x) > z(x, 0, \varepsilon) > z(x, k\tau, \varepsilon) > u(x, t_1 + k\tau; u_0), \quad x \in \Omega.$$

The latter inequality contradicts the way of determining $\varphi(x)$ as the limit in $C^{2+\alpha}(\bar{\Omega})$ of the sequence $u(x, t_1 + k\tau; u_0)$. The obtained contradiction shows that

$$u_t(x, t; \varphi(x)) \equiv 0;$$

therefore, $u_t(x, t; \varphi(x))$ is the stationary solution of (1.1)-(1.3). Note, that by the continuous dependence of solution $u(x, t; u_0)$ in the interval $[(k-1)\tau, k\tau]$, we have

$$\| u(x, t; u_0) - \varphi(x) \|_{2+\alpha}^{Q_{k\tau}^{(k-1)\tau}} \to 0 \quad \text{for} \quad k \to \infty,$$

where

$$Q_{k\tau}^{(k-1)\tau} = \Omega \times \left((k-1)\tau, k\tau\right).$$

Therefore, the solution $u(x, t; u_0)$ is stabilizing to the stationary solution $\varphi(x)$.

The first part of the theorem is proved. To complete the proof of the theorem, we have to show that $\varphi(x)$ is an asymptotically stable from below stationary solution.

Note, that

$$u(x, t_1 + \tau; u_0) < \varphi(x) \quad \text{for} \quad x \in \Omega,$$

and for each boundary-value problem different from the first boundary-value problem the latter inequality holds for $x \in \partial\Omega$. For the first boundary-value problem, by the Zhiro - Zaremba theorem (Protter and Weinberger, 1984), the inequality

$$\frac{\partial u}{\partial n}(x, t_1 + \tau; u_0) > \frac{\partial \varphi}{\partial n}(x), \quad x \in \partial\Omega$$

holds.

Therefore, such positive $\delta > 0$ exists that if

$$\| u_0^1(x) - \varphi(x) \|_1^\Omega < \delta, \quad u_0^1(x) \leq \varphi(x), \quad x \in \partial\Omega,$$

then

$$\varphi(x) \geq u_0^1(x) \geq u(x, t_1 + \tau; u_0), \quad x \in \Omega.$$

Applying the comparison theorem, and taking into account that for the problem the condition (3.5) (from chapter 4) holds, we obtain that $u(x, t; u_0^1)$ converges to the stationary solution $\varphi(x)$ in $C^{2+\alpha}(\bar{\Omega})$. Therefore, the stationary solution $\varphi(x)$ is asymptotically stable from below. The theorem is proved.

We must introduce some notations. Denote by S the set of all stationary solutions of problem (1.1)-(1.3), and by S_-, the set of all stationary solutions of problem (1.1)-(1.3) which have the negative first eigenvalue of the linearized on the stationary solution spectral problem. (If we use the notations of Belonosov (1983, 1984, 1985), we may say that the instability index is equal to zero). Denote by S_0 the set of stationary solutions of (1.1)-(1.3) for which the first eigenvalue of the linearized spectral problem is equal to zero. Denote by S_+ the set of all stationary solutions of (1.1)-(1.3) for which the greatest eigenvalue of the linearized spectral problem is positive (or, following Belonosov, 1983, 1984, 1985, the index of instability of stationary solution is not zero).

Note, that

$$S = S_+ \bigcup S_0 \bigcup S_-, \quad \varphi(x) \in S_- \bigcup S_0$$

where $\varphi(x)$ is the steady state from theorem 2.1.

We say that a subset $A \subset E$ is ω-*stable* if there exists such positive ε, that from

$$u_0(x) \in E, \quad \inf_{\varphi(x) \in A} \{ \| u_0(x) - \varphi(x) \|_1^\Omega \} < \varepsilon$$

it follows that $\omega(u_0) \subset A$.

Theorem 2.2. If A is an ω-stable connected set for (1.1)-(1.3), then A consists of a unique stationary solution.

Proof. Let

$$u_0(x) > \varphi(x), \quad \| u_0(x) - \varphi(x) \|_1^\Omega < \varepsilon.$$

Then such $\tau > 0$ exists that

$$u(x, \tau; u_0) \le u_0(x), \quad x \in \Omega.$$

By theorem 2.1, $\varphi(x)$ is an asymptotically stable stationary solution; therefore, $\varphi(x)$ is isolated. If A contains another function, then this function is to be (by the above established statement) the asymptotically stable stationary solution of the problem. But then, this solution, by the connectedness of the set A, must coincide with the solution $\varphi(x)$. The theorem is proved.

§6.3 Unstable ω-limit sets of solutions for the autonomous quasilinear parabolic equation.

Consider two solutions of (1.1)-(1.3): $u(x,t;u_0)$, $u(x,t;v_0)$ with initial data $u_0(x), v_0(x) \in E$. Assume that

$$u_0(x) > v_0(x); \quad x \in \Omega.$$

Then for $t > 0$, by the comparison theorem, we have

$$u(x,t;u_0) > u(x,t;v_0), \quad x \in \Omega.$$

Let $\varphi(x) \in \omega(u_0)$ and denote by $u(x,t;\varphi(x))$ the solution of (1.1)-(1.3) equal to $\varphi(x)$ for $t = 0$. In §1 it was shown that such a sequence $t_i \to +\infty$ exists that for $i \to \infty$

$$u(x,t_i + t;u_0) \to u(x,t;\varphi(x)), \quad t_i \to +\infty.$$

Choosing, if necessary, a subsequence of t_i, we obtain that $u(x,t_i + t;v_0)$ converges in $C^{2+\alpha}(\bar{\Omega})$ to a solution $u(x,t;\psi(x))$ of problem (1.1)-(1.3) for all t.

Note,that

$$u(x,t;\varphi(x)) \geq u(x,t;\psi(x)), \quad t \in R.$$

Denote the first solution by $\varphi(x,t)$ and the second - by $\psi(x,t)$.

Lemma 3.1. Either

$$\varphi(x,t) \equiv \psi(x,t), \quad (x,t) \in Q_0 = \Omega \times (-\infty, +\infty),$$

or

$$\varphi(x,t) > \psi(x,t), \quad (x,t) \in Q_0.$$

Proof. Prove that if such a point $(x_0, t_0) \in Q_0$ exists that

$$\varphi(x_0, t_0) = \psi(x_0, t_0),$$

then

$$\varphi(x, t) \equiv \psi(x, t), \quad (x, t) \in Q_0.$$

Denote

$$w(x, t) = \varphi(x, t) - \psi(x, t).$$

The function $w(x, t)$ is positive in Q_0 and solves in Q_0 the linear mixed problem

$$w_t = \sum_{i,j=1}^{n} a_{ij}^u(x, t) \frac{\partial^2 w}{\partial x_i \partial x_j} + \sum_{i=1}^{n} a_i^u(x, t) \frac{\partial w}{\partial x_i} + a^u(x, t)w, \quad (x, t) \in Q_0;$$

$$\alpha_1 \left(\sum_{i=1}^{n} b_i(x, t) \frac{\partial w}{\partial x_i} + b_0^u(x, t)w \right) + \alpha_0 w = 0,$$

$$(x, t) \in \Gamma_0 = \partial \Omega \times (-\infty, +\infty).$$

If λ is a sufficiently large positive number, then by theorem 2.1 the following problem

$$v_t = \sum_{i,j=1}^{n} a_{ij}^u(x, t) \frac{\partial^2 v}{\partial x_i \partial x_j} + \sum_{i=1}^{n} a_i^u(x, t) \frac{\partial v}{\partial x_i}$$

$$+ a^u(x, t)v - \lambda v, \quad (x, t) \in Q_0;$$

$$\alpha_1 \left(\sum_{i=1}^{n} b_i(x, t) \frac{\partial v}{\partial x_i} + b_0^u(x, t)v \right) + \alpha_0 v = 1, \quad (x, t) \in \Gamma_0$$

has a positive solution $\rho(x, t)$ determined in the whole axis.
Make the change

$$z(x, t)\rho(x, t) = w(x, t).$$

As a result, we obtain

$$z_t = \sum_{i,j=1}^{n} \tilde{a}_{ij}^u(x,t) \frac{\partial^2 z}{\partial x_i \partial x_j} + \sum_{i=1}^{n} \tilde{a}_i^u(x,t) \frac{\partial z}{\partial x_i} + \tilde{a}^u(x,t)z, \quad (x,t) \in Q_0;$$

$$\sum_{i=1}^{n} \tilde{b}_i(x,t) \frac{\partial z}{\partial x_i} - z, \quad (x,t) \in \Gamma_0.$$

Note, that for $t = t_0 - 1$ $z(x, t_0 - 1) \geq 0$, $x \in \Omega$. Therefore, if $z(x_0, t_0) = 0$ then $z(x, t) \equiv 0$ which proves the lemma.

Lemma 3.2. Under the lemma 3.1 conditions, if

$$\varphi(x,t) \equiv \psi(x,t), \quad (x,t) \in Q_0,$$

then $\varphi(x, t)$ is a stationary solution of (1.1)-(1.3).

Proof. As we have noticed above, for $t > 0$ the inequality

$$u(x, t; u_0) > u(x, t; v_0), \quad x \in \Omega$$

holds. In the case of each boundary-value problem differing from the first one, the inequality holds also for $x \in \partial\Omega$. For the first boundary-value problem, in addition, we have

$$\frac{\partial u}{\partial n}(x, t; u_0) < \frac{\partial u}{\partial n}(x, t; v_0), \quad x \in \partial\Omega.$$

Fix a positive t_0. Then there exists such positive τ_0 that for $|\tau| < \tau_0$ we have

$$u(x, t_0 + \tau; u_0) > u(x, t_0; v_0), \quad x \in \Omega.$$

Therefore,

$$u(x, t_i + t_0 + \tau; u_0) > u(x, t_i + t_0; v_0), \quad x \in \Omega,$$

hence follows

$$\varphi(x, t_0 + \tau) \geq \psi(x, t_0).$$

Similarly, we may prove that

$$\varphi(x, t_0) \geq \psi(x, t_0 + \tau).$$

Combining these inequalities, we obtain

$$\varphi(x, t_0 + \tau) \geq \varphi(x, t_0) \geq \varphi(x, t_0 + \tau),$$

which is possible only if $\varphi(x, t)$ does not depend on time. The lemma is proved.

Theorem 3.3. Let $\varphi(x, t), \psi(x, t)$ be solutions of (1.1)-(1.3) defined in the beginning of the paragraph. Then, if

$$\varphi(x, t) > \psi(x, t), \quad (x, t) \in Q_0,$$

then

$$\varphi(x, t) > \psi(x, \tau), \quad (x, t) \in Q_0, \quad (x, \tau) \in Q_0.$$

Proof. If at least one of solutions $\varphi(x, t), \psi(x, t)$ is a stationary solution of (1.1)-(1.3), then the statement of the theorem is evident.

Assume that $\varphi(x, t), \psi(x, t)$ are nonstationary solutions of (1.1)-(1.3). Doing the same as in the proof of lemma 3.2, we obtain that for $|\tau| < \tau_0$

$$\varphi(x, t + \tau) \geq \psi(x, t), \quad (x, t) \in Q_0.$$

Taking into account lemma 3.1 and that $\varphi(x, t), \psi(x, t)$ are nonstationary solutions, we obtain

$$\varphi(x, t + \tau) > \psi(x, t), \quad (x, t) \in Q_0, \quad |\tau| < \tau_0.$$

Let

$$T = \sup \{\tau_0 : |\tau| < \tau_0, \ \varphi(x, t + \tau) > \psi(x, t), \ (x, t) \in Q_0\}.$$

The theorem will be proved, if we show that $T = \infty$. Assume the contrary. Then, for $|\tau| = T$, in at least one point (x_0, t_0), either the equality

$$\varphi(x_0, t_0 + T) = \psi(x_0, t_0),$$

or the equality

$$\varphi(x_0, t_0 - T) = \psi(x_0, t_0)$$

will hold. Then, by lemma 3.2, $\varphi(x, t + T) \equiv \psi(x, t)$ is the stationary solution of (1.1)-(1.3) which contradicts the initial assumption. Therefore, $T = +\infty$ and the theorem is proved.

Theorem 3.4. Let the set $\omega(u_0)$ contain at least one nonstationary solution of (1.1)-(1.3), or an unstable stationary solution from S_+. Then the set $\omega(u_0)$ is unstable in the following sense. There exist unique stationary solutions $m_+(u_0), m_-(u_0)$ such that

1) $m_+(u_0) > m_-(u_0)$, $x \in \Omega$. The solution $m_+(u_0)$ is asymptotically stable from below; $m_-(u_0)$ is asymptotically stable from above.

2) Each function from $\omega(u_0)$, or each function $u(x, t; u_0)$, for any fixed $t \geq 0$ lies between $m_+(u_0)$ and $m_-(u_0)$ and divides the attraction domains of $m_+(u_0), m_-(u_0)$.

Proof. The proof is analogous to the case when $\omega(u_0)$ contains a nonstationary solution or a stationary solution from S_+.

Let a nonstationary solution $\varphi(x, t)$ for all t belong to $\omega(u_0)$ and the sequence $u(x, t_i + t; u_0)$ for all t converge to $\varphi(x, t)$ in $C^{2+\alpha}(\bar{\Omega})$.

We must show that for all $t_1, t_2 \in R$ the functions $\varphi(x, t_1)$ and $\varphi(x, t_2)$ intersect in Ω.

Assume the contrary. There exist such $t_1, t_2 \in R$ that $\varphi(x, t_1)$ and $\varphi(x, t_2)$ do not intersect in Ω. Let $\varphi(x, t_1) > \varphi(x, t_2)$ and for definiteness $t_2 > t_1$. Denote $\tau = t_2 - t_1$. Consider the function

$$w(x, t) = \varphi(x, t + \tau) - \varphi(x, t)$$

as a solution of a homogeneous linear mixed parabolic problem with nonnegative initial data for $t = t_1$. Then, we obtain

$$\varphi(x, t_1 + 2\tau) > \varphi(x, t_1 + \tau), \quad x \in \Omega.$$

Therefore, for sufficiently large t_i, we have

$$u(x, t_i + t_1 + 2\tau; u_0) \geq u(x, t_i + t_1 + \tau; u_0).$$

Applying theorem 2.1, we obtain that the solution $u(x, t; u_0)$ is stabilized to a unique stationary solution, which contradicts the above assumption. The case $t_2 < t_1$ may be considered analogously.

Therefore, functions $\varphi(x, t_1), \varphi(x, t_2)$ intersect in Ω for each $t_1, t_2 \in R$.

Consider the function $\eta_0(x) \neq 0$ in Ω, where

$$0 \leq \eta_0(x) \leq \max\{\varphi(x, \tau) - \varphi(x, 0), \ x \in \Omega\}, \tau \in R.$$

As the functions $\varphi(x, \tau)$ and $\varphi(x, 0)$ intersect in Ω, we may always take such function $\eta_0(x)$. For each $\sigma \in (0, 1)$ the inequality

$$\varphi(x, 0) \leq \varphi(x, 0) + \sigma\eta_0(x) \leq \max\{\varphi(x, \tau), 0, \}, \quad x \in \Omega$$

holds.

Denote by $\eta(x, t; \sigma)$ the solution of (1.1)-(1.3) with the initial data $\varphi(x, 0) + \sigma\eta_0(x)$. If $\sigma \in (0, 1)$, then $\varphi(x, 0) \leq \eta(x, 0; \sigma)$. By the comparison theorem, we have

$$\varphi(x, t) < \eta(x, t; \sigma), \quad x \in \Omega, \quad t > 0.$$

If t_i is sufficiently large, then the latter inequality yields

$$u(x, t_i + t; u_0) < \eta(x, t; \sigma), \quad x \in \Omega.$$

In particular, for $j > i$, we have

$$u(x, t_i + t; u_0) < \eta(x, t - t_i + t_j; \sigma), \quad x \in \Omega.$$

As in lemma 1.1, choose from the sequence $\eta(x, t - t_i + t_j; \sigma)$ a subsequence which converges in $C^{2+\alpha}(\bar{\Omega})$ for each t. So, we may assume that

$$\eta(x, t - t_i + t_j; \sigma) \to \psi(x, t, \sigma) \quad \text{for} \quad j \to \infty.$$

Evidently, $\psi(x, t, \sigma)$ solves (1.1)-(1.3). Lemmas 3.1, 3.2 and the theorem suppositions yield

$$\psi(x, t, \sigma) > \varphi(x, \tau); \quad (x, t), (x, \tau) \in Q_0.$$

In particular, hence follows that

$$\psi(x, 0, \sigma) > \varphi(x, 0), \quad \psi(x, 0, \sigma) > \varphi(x, \tau).$$

Therefore, for sufficiently large t, we shall have

$$\eta(x, t_j - t_i; \sigma) > \varphi(x, 0); \quad \eta(x, t_j - t_i; \sigma) > \varphi(x, \tau); \quad x \in \Omega.$$

Taking into account the method of constructing the function $\eta(x, 0; \sigma)$, we obtain that

$$\eta(x, t_j - t_i; \sigma) > \eta(x, 0, \sigma), \quad x \in \Omega.$$

Therefore, by theorem 2.1, for each $\sigma \in (0, 1)$, the solution $\eta(x, t; \sigma)$ is stabilized to a stable from below stationary solution $\psi(x, \sigma)$, where

$$\eta(x, t; \sigma) < \psi(x, \sigma), \quad (x, t) \in Q.$$

Now we show that the solution $\psi(x, \sigma)$ does not depend on σ. Really, for each $\sigma_1, \sigma_2 \in (0, 1)$, we have

$$\psi(x, \sigma_1) > \eta(x, 0; \sigma_2); \quad \psi(x, \sigma_1) > \eta(x, t; \sigma_2), \quad x \in \Omega, \ t > 0.$$

Therefore, by the comparison theorem, $\psi(x, \sigma_1) > \psi(x, \sigma_2)$. Since σ_1, σ_2 are arbitrary, we obtain that $\psi(x, \sigma)$ does not depend on σ. Denote this stable from below stationary solution $\psi(x, \sigma) = \psi(x)$ by $m_+(u_0)$.

Show that the set of functions $u(x, t; u_0)$ for each $t > 0$ belongs to the boundary of the attraction domain $A(m_+(u_0))$. For this purpose it suffices to show that if

$$v_0(x) \in E, \quad m_+(u_0) \geq v_0 \geq u(x, t; u_0), \quad v_0 \not\equiv m_+(u_0)$$

for a certain $t \geq 0$, $x \in \Omega$, then $u(x, t; v_0)$ converges to $m_+(u_0)$ for $t \to +\infty$.

It is clear, that we always may consider $t = 0$. As we have assumed that

$$m_+(u_0) \geq v_0 \geq u_0, \quad x \in \Omega,$$

then, by the comparison theorem, we obtain that

$$m_+(u_0) \geq u(x, t; v_0) \geq u(x, t; u_0), \quad t \geq 0, \quad x \in \Omega.$$

Therefore, taking account of restrictions on the growth of nonlinearity by the gradient, the solution $u(x, t; v_0)$ is determined for all $t > 0$. As the sequence $u(x, t; v_0)$ converges in $C^{2+\alpha}(\bar{\Omega})$ to $\varphi(x, 0)$, then for each $\sigma \in (0, 1)$ there exists such a moment $\hat{t}$ that

$$m_+(u_0) \geq u(x, \hat{t}; v_0) \geq \eta(x, 0; \sigma), \quad \hat{t} \geq 0, \quad x \in \Omega.$$

Therefore,

$$m_+(u_0) \geq u(x, \hat{t} + t; v_0) \geq \eta(x, t; \sigma), \quad \hat{t}, t \geq 0, \quad x \in \Omega.$$

Hence follows that since

$$\lim_{t \to +\infty} \| \eta(x, t; \sigma) - m_+(u_0) \|_{2+\alpha}^{\Omega} = 0,$$

then

$$\lim_{t \to +\infty} \| u(x, t + \hat{t}; \sigma) - m_+(u_0) \|_0^{\Omega} = 0.$$

Using condition (1.4) from chapter 5, finally, we obtain

$$\lim_{t \to +\infty} \| u(x, t + \hat{t}; \sigma) - m_+(u_0) \|_{2+\alpha}^{\Omega} = 0.$$

Solution $m_-(u_0)$ and its basis properties are obtained analogously. The theorem is proved.

Theorem 3.5. Let the set $\omega(u_0)$ contain at least one stationary solution from S_+; then $\omega(u_0)$ and $S_- \bigcup S_0$ do not interest.

First, we shall prove two lemmas.

Lemma 3.6. Let M be a connected and bounded subset of S_+. Then the closure of M is also contained in S_+. (The closure and boundedness of M are taken in $C^{2+\alpha}(\bar{\Omega})$).

Proof. Let $\varphi_m \in M \subset S_+$ and the sequence of these stationary solutions of (1.1)-(1.3) converge in $C^{2+\alpha}(\bar{\Omega})$ to a stationary solution $\varphi(x)$ of problem (1.1)-(1.3).

Let $\psi(x)$ be an arbitrary stationary solution of (1.1)-(1.3). As earlier, denote by $\lambda_1(\psi)$ the first eigenvalue of the spectral problem

$$\lambda w = L(x, \psi)w, \quad x \in \Omega, \quad B(x, \psi)w = 0, \quad x \in \partial\Omega,$$

which has, by the Krein-Rutman theorem (Krein and Rutman, 1948), the maximal real part.

Show that $\lambda_1(\varphi) > 0$. As M is a connected set, then for each m there exists a sequence $\varphi_{ml}(x)$ consisting of stationary solutions which belong to the set M and for $l \to +\infty$ converge in $C^{2+\alpha}(\bar{\Omega})$ to the stationary solution $\varphi_m(x), \varphi_{ml} \not\equiv \varphi_m(x)$. Denote

$$\psi_{ml}(x) = \varphi_{ml}(x) - \lambda\varphi_m(x).$$

Then $\psi_{ml}(x)$ is a nontrivial solution of the linear homogeneous problem

$$0 = \sum_{i,j=1}^{n} a_{ij}^{ml}(x)\frac{\partial^2 \psi_{ml}}{\partial x_i \partial x_j} + \sum_{i=1}^{n} a_i^{ml}(x)\frac{\partial \psi_{ml}}{\partial x_i} + a^{ml}(x)\psi_{ml}$$

$$= L^{ml}(x)\psi_{ml}, \quad x \in \Omega;$$

$$B^{ml}(x)\psi_{ml} = 0, \qquad x \in \partial\Omega. \tag{3.1}$$

The coefficients of (3.1) converge for $l \to \infty$ to the coefficients of the problem

$$\lambda\psi = L(x, \varphi_m)\psi, \quad x \in \Omega, \quad B(x, \varphi_m)\psi = 0, \quad x \in \partial\Omega. \tag{3.2}$$

The function $\psi_{ml}(x)$ is the eigenfunction corresponding to the trivial eigenvalue for a certain linear problem

$$\lambda\psi = L^{ml}(x)\psi, \quad x \in \Omega, \quad B^{ml}(x)\psi = 0, \quad x \in \partial\Omega.$$

Note, that $\varphi_m(x) \in S_+$; therefore, problem (3.2) always has positive eigenvalues. On the other hand, passing in the latter problem to the limit for $l \to \infty$, we obtain that the zero is the eigenvalue of (3.2).

Passing to the limit, in (3.2) for $m \to \infty$, we obtain that the zero is the eigenvalue of the problem

$$0 = L(x, \varphi)\psi, \quad x \in \Omega, \quad B(x, \varphi)\psi = 0, \quad x \in \partial\Omega. \tag{3.3}$$

If zero is the maximal eigenvalue of (3.3), then $\lambda_1(\varphi_m)$ must converge to zero for $m \to \infty$. Applying the disturbance for spectral problems (Kato, 1972), we see that the multiplicity of the zero eigenvalue of (3.3) is greater than unity. On the other hand, by the Krein-Rutman theorem, the first eigenvalue of the problem has multiplicity equal to unity. The obtained contradiction shows that $\lambda_1(\varphi) > 0$ which proves the lemma.

Lemma 3.7. Let $\omega(u_0)$ have a nonempty intersection with S_0; then the solution is stabilized to the stationary solution $\varphi(x) = \omega(u_0) \in S_0$.

Proof.   Let $u(x, t_i; u_0)$ converge to the stationary solution $\varphi(x) \in S_0$ in $C^{2+\alpha}(\bar{\Omega})$.   Begining with a certain number t_i the function $u(x, t_i; u_0)$ is contained in such neighborhood $\varphi(x)$, where we may use the results of theorem 1.6.   In particular, we may apply the results of this theorem to $u(x, t_i + t; u_0)$ for those $t \in [0, \tau_i]$, for which this solution is contained in the selected neighborhood of the stationary solution. The theorem on continuous dependence of the boundary-value problem solution on initial data yields $\tau_i \to +\infty$, $i \to \infty$. Taking into account the formula (1.38) chapter 1, we see that such $\lambda > 0$ exists that

$$\| u(x, t_i + t; u_0) - \varphi(x) - \xi_i(x, t) \|_{2+\alpha}^{\Omega} \tag{3.4}$$

$$\leq C(\lambda)e^{-\lambda t} \times \| u(x, t; u_0) - \xi_i(x, 0) \|_{2+\alpha}^{\Omega}, \quad t \in [0, \tau_i].$$

The function

$$\xi_i(x, t) = \zeta_i(t)\xi_1(x) + \Psi_0\Big(\zeta_i(t)\xi_1(x)\Big)$$

is the solution of (1.1)-(1.3) of a special form. The functions $\zeta_i(t)$ for all natural and sufficiently large i solve the ordinary differential equation of the first order with one unknown right side

$$\zeta'(t) = \Phi_i(\xi_i). \tag{3.5}$$

The solution of (3.5) either monotonically increases, or monotonically decreases, or is a constant. Relation (1.31) yields that $\xi_i(x, t)$ has the same properties.

Consider the case when $\xi_i(x, 0) > 0$. There are two possibilities: the solution either monotonically increases, or monotonically decreases (if $\xi_i(x)$ is stationary solution, then the theorem statement immediately follows from (3.5)).

Assume, firstly, that $\xi_i(x, t)$ monotonically increases. Inequality (3.4) yields, that for sufficiently large τ_i, the inequality

$$u(x, t_i + \tau_i; u_0) - \varphi(x) > \xi_i(x, 0)$$

holds.

As

$$\xi_i(x,t) > \xi_i(x,0), \qquad x \in \Omega,$$

then

$$u(x, t + t_i + \tau_i; u_0) - \varphi(x) > \xi_i(x,t), \qquad x \in \Omega, \quad t > 0.$$

Hence it follows that $\varphi(x)$ can not belong to $\omega(u_0)$. This contradicts the initial assumption.

If $\xi_i(x,0) > 0$ and $\xi_i(x,t)$ monotonically decreases then

$$0 < \xi_i(x,t) < \xi_i(x,0), \qquad x \in \Omega.$$

Taking into account condition (3.5), we obtain that for sufficiently large i the solution $\xi_i(x,t)$ is contained in the selected neighborhood of the stationary solution. Therefore, for a certain i, $\quad \tau_i = \infty$ and the sequence is stabilizing. Then, (3.4) yields that $u(x,t;u_0)$ also is stabilizing. The lemma is proved.

Theorem 3.5 is an evident corollary of lemmas 3.6 and 3.7. In lemma 3.7 we have considered the case when $\lambda_1(\varphi) \geq 0$, i.e., $\varphi \in S_0$. Assume that

$$\varphi(x) \in \omega(u_0),$$

where $\varphi \in S_+$, i.e., $\lambda_1(\varphi) > 0$. In this case the following theorem holds.

Theorem 3.8. Let the set $\omega(u_0)$ contain at least one stationary solution of (1.1)-(1.3) from S_+. Then the set $\omega(u_0)$ is unstable in the following sense: there exist unique stationary solutions $m_+(u_0), m_-(u_0)$ such that

1. $m_+(u_0) > m_-(u_0)$, $x \in \Omega$. The solution $m_+(u_0)$ is asymptotically stable from below. The solution $m_-(u_0)$ is asymptotically stable from above.

2. Each function from the set $\omega(u_0)$, or each function $u(x,t;u_0)$ for $t \geq 0$ lies between the stationary solutions $m_+(u_0), m_-(u_0)$ and divides the attraction domains of $m_+(u_0), m_-(u_0)$.

Proof. This proof may be established in a similar way to the proof of theorem 3.4. The necessary auxiliary statements are proved in the lemmas.

§6.4   Stabilization of solutions of boundary-value problems and monotone solutions of boundary-value problems.

In this paragraph we shall assume problem (1.1)-(1.3) to be dissipative. One of the main results of this chapter is the following theorem.

Theorem 4.1. There exists such an open and dense in $C^1(\bar{\Omega})$ set $M \subset E$ that if initial data $u_0 \in M$, then the solution $u(x,t;u_0)$ of problem (1.1)-(1.3) is stabilized for $t \to +\infty$.

Proof. For each $u_0 \in E$, construct an open set $V_{u_0} \subset E$ such that a) the closure of V_{u_0} contains the function u_0, and b) if $v_0 \in V_{u_0}$ then $u(x,t;u_0)$ is stabilized for $t \to +\infty$. Then, setting $M = \bigcup_{u_0 \in E} V_{u_0}$, we shall prove the theorem.

Consider the possible cases.

1. The set $\omega(u_0)$ is not empty and contains the stationary solution $\varphi \in S_-$. Then, taking into account the chapter 1 results, we see that the solution $u(x,t;u_0)$ being in a sufficiently small neighborhood of the stationary solution, always remains there. Therefore, $\omega(u_0) = \varphi$, and as V_{u_0} we may consider any sufficiently small neighborhood of the stationary solution φ.

2. The set $\omega(u_0)$ is not empty and contains a stationary solution $\varphi \in S_+$, or a nonstationary solution. In both cases the set V_{u_0} is constructed by means of theorem 3.4. For example, we may take the set

$$\{v_0 \in E, \quad m_+(x) \geq v_0(x) \geq u_0(x), \quad x \in \Omega; \quad v_0(x) \not\equiv m_+(x)\},$$

or the set

$$\{v_0 \in E, \quad m_-(x) \ge v_0(x) \ge u_0(x), \quad x \in \Omega; \quad v_0(x) \not\equiv m_-(x)\}.$$

3. The set $\omega(u_0)$ contains a stationary solution from S_0. Two situations are possible.

a) There exists a neighborhood of u_0 in E such that all solutions from this neighborhood are stabilizing (the solution $u(x, t; u_0)$ is stabilizing by lemma 3.7).

b) There exists a sequence of initial data $u_{n0}(x) \in E$ and

$$u_{n0}(x) \to u_0(x) \quad \text{in} \quad C^1(\bar{\Omega}),$$

such that the solution $u(x, t; u_{n0})$ is not stabilized for $t \to +\infty$. Then, the nonempty set $\omega(u_{n0})$ must contain (as was shown in §2, §3 of this chapter) either a nonstationary solution of the problem, or a stationary solution from S_+. Using the second point, construct the set $V_{u_{n0}}$ and take

$$V_{u_0} = \bigcup_{n=1}^{\infty} V_{u_{n0}}.$$

The theorem is proved.

Let the functions $a_{ij}(x, u, p), a(x, u, p)$ be analytic by (u, p) and be twice differentiable by x. Assume, also, for simplicity of presentation, that boundary conditions (1.2) are linear, i.e.,

$$g(x, u) = b_0(x)u.$$

It is easy to remove this assumption, but then, some technical difficulties arise when proving the theorem. Then there exists such open and dense in E set M_1, that if $u_0 \in M_1$ then $u(x, t; u_0)$ becomes monotone for $t \ge \tau(u_0)$. This statement will follow from the theorems below.

Theorem 4.2. Suppose that for (1.1)-(1.3) the functions $a_{ij}(x, u, p), a(x, u, p)$ are analytic by (u, p) and are twice differentiable by x; $\varphi(x)$ is an unstable solution of the problem. Let

$$u_0, v_0 \in A(\varphi), \quad u_0 \geq v_0, \quad u_0 \not\equiv v_0$$

and $u(x, t; v_0)$ intersect the stationary solution $\varphi(x)$ for all $t \geq 0$. Then such positive $\tau(u_0)$ exists that

$$u(x, t; u_0) > \varphi(x), \qquad t \geq \tau(u_0),$$

and $u(x, t; u_0)$ monotonically decreases.

The analogous statement holds in the case $u_0 \leq v_0$.

In order to formulate the statements below we shall introduce some new definitions. Denote by $\hat{S}$ the set of stationary solutions to problem (1.1)-(1.3) with the following properties:

1. Let $\varphi \in \hat{S}$; then the spectral problem

$$\lambda \xi = L(\varphi)\xi, \quad x \in \Omega, \quad B(\varphi)\xi = 0, \quad x \in \partial\Omega \qquad (4.1)$$

(by $L(\varphi)$ in (4.1) we have denoted the operator from the right side of (1.1) linearized on the stationary solution $\varphi(x)$) has no more than one eigenvalue in the right half-plane of the complex plane. (Here, as earlier, we denote by $L(\varphi)$ and $B(\varphi)$ parabolic equation (1.1) and boundary conditions (1.2) linearized on the stationary solution $\varphi(x)$).

2. Denote by $\lambda_2(\varphi)$ the second eigenvalue of problem (4.1). By the second eigenvalues here is meant all eigenvalues $\lambda_2(\varphi)$ such that $\mathrm{Re}\lambda_2(\varphi) < \lambda_1(\varphi)$, and in the band $\mathrm{Re}\lambda_2(\varphi) < z < \lambda_1(\varphi)$ of the complex plane there are no other eigenvalues of problem (4.1). Note, that such eigenvalue may not be unique, for example, if $\lambda_2(\varphi)$ is a complex number, as in the example in §5 chapter 5. Besides, the complex numbers $\lambda_2(\varphi)$ may be multiple. Suppose that the set $\{\lambda_2(\varphi), \varphi \in \hat{S}\}$ has no limit points in the imaginary axis of the complex plane.

Theorem 4.3. Suppose that all the suppositions of theorem 4.2 hold true. Fix an arbitrary positive constant K. Then problem (1.1)-(1.3) has no more than a finite number of solutions from $\hat{S}$ which are less by the norm of $C^1(\bar{\Omega})$ than K.

Denote by M_1 the subset of the set E consisting of the functions for which the solution $u(x, t; u)$ constructed by the initial data $u_0 \in M_1$ is strictly monotone for $t \geq \tau(u_0)$.

Theorem 4.4. Let all suppositions of theorem 4.2 hold. Then the set M_1 contains an open and dense in E subset.

We can now prove theorems 4.2-4.4 formulated above, but we must preface the proofs of the theorems with some lemmas.

Consider two functions $u_0, v_0 \in E$. We shall say that $u_0 \gg v_0$, if $u_0 > v_0$, $x \in \bar{\Omega}$ for each boundary-value problem different from the first boundary-value problem. For the first boundary-value problem:

$$u_0 > v_0, \quad x \in \Omega$$

and, in addition,

$$\frac{\partial u_0}{\partial n} < \frac{\partial v_0}{\partial n} \quad x \in \partial\Omega.$$

Lemma 4.5. Let initial data u_0, v_0 belong to E, $u_0 \geq v_0$, $u_0 \not\equiv v_0$. Then

$$u(x, t; u_0) \gg u(x, t; v_0) \quad \text{for} \quad t > 0.$$

Proof. Consider the difference

$$w(x, t) = u(x, t; u_0) - u(x, t; v_0).$$

This function may be considered as a solution of a linear parabolic equation with positive initial data. Applying the strict maximum principle, or, in the case of the first boundary-value problem the Zhiro-Zaremba theorem (see, for example, Protter and Weinberger, 1984), we obtain

$$w(x, t) \gg 0.$$

This proves the lemma.
   Make the change of variables

$$w(x, t; w_0) = u(x, t; u_0) - \varphi(x).$$

Assume, for simplicity, that boundary conditions (1.2) are homogeneous. The inhomogeneous case may be considered analogously. Then $w(x, t; w_0)$ solves the problem

$$w_t = L(\varphi)w + F(w), \qquad (x, t) \in Q;$$

$$B(\varphi)w = 0, \quad (x, t) \in \Gamma; \quad w(x, 0) = w_0(x) = u_0(x) - \varphi(x). \quad (4.2)$$

Here, as earlier, we denote by $L(\varphi)$ the linearized on the stationary solution $\varphi(x)$ operator from the right side of (1.1); $F(w)$ are the nonlinear terms in the Taylor expansion in the $\varphi(x)$ neighborhood. Here F is a nonlinear operator acting from $H_1^{2+\alpha}(Q)$ into $H_{-1}^{\alpha}(Q)$, where (see chapter 1) the inequality

$$\| F(w_1) - F(w_2) \|_{-1,\alpha}^Q \leq q \left(\| w_1 \|_{1,\alpha}^Q, \| w_2 \|_{1,\alpha}^Q \right) \| w_1 - w_2 \|_{1,2+\alpha}^Q \quad (4.3)$$

holds.

In this case the value $q(\varepsilon_1, \varepsilon_2)$ tends to zero for $\varepsilon_1, \varepsilon_2 \to 0$.

Let ε be a sufficiently small positive number. Denote by q the value $q(2\varepsilon, 2\varepsilon)$ and assume that $q < l_{01}$, where l_{01} is determined as in chapter 1.

Modify the mapping F extending it out of an ε-neighborhood in $C^1(\bar{\Omega})$, so that for the extended mapping the equality (4.3) would hold for all $w_i(x, t) \in H_1^{2+\alpha}(Q)$. Such extension was constructed, for example, in Hartman (1970). Denote this extension by $\tilde{F}$ and by $w(x, t; w_0)$ denote the solution of the problem

$$\tilde{w}_t = L(\varphi)\tilde{w} + \tilde{F}(\tilde{w}), \qquad (x, t) \in Q; \quad (4.4)$$

$$B\tilde{w} = 0, \quad (x, t) \in \Gamma; \quad \tilde{w}(x, t; \tilde{w}_0) = \tilde{w}_0(x).$$

Solutions $u(x, t; u_0)$ and $u(x, t; v_0)$ do not leave the ε-neighborhood of φ in $C^1(\bar{\Omega})$. Therefore, in the proof of theorem (4.7) we may consider that $u_0 \gg v_0$ and that

$$w(x, t; u_0 - \varphi) = \tilde{w}(x, t; u_0 - \varphi),$$

$$w(x, t; v_0 - \varphi) = \tilde{w}(x, t; v_0 - \varphi).$$

For problem (4.4) all theorems proved in the first chapter hold. Denote by E_1 the one-dimensional space stretched on the first eigenfunction $\xi_1(\varphi)$.

We have constructed the invariant set in the first chapter, which we have denoted by $N = \{\tilde{w}_- + \Psi_-(\tilde{w}_-))$. Here Ψ_- is the mapping from E_- into E_1 such that the solution $\tilde{w}(x, t; \tilde{w}_0)$ with the initial data $\tilde{w}_0 \in N$ converges to zero faster than

$$e^{(\mathrm{Re}\lambda_2(\varphi) + C_1 q)t}.$$

The constant C_1 here does not depend on initial data. Otherwise, the following inequality

$$\| w(x, t; w_- + \Psi_-(w_-)) \|_{2+\alpha}^{\Omega} \leq C_2 e^{(\mathrm{Re}\lambda_2(\varphi) + C_1 q)t} \| w_- \|_1^{\Omega} \qquad (4.5)$$

holds.

The mapping Ψ_- satisfies the inequality

$$\| \Psi_-(w_-^1) - \Psi_-(w_-^2) \|_{2+\alpha}^{\Omega} \leq C_3 q \| w_-^1 - w_-^2 \|_1^{\Omega}. \qquad (4.6)$$

The set N is the integral set of (4.4); therefore, if $\tilde{w}_0(x) \in N$, then $\tilde{w}(x, t; \tilde{w}_0) \in N$ for all t from the domain of definition of the problem solution.

Lemma 4.6. Assume that initial data of (4.4) satisfy the inequality

$$\tilde{w}_0(x) \geq w_-(x) + \Psi_-(w_-(x)),$$

$$w_0(x) \not\equiv w_-(x) + \Psi_-(w_-(x)).$$

Then there exists such positive τ_2 that for $t \geq \tau_2$ the inequality

$$\tilde{w}(x,t;\tilde{w}_0) \gg 0$$

holds.

Proof. Consider, first, the spectral case

$$\tilde{w}_0 = \beta\xi_1(x) + w_-(x) + \Psi_-(w_-), \quad \beta > 0;$$

$\xi_1(x)$ is the eigenfunction of (4.1) corresponding to the first eigenvalue $\lambda_1(\varphi)$; $w_-(x) \in E_-$ is arbitrary. Denote

$$z(x,t) = \tilde{w}(x,t;\tilde{w}_0) - w(x,t;w_- + \Psi_-(w_-)).$$

By lemma 4.5, we obtain that $z(x,t) \gg 0$ for $t \geq 0$. Thus, taking into account (4.3), we obtain

$$z_t \geq L(\varphi)z - qz, \qquad (x,t) \in Q,$$

$$B(\varphi)z = 0, \qquad (x,t) \in \Gamma.$$

By the comparison theorem, we have

$$z(x,t) \geq \beta\xi_1(x)e^{(\lambda_1(\varphi)-q)t}. \tag{4.7}$$

Suppose, ε is chosen so that such μ exists that

$$\mathrm{Re}\lambda_2(\varphi) + C_1 q < \mu < \lambda_1(\varphi) - q.$$

Then, the statement of the lemma follows from inequalities (4.5) and (4.7).

Now, consider the case when $\tilde{w}_0$ is an arbitrary function from E_-. By lemma 4.5, for each $t > 0$, we have

$$w(x, t; w_0) \gg w(x, t, w_- + \Psi_-(w_-)).$$

Therefore, there exists such positive β that

$$w(x, t; w_0) \gg w(x, t, \beta \xi_1(x) + w_- + \Psi_-(w_-))$$

$$\gg w(x, t, w_- + \Psi_-(w_-)).$$

Using the statement we proved before and the comparison theorem, we obtain that the lemma holds also in the general case.

Lemma 4.7. Let initial data w_0 of problem (4.4) lie in N. Then the solution $w(x, t; w_0)$ of (4.4) with the initial data w_0 intersects the trivial solution for all $t \geq 0$.

Proof. Assume that for a certain $t_1 \geq 0$ the solution of the problem becomes nonnegative, i.e., $w(x, t; w_0) \geq 0$, $x \in \Omega$. Then $w(x, t; w_0) \geq 0$ for $t > t_1$. There exists such $\beta > 0$, that

$$w(x, t; w_0) > \beta \xi_1(\varphi).$$

By arguments similar to those in lemma 4.6, we see that

$$w(x, t; w_0) \gg \beta \xi_1(x) e^{(\lambda_1(\varphi) - q)(t - t_1)} \quad \text{for} \quad t > t_1.$$

This contradicts (4.5) which has to be true for the solution $w(x, t; w_0)$ if the initial data w_0 belongs to the integral set N. This contradiction shows that the initial assumption is false, and the lemma is proved.

In the first chapter, in theorem 3.2 and its corollary it was proved that the set of solutions of the special form

$$\eta(\gamma(t)) = \gamma(t) \xi_1(\varphi) + \Psi(\gamma(t)).$$

Here $\gamma(t)$ solves the ordinary differential equation

$$\gamma'(t) = \lambda_1(\varphi)\gamma(t) + P_1\tilde{F}\big(\eta(\gamma(t))\big), \quad \gamma(0) = \gamma_0. \tag{4.8}$$

The mapping $\Psi(\gamma)$ acts from R into E_2 and

$$\| \ \Psi(\gamma_1) - \Psi(\gamma_2) \ \|_{2+\alpha}^{\Omega} \leq Cq|\gamma_1 - \gamma_2|, \quad \Psi(0) = 0. \tag{4.9}$$

If $w(x, t; w_0)$ solves (4.4), then, by theorem 1.9, there exists a unique γ_0 such that

$$\| \ (w(x,t;w_0) - \eta(\gamma(t)))e^{-(\mathrm{Re}(\lambda_2)+C_1 q)t} \ \|_{1,2+\alpha}^{Q} \leq C \ \| \ w_0 - \eta(\gamma_0) \ \|_1^{\Omega}, \tag{4.10}$$

$$\| \ w_0 - \eta(\gamma_0) \ \|_1^{\Omega} \leq \min_{\gamma \in R} \| \ w_0 - \eta(\gamma) \ \|_1^{\Omega}. \tag{4.11}$$

In particular, for all solutions lying in N, $\gamma_0 = 0$ and $\eta(\gamma(t)) \equiv 0$ is the stationary trivial solution of the problem.

Note, that due to the suppositions, the trivial solution of (4.4) is asymptotically stable from above and, therefore, the trivial stationary solution of (4.8) is stable from above also. Consequently, for sufficiently small positive $\gamma_0 > 0$ the solution $\gamma(t)$, monotonically decreasing, tends to zero.

Now we estimate the rate of convergence for $\lambda_1(\varphi) = 0$.

Lemma 4.8. Let $\lambda_1(\varphi) = 0$ and for problem (4.4) all the above assumptions hold. Let a positive number γ_0 be such that for all $t \in [0, \delta]$ for a certain $\delta > 0$, solution $\eta(\gamma(t))$ does not go out from the ε-neighborhood of the trivial solution in the $C^1(\bar{\Omega})$ norm. Then, such a positive integer m, a positive number $\tilde{d}_m$, and a bounded function $f(\gamma, t)$ exist, that for $t \in [0, \delta]$ the function $\gamma(t)$ solves the special differential equation

$$\gamma'(t) = -\tilde{d}_m\gamma^m(t) + \gamma^{(m+1)}f(\gamma(t), t), \ \gamma(0) = \gamma_0. \tag{4.12}$$

Proof. We shall seek the solution of (4.12) in the form of the following one-parameter family

$$w(x, t; \tau) = \tau \xi_1(x) + \ldots + \tau^{m-1} \xi_{m-1}(x)$$

$$+ \tau^m \xi_{m+1}(x, t) + \tau^{m+1} \xi_{m+2}(x, t; \tau). \qquad (4.13)$$

Determine the functions ξ_i. Set $\xi_1(x) = \xi_1(\varphi)$. Denote by ξ_1^* the first eigenfunction of the problem formally conjugate with (4.1). Since the mapping is analytic $F(w(x, t; \tau)) = \tilde{F}(w(x, t; \tau))$, and $w(x, t; \tau)$ has the form (4.13), we have

$$F(w(x, t; \tau)) = \tau^2 d_2(x) + \ldots + \tau^m d_m(x, t) +$$

$$+ \tau^{m+1} d_{m+1}(x, t, \tau, \xi_{m+1}). \qquad (4.14)$$

We must show how to find the functions d_i, ξ_i if we know $\xi_1(x), \ldots,$ $\xi_{i-1}(x)$. The function ξ_1 is the known function; therefore, substituting $\tau \xi_1$ into (4.14), we obtain the function $d_2(x)$.

If $d_2(x)$ is orthogonal to the first eigenfunction of the conjugate problem, which corresponds to the trivial eigenvalue, then the inhomogeneous problem

$$\lambda \xi = L(\varphi) \xi + d_2(x), \ x \in \Omega; \ B(\varphi) \xi = 0, \ x \in \partial \Omega \qquad (4.15)$$

is solvable.

In this case, (4.15) has a unique solution orthogonal to the first eigenfunction of the conjugate problem. Denote this solution by $\xi_2(x)$. Further, by ξ_1, and ξ_2, we find the function d_3. If d_3 is orthogonal to the first eigenfunction of the conjugate problem, then, as earlier, denote by ξ_3 the solution of the problem

$$\lambda \xi = L(\varphi) \xi + d_3(x), \ x \in \Omega,$$

$$B(\varphi) \xi = 0, \ x \in \partial \Omega,$$

which is orthogonal to the first eigenfunction of the conjugate problem.
Assume that for a certain m

$$\int_\Omega d_m(x)\xi_1^*(x)dx = \tilde{d}_m \neq 0. \tag{4.16}$$

Define

$$\xi_m(x,t) = t\tilde{d}_m\xi_1^*(x)\tilde{\xi}_m(x),$$

where $\tilde{\xi}_m$ solves the problem

$$\lambda\xi = L(\varphi)\xi + d_m(x) - \tilde{d}_m\xi_1^*(x), \quad x \in \Omega;$$

$$B(\varphi)\xi = 0, \quad x \in \partial\Omega,$$

and is orthogonal to the firste eigenfunction of the conjugate problem.
To determine the function $\xi_{m+1}(x,t,\tau)$ consider the following problem

$$(\xi_{m+1})_t = L(\varphi)\xi_{m+1} + d_{m+1}(x) + \tau d_{m+2}(x,t,\tau,\xi_{m+1}), \quad (x,t) \in Q_T,$$

$$B(\varphi)\xi_{m+1} = 0, \quad (x,t) \in \Gamma_T, \quad \xi_{m+1}(x,0,\tau) = 0. \tag{4.17}$$

It was shown in Vishnevskii (1981b), by the disturbance method,
that such positive τ_0 exists, that for $|\tau| < \tau_0$, problem (4.17) has a
unique solution in Q_T from the class $H_1^{2+\alpha}(Q_T)$. Moreover, for the
function $w(x,t,\tau)$ from (4.13) for $t = 0$, we have

$$w_t(x,0;\tau) = \tau^m(\tilde{d}_m\xi_1(x) + \tau\xi_{m+1}(x,0,\tau)). \tag{4.18}$$

Consider the case when the described process will not stop for any
m. Then, we obtain the formal series

$$w(x,\tau) = \tau\xi_1(x) + \ldots + \tau^m\xi_m(x) + \ldots. \tag{4.19}$$

Now we must prove that for the small τ this series converges in

$C^{2+\alpha}(\bar{\Omega})$ and $w(x;\tau)$ satisfies (4.4), (1.1)-(1.3). Assume that $z(x)$ solves the problem

$$0 = L(\varphi)z + \tilde{F}(z + \tau\xi_1) - F_1(z + \tau\xi_1), \quad x \in \Omega,$$

$$B(\varphi)\xi = 0, \ x \in \partial\Omega, \quad F_1(z + \tau\xi_1) = \int \tilde{F}(z + \tau\xi_1)\xi_1^*(x)dx, \quad (4.20)$$

and the function $z(x)$ is orthogonal to the first eigenfunction of the conjugate problem.

It was shown in Vishnevskii (1981) that problem (4.20) has for $|\tau| < \tau_0$ a unique solution which analytically depends on τ. Therefore,

$$z(x,\tau) = \tau^2 z_2(x) + \dots.$$

Substituting the series for $z(x,\tau)$ into (4.4), we note that for $z_i(x)$ we obtain the same relations as for $\xi_i(x,\tau)$. Therefore,

$$\xi_i(x) = z_i(x); \ F_1(z + \tau\xi_1) \equiv 0,$$

and $w(x,\tau) \equiv z(x,\tau)$ is a one-parameter family of problem (4.4) solutions.

Thus, we have shown that the set $w(x,\tau)$ is a one-parameter set of stationary solutions of problem (4.4) which lies in a neighborhood of the trivial stationary solution.

This contradicts the initial assumption, that the trivial solution of (4.4) is asymptotically stable from above. Consequently, there always exists such m that (4.16) will hold true. As the trivial solution is stable from above, we obtain $-\tilde{d}_m < 0$, or $\tilde{d}_m > 0$.

Note, that (see (4.16))

$$\gamma = \tau - \tilde{d}_m \tau^m + \tau^{m+1} P_1 \xi_{m+1}(x,t,\tau).$$

Expressing from this formula τ by γ and t, we obtain (4.12). The lemma is proved.

Lemma 4.9.   Let all the suppositions of lemma 4.8 hold. Then there exists such $\delta > 0$ that for $t \in [0, \delta]$, $\gamma > 0$ the inequality

$$\frac{\partial}{\partial t}\eta(\gamma(t)) \leq -C_4(\gamma_0)e^{(-qm(C+1))t}\xi_1(\varphi). \qquad (4.21)$$

holds.

Proof.   Note, that for $\gamma_0 > 0$ the function $\gamma(t)$ is positive, Therefore, (4.3), (4.6) yield that

$$P_1 F(\eta(\gamma(t))) \geq -Cq\gamma - \gamma,$$

i.e.

$$\gamma(t) \geq \gamma_0 e^{-q(C+1)t}.$$

Substituting the latter inequality into (4.12) and taking positive ε so that

$$\varepsilon f(\gamma, t) < 1/2\tilde{d}_m,$$

we obtain the required inequality (4.21). The lemma is proved.

Lemma 4.10.   Let $\lambda_1(\varphi) < 0$. Then there exists such $\delta > 0$ that $t \in [0, \delta]$, $\gamma > 0$ and the inequality

$$\frac{\partial}{\partial t}\eta(\gamma(t)) \leq -\frac{1}{2}\lambda_1(\varphi)e^{(\lambda_1(\varphi)-q(C+1))t}\xi_1(\varphi). \qquad (4.22)$$

holds.

Proof   of this lemma is just the same as the proof of lemma 4.9, and is therefore omitted.

Proof of theorem 4.2. Choose a positive ε so that

$$\mathrm{Re}\lambda_2(\varphi) + C_1 q(2\varepsilon, 2\varepsilon) < -(C+1)mq(2\varepsilon, 2\varepsilon).$$

Write the solution $w(x, t; u_0 - \varphi)$ as follows

$$\eta(\gamma(t)) + \zeta(x,t)e^{(\mathrm{Re}\lambda_2(\varphi) + Cq(2\varepsilon, 2\varepsilon))t} \qquad (4.23)$$

By (4.10), the function $\zeta(x,t)$ is bounded in $H_1^{2+\alpha}(Q)$. Relation (4.11) yields

$$\| w_0 - \eta(\gamma_0) \|_1^\Omega < Cl\varepsilon.$$

The monotonicity of the solution gives

$$\| \eta(\gamma(t)) \|_{1,2+\alpha}^\Omega < Cl\varepsilon.$$

Choose l so that $Cl < 2$. Applying (4.21) or (4.22), we obtain

$$\frac{\partial w}{\partial t}(x,t; u_0 - \varphi) \le C_4\xi_1(\varphi)e^{-(C+1)mqt} + \zeta(x,t)e^{(\mathrm{Re}\lambda_2 + C_1 q)t}$$

where $w(x, t, u_0 - \varphi)$ solves (4.2).

Therefore, for $t > \tau(u_0)$, $\partial w/\partial t\, (x, t; u_0 - \varphi)$ becomes negative. To complete the proof note that

$$\frac{\partial u}{\partial t}(x,t; u_0) = \frac{\partial w}{\partial t}(x,t; u_0 - \varphi),$$

where $u(x, t; u_0)$ solves (1.1)-(1.3). Theorem 4.2 is proved.

Proof of theorem 4.3. Let φ_n be a sequence of stationary solutions of problem (1.1)-(1.3) from $\hat{S}$. This sequence is bounded in $C^{2+\alpha'}(\bar\Omega)$; therefore, we may choose from it the converging subsequence ($0 < \alpha < \alpha' < 1$).

Denote by φ the limit of this sequence. Evidently φ is a stationary solution of the problem. Note, that φ has a trivial eigenvalue, since it is not isolated. As the second eigenvalues of the stationary solutions φ_n are strictly separated from the imaginary axis, then the trivial eigenvalue may be only the first eigenvalue (Lions, 1984; Vishnevskii, 1987, 1992). Recall the proof of lemma 4.8. The stationary solution φ is not isolated; therefore, (4.16) is false for all m. Consequently, in a neighborhood of stationary solution φ there is the one-parameter family of the stationary solutions of the problem which contains the family (4.19). This component is ordered on the one hand, and on the other hand, by the analycity, this component could not contain the maximal element, i.e., it must be unbounded. This contradicts the dissipativity of the problem. Therefore, our initial supposition, that there exists a bounded sequence of stationary solutions from $\hat{S}$, is false. This means that in a bounded domain, problem (1.1)-(1.3) cannot have infinitely many stationary solutions from $\hat{S}$. But since the problem is dissipative, all stationary solutions lie in a certain bounded domain. Therefore, the problem has no more than a finite number of stationary solutions from $\hat{S}$. The theorem is proved.

Proof of theorem 4.16. Numerate the finite number of stationary solutions from $\hat{S}$. Select from this set of stationary solutions those which are asymptotically stable from below, from above, or simply asymptotically stable. Denote them by $\varphi_1, \ldots, \varphi_m$. Recall, that if φ is an asymptotically stable from above stationary solution, then, by $A^+(\varphi)$, we have denoted the subset consisting of such functions u_0, that $u(x, t; u_0)$ for $t > \tau(u_0)$ becomes strictly monotone decreasing. Analogously we may define $A^-(\varphi)$.

Consider the following set

$$M_1 = \left(\bigcup_{i=1}^{m} A^+(\varphi_i) \right) \bigcup \left(\sum_{i=1}^{m} A^-(\varphi_i) \right).$$

Prove that this set satisfies all conditions of theorem 4.4. For this purpose take any $v_0 \in E$ and show that such an open set V_{v_0} exists that $v_0 \in \bar{V}_{v_0}$, $V_{v_0} \subset M_1$.

From the theorem on the continuous dependence on initial data it follows that M_1 is open. Therefore, if $v_0 \in M_1$ then our statement is evident. Let $v_0 \notin M_1$. As the problem is dissipative, $w(u_0)$ is not empty. Assume that $w \in w(u_0)$.

Three cases are possible.

1). The function w is an asymptotically stable from above or from below stationary solution of the problem. In this case, as V_{v_0} we may take either $A^+(w)$ or $A^-(w)$ depending on the solution being asymptotically stable from below or from above.

2). The function w is a stationary solution, which is not asymptotically stable from above or from below. In this case, if $w \in \hat{S}$, then it is isolated, and therefore, is unstable. If $w \notin \hat{S}$ then it is also unstable. Assume for definiteness that the stationary solution w is unstable from above. Then, by theorem 3.8, there exists such an asymptotically stable from below stationary solution w_1, that the closure of $A^-(w_1)$ contains the stationary solution w. Let

$$w_1 \geq u_0 \geq v_0, \quad u_0 \not\equiv v_0.$$

Then, beginning with a certain moment of time, as was shown in lemma 4.5, the solution $u(x, t; u_0)$ becomes greater then w, and, therefore, u_0 belongs to $A^-(w_1)$. Consequently, in this case, as V_{v_0} we may take the set $A^-(w_1)$.

3). The function w is not a stationary solution of the problem. Then, we must construct V_{v_0} as in the point 2), but instead of theorem 3.8 we have to use theorem 3.4. The theorem is proved.

A

Appendix.

§A.1 Setting of the problem.

Here we shall consider the following problem

$$u_t = a(u_x)u_{xx} + 2\mu u u_x, \qquad (x,t) \in Q_T, \qquad (1.1)$$

$$u(0,t) = u(1,t) = 0, \qquad (1.2)$$

$$u(x,0) = u_0(x). \qquad (1.3)$$

Here $Q_T = [0,1] \times [0,T]$, and μ is a constant. Moreover, the smooth coefficient $a(\xi)$ satisfies the inequality

$$a(\xi) \geq \delta > 0$$

for $|\xi| > N$, and its sign may change for $|\xi| < N$.

Our interest in the problems similar to (1.1)-(1.3) is mainly due to Yanenko (Yanenko and Novikov, 1973; Yanenko, 1980). For the physical interpretation of the model and the applications we refer the reader to Lar'kin *et al.*(1983) and the references therein.

In Lar'kin *et al.*(1983) equations similar to (1.1) are named changing type equations (or equations of variable type). These equations also arise as the first differential approximation of some numerical schemes for the wave equation (Lar'kin *et al.*, 1983). Without entering into details one can mention, e.g., Starr (1968) and Gokhman *et al.*, (1981), where some physical motivations can be found.

Consider (1.1) as a one-dimensional Navier-Stokes equation, then the term $a(u_x)$ is related to the kinematic viscosity. The Navier-Stokes system with viscosity coefficient dependent on the gradient of the solution is considered in Golovkin (1967) and in Ladyzhenskaya (1967, 1968), and it is proved that a special type of this dependence makes it possible to get some existence theorems.

Qualitative investigation of problem (1.1)-(1.3) appears to be not trivial. In the papers of Zelenyak *et al.*(1974) and Lavrentiev, Jr. (1980, 1982) the following estimate (see also chapter 2)

$$\iint\limits_{Q_T} \left[u_t^2 + a^2 u_{xx}^2 \right] dx dt + \sup\left(|u| + |u_x|\right) \le K(u_0, \delta, N) \qquad (1.4)$$

was proved. Here $u(x,t)$ is a smooth solution to problem (1.1)-(1.3).

Inequalities similar to (1.4) hold for solutions of different types of approximated problems, such as those obtained by adding to (1.1) the terms like, e.g, $\varepsilon \partial^4 u / \partial x^4$ (or - $\varepsilon \partial^5 u / \partial x^4 \partial t$), or by substituting differential operators acting on a space variable by the related finite-difference operators. From these estimates we can conclude that the families of approximate solutions are weakly compact in suitable functional spaces. Moreover, it suffices to prove an existence theorem if $a(\xi) \ge 0, \forall \xi \in R$ (Lavrentiev, Jr., 1980). But we do not know as yet how to treat the limit of the term $a(u_x)u_{xx}$.

In certain cases however we can pass the limit by taking into consideration a special probability measure. In this way, we can obtain the so-called measure valued solution (Slemrod, 1991, and references therein).

The main difficulty results from the fact that we do not know *a priori* the set of degeneration, i.e., the set on which $a(u_x)$ vanishes. There are many investigations devoted to the equations with a definite line of sign changing of the leading coefficient u_{xx} (see, for example, Podgaev, 1982; Tersenov, 1982). It has been proved that sometimes the correct problem can be obtained by setting so-called nonlocal initial conditions. Roughly speaking, it means that we must set the initial data at $t = 0$ on the domain in which the coefficient is positive, and set this initial data at $t = T$ on the domain in which the coefficient is negative. Taking account of the effect of a bounded speed of propagation of the perturbation for certain nonlinear parabolic equations (Kalashnicov, 1981; Samarskii *et al.*, 1987; Herrego and Vazquez, 1987) yields several similar results for equations with an unknown *a priori* line of degeneration (Grebnev and Novikov, 1987), but in these latter cases the problem actually splits into different problems. For each of these problems, the coefficient has a definite sign on the corresponding domains and vanishes on the line dividing such domains (Lavrentiev, Jr., 1987).

Regarding problem (1.1)-(1.3), many other aspects have been investigated. Among them, let us recall numerical computations (Voronko, 1977; Maslova and Novikov, 1980; Kim and Zelinskaya, 1981), analysis of the stationary equation (Grebnev and Novikov, 1980; Akhmerov, 1984, 1985), stability (Maslova *et al.*, 1984; Novikov and Shvab, 1985) and so on (see also Belonosov and Zelenyak, 1975; Pyatkov, 1987; Pyatkov and Podgaev, 1987; Lavrentiev, Jr., 1989). Moreover, in Hollig (1983), a class of special solutions is constructed. The solutions belonging to this class uniformly converge to certain initial data as $t \to 0$, but do not lie in the domain on which the principal coefficient is negative if $t > 0$.

Here we shall prove *a priori* estimates for smooth solutions of (1.1)-(1.3) similar to

$$\| u_{xx}(x,t) \|_{L_2(Q_{T-\epsilon})} \le K, \tag{1.5}$$

where K is a positive constant, and $Q_{T-\epsilon} = [0,1] \times [0, T-\varepsilon]$, $\varepsilon \in (0,T)$.

They are quite unexpected for equation (1.1) with a degenerate leading coefficient.

It should be noted that if one could prove (1.5) for the set of approximate solutions, then it would be easy to get the existence theorem for (1.1)-(1.3) by using a compactness argument.

To get (1.5) the Liapunov functional method is used.

§A.2   The basic estimates.

Here, we shall investigate equation (1.1) with $\mu = 0$, assuming the function $a(\xi)$ to be of special form. We shall show the basic idea obtaining the proof of the desired estimate. Further, we shall prove our estimates for other cases in similar fashion.

Let us consider the following problem

$$u_t = a(u_x)u_{xx}, \tag{2.1}$$

$$u(0,t) = u(1,t) = 0, \tag{2.2}$$

$$u(x,0) = u_0(x), \tag{2.3}$$

where $(x,t) \in Q_T = [0,1] \times [0,T]$, and $a(\xi)$ is defined as follows

$$a(\xi) = \begin{cases} -\cos\xi & \text{for} \quad |\xi| \le \pi/2, \\ |\xi| - \pi/2 & \text{for} \quad |\xi| > \pi/2. \end{cases} \tag{2.4}$$

Notice that $a(\xi) \in C^2$ and $a(\xi) \ge \pi/2$ for $|\xi| \ge \pi$, $a(\xi) < 0$ for $|\xi| < \pi/2$. Moreover, note that (see the next remark)

$$a(\xi)a''(\xi) \le 0 \quad \text{for} \quad \xi \in R. \tag{2.5}$$

Remark 2.1.   All the following results remain true for any $a(\xi) \in C^2$ such that it satisfies (2.5), has no multiple roots, and $a(\xi) \ge \delta > 0$ for $|\xi| > N$.

Theorem 2.1.   Let $u(x,t) \in C^3(Q_T)$ be a solution of (1.1)-(1.3) with $a(\xi)$ given by (2.4).

Then for any $\varepsilon \in (0,T)$, the inequality

$$\| u_x(x,t) \|_{W_2^1(Q_{T-\varepsilon})} \le K\varepsilon^{-1} \tag{2.6}$$

holds.

Below we shall use the equation

$$u_{xt} = a(u_x)u_{xxx} + a'(u_x)u_{xx}^2 \tag{2.7}$$

and the boundary conditions

$$u_t\,|_{x=0} = u_t|_{x=1} = 0,$$

$$a(u_x)u_{xx}\,|_{x=0} = a(u_x)u_{xx}|_{x=1} = 0. \tag{2.8}$$

One can easily get (2.7), (2.8) from (2.1), (2.2) by using the smoothness of $u(x,t)$.

We start by proving some estimates necessary to prove the theorem.

Lemma 2.1. Under assumptions of theorem 2.1, the following estimates

$$\int\limits_0^T \iint\limits_{Q_t} \left(u_{x\tau}^2 + a^2 u_{xxx}^2\right) dx d\tau dt \le K, \tag{2.9}$$

$$\int\limits_0^T \iint\limits_{Q_t} \left(a'(u_x)u_{xx}^2\right)^2 dx d\tau dt \le K \tag{2.10}$$

hold.

Proof. Let us multiply (2.1) by u_{xx}. Using (2.8), write the identities

$$\int\limits_0^1 a(u_x)u_{xx}^2 dx = \int\limits_0^1 u_t u_{xx} dx = -\int\limits_0^1 u_x u_{xt} dx = -\frac{1}{2}\frac{d}{dt}\int\limits_0^1 u_x^2 dx.$$

According to (1.4) it follows from these identities that

$$\left|\int\limits_0^t\int\limits_0^1 a(u_x)u_{xx}^2 dx d\tau\right| \le K, \quad \forall t \in [0,T]. \tag{2.11}$$

Let us consider the equality

$$\frac{d}{dt}\int_0^1 a(u_x)u_{xx}^2 dx = 2\int_0^1 a(u_x)u_{xx}u_{xxt}dx + \int_0^1 a'(u_x)u_{xx}^2 u_{xt}dx. \quad (2.12)$$

Integrating by parts in the right side of (2.12), and using (2.7), (2.8), we obtain

$$\frac{d}{dt}\int_0^1 a(u_x)u_{xx}^2 dx = -2\int_0^1 a(u_x)u_{xxx}u_{xt}dx - \int_0^1 a'(u_x)u_{xx}^2 u_{xt}dx$$

$$= -\int_0^1 u_{xt}^2 dx - \int_0^1 a(u_x)u_{xxx}\left(au_{xxx} + a'u_{xx}^2\right)dx$$

$$= -\int_0^1 \left[u_{xt}^2 + a^2 u_{xxx}^2\right]dx - \int_0^1 aa'u_{xx}^2 u_{xxx}dx. \quad (2.13)$$

It is obvious that

$$-\int_0^1 a(u_x)a'(u_x)u_{xx}^2 u_{xxx}dx = -\frac{1}{3}\int_0^1 aa'\frac{d}{dx}u_{xx}^3 dx$$

$$= \frac{1}{3}\int_0^1 (a')^2 u_{xx}^4 dx + \frac{1}{3}\int_0^1 a(u_x)a''(u_x)u_{xx}^4 dx. \quad (2.14)$$

Thus, from (2.13), we can get

$$\frac{d}{dt}\int_0^1 au_{xx}^2 dx = -\int_0^1 \left(u_{xt}^2 + a^2 u_{xxx}^2\right)dx + \frac{1}{3}\int_0^1 \left(a'u_{xx}^2\right)^2 dx$$

$$+ \frac{1}{3}\int_0^1 a(u_x)a''(u_x)u_{xx}^4 dx. \quad (2.15)$$

From equation (2.7) it is easy to see that

$$\int_0^1 \left(a'u_{xx}^2\right)^2 dx \le 2\int_0^1 \left(u_{xt}^2 + a^2 u_{xxx}^2\right)dx. \quad (2.16)$$

On the other hand, because of (2.5), the inequality

$$\int_0^1 a(u_x)a''(u_x)u_{xx}^4 dx \leq 0 \tag{2.17}$$

holds.

Using (2.14)-(2.17) one can see that

$$3\int_0^1 au_{xx}^2 dx + \iint_{Q_T} \left(u_{xt}^2 + a^2 u_{xxx}^2\right) dxdt \leq K(u_0).$$

From this and (2.11) we get (2.9). Besides, from (2.14), we also have (2.10). Lemma 2.1 is completely proved.

This lemma will be used further to prove theorem 2.1.

§A.3 Proof of theorem 2.1.

It is well-known that

$$\int_0^t \int_0^\tau f(\xi)d\xi d\tau = \int_0^t (t - \tau)f(\tau)d\tau. \tag{3.1}$$

Therefore, it follows from lemma 2.1 that for any $\varepsilon \in (0, T)$

$$\iint_{Q_{T-\epsilon}} \left(u_{xt}^2 + a^2 u_{xxx}^2\right) dxdt \leq K\varepsilon^{-1}, \tag{3.2}$$

$$\iint_{Q_{T-\epsilon}} \left(a'u_{xx}^2\right)^2 dxdt \leq K\varepsilon^{-1}. \tag{3.3}$$

From (1.4), (3.3) we can write the following inequality

$$\iint_{Q_{T-\epsilon}} \left[a^2(u_x)u_{xx}^2 + (a'(u_x))^2 u_{xx}^4\right] dxdt \leq K\varepsilon^{-1}. \tag{3.4}$$

It is easy to see that the function $a(\xi)$ (see (2.4)) is such that the following condition holds

i)   if $|a(\xi)| < 1/4$ then $|a'(\xi)| > 1/4$.

Let us set for any fixed $\varepsilon > 0$

$$\Omega_1 = \left\{(x,t) \in Q_{T-\varepsilon} : |a(u_x(x,t))| \geq 1/4\right\},$$

$$\Omega_2 = \left\{(x,t) \in Q_{T-\varepsilon} : |a(u_x(x,t))| < 1/4\right\},$$

$$\Omega_3 = \left\{(x,t) \in \Omega_2 : |u_{xx}(x,t)| < 1\right\},$$

$$\Omega_4 = \left\{(x,t) \in \Omega_2 : |u_{xx}(x,t)| \geq 1\right\}.$$

One can easily see that $Q_{T-\varepsilon} = \Omega_1 \bigcup \Omega_2 = \Omega_1 \bigcup \Omega_3 \bigcup \Omega_4$. So, because of the property (i), we can write the following chain of inequalities

$$\iint\limits_{Q_{T-\varepsilon}} \left(a^2 u_{xx}^2 + a'^2 u_{xx}^4\right) dxdt \geq \iint\limits_{\Omega_1} a^2 u_{xx}^2 dxdt$$

$$+ \iint\limits_{\Omega_2} a'^2 u_{xx}^4 dxdt \geq \frac{1}{16}\iint\limits_{\Omega_1} u_{xx}^2 dxdt + \frac{1}{16}\iint\limits_{\Omega_2} u_{xx}^4 dxdt$$

$$\frac{1}{16}\iint\limits_{\Omega_1 \cup \Omega_2} u_{xx}^2 dxdt \geq \frac{1}{16}\iint\limits_{Q_{T-\varepsilon}} u_{xx}^2 dxdt - \frac{1}{16}T.$$

It means that (see (3.4))

$$\iint\limits_{Q_{T-\varepsilon}} u_{xx}^2 dxdt \leq K\varepsilon^{-1}. \tag{3.5}$$

Finally, the proof follows from (3.2), (3.5).

§A.4   Estimate for the polynomial function $a(\xi)$.

Here, we are going to prove an analog of theorem 2.1 for

$$a(\xi) = \gamma_2 \xi^2 + \gamma_1 \xi + \gamma_0 \qquad (\gamma_2 > 0).$$

The quadratic function $a(\xi)$ is related to the above mentioned numerical computations (Ladyzhenskaya, 1967, 1968; Lar'kin et al, 1983) and to the Cahn-Hilliard equation, see chapter 3 (Starr, 1968; Voronko, 1977; Tersenov, 1982; Pyatkov and Podgaev, 1987; Samarskii et al, 1987; Slemrod, 1991).

Let us consider problem (1.1)-(1.3) with

$$a(\xi) = \gamma_2 \xi^2 + \gamma_1 \xi + \gamma_0, \tag{4.1}$$

where $\gamma_2 > 0$ and

$$\gamma_1^2 - 4\gamma_0\gamma_2 > 0. \tag{4.2}$$

Condition (4.2) implies $a(\xi) < 0$ for $\xi \in (\gamma_+, \gamma_-)$, where

$$\gamma_\pm = \left(-\gamma_1 \pm \sqrt{\gamma_1^2 - 4\gamma_0\gamma_0}\right)/(2\gamma_2).$$

Then we have

Theorem 4.1. Let $u(x,t) \in C^3(Q_T)$ be a solution to (2.1)-(2.3) with function $a(\xi)$ given by (4.1) and let inequality (4.2) hold.

Then for any $\varepsilon \in (0,T)$, inequality (2.6) holds.

Proof. By using transformations quite similar to those used in §A.3 one can get the analog of identity (2.13)

$$\frac{d}{dt} \int_0^1 a(u_x)u_{xx}^2 dx = -\int_0^1 \left(u_{xt}^2 + a(u_x)u_{xxx}^2\right) dx +$$

$$\frac{1}{3}\int_0^1 a'^2 u_{xx}^4 dx + \frac{1}{3}\int_0^1 a(u_x)a''(u_x)u_{xx}^4 dx. \tag{4.3}$$

From (4.1), we obtain

$$a(\xi)a''(\xi) = \frac{1}{2}(a'(\xi))^2 - \frac{1}{2}(\gamma_1^2 - 4\gamma_0\gamma_2). \tag{4.4}$$

Since $|u_x(x,t)| < K$ (see (1.4)), then there exists $\delta > 0$ such that, (taking into account (4.2), (4.4))

$$a(u_x)a''(u_x) \leq (1/2 - 3\delta)(a'(u_x))^2. \tag{4.5}$$

By using (4.3), (4.5), we can write the following estimate

$$\int_0^1 a(u_x)u_{xx}^2 dx - \int_0^1 a(u_0')(u_0'')^2 dx$$

$$\leq \iint_{Q_t} \left(u_{xt}^2 + a^2 u_{xxx}^2 \right) dx\,dt + (1/2 - \delta) \iint_{Q_t} \left(a' u_{xx}^2 \right)^2 dx\,dt.$$

Thus, owing to (2.14), (2.8) we see that

$$\int_0^T \iint_{Q_t} \left[u_{x\tau}^2 + a^2 u_{xxx}^2 + \left(a' u_{xx}^2 \right)^2 \right] dx\,d\tau\,dt \leq K. \tag{4.6}$$

Reasons given in §A.3 show that the proof of theorem 4.1 follows from the inequality (4.6).

§A.5   Estimate for the case $\mu \neq 0$.

In this paragraph, we consider the most involved case, i.e., problem (1.1)-(1.3) with $\mu \neq 0$ and $a(\xi)$ given by (2.4).

More precisely, we consider the following problem

$$u_t = a(u_x)u_{xx} + 2\mu u u_x, \tag{5.1}$$

$$u(0,t) = u(1,t) = 0, \tag{5.2}$$

$$u(x,0) = u_0(x). \tag{5.3}$$

Here $\mu = $ const and the coefficient $a(\xi) \in C^2$ (without multiple roots) is such that

$$a(\xi)a''(\xi) \le 0, \qquad \forall \xi \in R, \tag{5.4}$$

$$a(\xi) \ge \delta > 0 \quad \text{for} \quad |\xi| > N.$$

Theorem 5.1. Let $u(x,t) \in C^3(Q_T)$ be a solution to problem (5.1)-(5.4). Then for any $\varepsilon \in (0,T)$, inequality (2.6) holds.

Proof. As in §A.3, we use the following corollaries of equation (5.1) and boundary conditions (5.2)

$$u_{xt} = au_{xxx} + a'u_{xx}^2 + 2\mu u u_{xx} + 2\mu u_x^2;$$

$$u_t(0,t) = u_t(1,t) = 0, \tag{5.5}$$

$$a(u_x)u_{xx}\big|_{x=0} = a(u_x)u_{xx}\big|_{x=1} = 0.$$

Let us consider the following chain of inequalities (see (5.1), (5.2), (5.5))

$$\frac{d}{dt}\int_0^1 a(u_x)u_{xx}^2 dx = \int_0^1 \left(2au_{xx}u_{xxt} + a'u_{xx}^2 u_{xt}\right)dx$$

$$= 2\int_0^1 (au_{xx} + 2\mu u u_x)u_{xxt}dx - \int_0^1 4\mu u u_x u_{xxt}dx$$

$$+ \int_0^1 \left(u_{xt} - au_{xxx} - 2\mu u u_{xx} - 2\mu u_x^2\right)u_{xt}dx = 2\int_0^1 u_t u_{xxt}dx$$

$$- \int_0^1 4\mu u u_x u_{xxt}dx + \int_0^1 u_{xt}^2 dx - \int_0^1 \left(2\mu u u_{xx} + 2\mu u_x^2\right)u_{xt}dx$$

$$- \int_0^1 au_{xxx}\left(au_{xxx} + a'u_{xx}^2 + 2\mu u u_{xx} + 2\mu u_x^2\right)dx$$

$$= -\int_0^1 \left(u_{xt}^2 + a^2 u_{xxx}^2\right) dx - \int_0^1 4\mu u u_x u_{xxt} dx - \int_0^1 aa' u_{xx}^2 u_{xxx} dx$$

$$-\int_0^1 \left(2\mu u u_{xx} + 2\mu u_x^2\right)\left(u_{xt} + a u_{xxx}\right) dx. \qquad (5.6)$$

Notice that

$$-\int_0^1 aa' u_{xx}^2 u_{xxx} dx = -\frac{1}{2}\int_0^1 (a u_{xx}) a'(u_x)\frac{d}{dx} u_{xx}^2 dx$$

$$= \frac{1}{2}\int_0^1 \frac{d}{dx}(u_t - 2\mu u u_x) a' u_{xx}^2 dx + \frac{1}{2}\int_0^1 a u_{xx} a'' u_{xx} u_{xx}^2 dx$$

$$= \frac{1}{2}\int_0^1 aa'' u_{xx}^4 dx - \frac{1}{2}\int_0^1 \left(2\mu u u_{xx} + 2\mu u_x^2\right) a' u_{xx}^2 dx$$

$$+ \frac{1}{2}\int_0^1 u_{xt}\left(u_{xt} - a u_{xxx} - 2\mu u u_{xx} - 2\mu u_x^2\right) dx. \qquad (5.7)$$

On the other hand, from (5.6), (5.7) we obtain

$$\frac{d}{dt}\int_0^1 a u_{xx}^2 dx = -\frac{1}{2}\int_0^1 \left[u_{xt}^2 + u_{xt} a u_{xxx} + 2a^2 u_{xxx}^2\right] dx$$

$$-\int_0^1 4\mu u u_x u_{xxt} dx - \int_0^1 \left(3\mu u u_{xx} + 3\mu u_x^2\right) u_{xt} dx + \frac{1}{2}\int_0^1 aa'' u_{xx}^4 dx$$

$$-\int_0^1 \left(2\mu u u_{xx} + 2\mu u_x^2\right) a u_{xxx} dx - \frac{1}{2}\int_0^1 \left(2\mu u u_{xx} + 2\mu u_x^2\right) a' u_{xx}^2 dx. \qquad (5.8)$$

Moreover, we have

$$-\int_0^1 \left(2\mu u u_{xx} + 2\mu u_x^2\right) u_{xt} dx = 2\mu \int_0^1 u u_x u_{xxt} dx, \qquad (5.9)$$

$$-\frac{1}{2}\int_0^1 \left(2\mu u u_{xx} + 2\mu u_x^2\right) a' u_{xx}^2 dx$$

$$= -\frac{1}{2}\int_0^1 2\mu u u_{xx}^2 \frac{d}{dx} a(u_x) dx - \frac{1}{2}\int_0^1 2\mu u_x^2 a' u_{xx}^2 dx$$

$$= \int_0^1 2\mu u u_{xx} a u_{xxx} dx + \frac{1}{2}\int_0^1 2\mu u_x a u_{xx}^2 dx$$

$$-\frac{1}{2}\int_0^1 2\mu u_x^2 u_{xt} dx + \frac{1}{2}\int_0^1 2\mu u_x^2 a u_{xxx} dx$$

$$+\frac{1}{2}\int_0^1 2\mu u_x^2 \left(2\mu u u_{xx} + 2\mu u_x^2\right) dx, \tag{5.10}$$

$$\int_0^1 u u_{xx} u_x^2 dx = \frac{1}{3}\int_0^1 u \frac{d}{dx} u_x^3 dx = -\frac{1}{3}\int_0^1 u_x^4 dx. \tag{5.11}$$

By using (5.9)-(5.11) we find (taking into account (5.8))

$$\frac{d}{dt}\int_0^1 a u_{xx}^2 dx = -\frac{1}{2}\int_0^1 \left(u_{xt}^2 + u_{xt} a u_{xxx} + 2a^2 u_{xxx}^2\right) dx$$

$$+\frac{1}{2}\int_0^1 a a'' u_{xx}^4 dx - \int_0^1 \mu u u_x u_{xxt} dx - \frac{1}{2}\int_0^1 2\mu u_x^2 a u_{xxx} dx$$

$$+\frac{1}{2}\int_0^1 2\mu u_x a u_{xx}^2 dx - \int_0^1 \mu u_x^2 u_{xt} dx + \frac{4}{3}\int_0^1 \mu^2 u_x^4 dx. \tag{5.12}$$

Notice that

$$\frac{1}{2}\int_0^1 2\mu u_x u_{xx} a u_{xx} dx = \frac{1}{2}\int_0^1 2\mu u_x u_{xx} u_t dx$$

$$-\frac{1}{2}\int_0^1 4\mu^2 u u_x^2 u_{xx} dx = -\frac{1}{4}\int_0^1 2\mu u_x^2 u_{xt} dx + \frac{2}{3}\int_0^1 \mu^2 u_x^4 dx, \tag{5.13}$$

$$-\int_0^1 \mu u_x^2 a u_{xxx} dx \le \frac{1}{2}\int_0^1 a^2 u_{xxx}^2 dx + \frac{1}{2}\int_0^1 \mu^2 u_x^4 dx. \qquad (5.14)$$

Besides, from (5.4) it follows that

$$\int_0^1 a(u_x)a''(u_x)u_{xx}^4 dx \le 0. \qquad (5.15)$$

Summing up, by virtue of (5.12)-(5.15), we have the inequality

$$\int_0^1 a(u_x)u_{xx}^2 dx - \int_0^1 a(u_0')(u_0'')^2 dx$$

$$\le -\frac{1}{2}\iint_{Q_t}\left(u_{x\tau}^2 + u_{x\tau}a u_{xxx} + a^2 u_{xxx}^2\right)dx d\tau$$

$$-\iint_{Q_t}\mu u u_x u_{xx\tau} dx d\tau + \frac{5}{2}\iint_{Q_t}\mu^2 u_x^4 dx d\tau$$

$$-\frac{1}{2}\int_0^1 \mu\left(u_x^3 - (u_0')^3\right)dx. \qquad (5.16)$$

Recalling (5.1), (5.2), (5.5) we get

$$-\iint_{Q_t}\mu u u_x u_{xx\tau} dx d\tau = -\frac{1}{2}\iint_{Q_t}\left(u_\tau - a u_{xx}\right)u_{xx\tau} dx d\tau$$

$$-\frac{1}{2}\iint_{Q_t}u_\tau u_{xx\tau} dx d\tau + \frac{1}{4}\iint_{Q_t}a(u_x)\frac{d}{d\tau}u_{xx}^2 dx d\tau$$

$$= \frac{1}{2}\iint_{Q_t}u_{x\tau}^2 dx d\tau + \frac{1}{4}\int_0^1 a u_{xx}^2|_{\tau=0}^t dx - \frac{1}{4}\iint_{Q_t}a' u_{xx}^2 u_{x\tau} dx d\tau$$

$$= \frac{1}{2}\iint_{Q_t}u_{x\tau}^2 dx d\tau + \frac{1}{4}\int_0^1 a u_{xx}^2|_{\tau=0}^t dx$$

$$-\frac{1}{4} \iint\limits_{Q_t} \left(u_{x\tau} - a u_{xxx} - 2\mu u u_{xx} - 2\mu u_x^2 \right) u_{x\tau} dx d\tau$$

$$= \frac{1}{4} \iint\limits_{Q_t} \left(u_{x\tau}^2 + u_{x\tau} a u_{xxx} \right) dx d\tau$$

$$+ \frac{1}{4} \int\limits_0^1 a u_{xx} \big|_{\tau=0}^t dx - \frac{1}{2} \iint\limits_{Q_t} \mu u u_x u_{xx\tau} dx d\tau.$$

Hence, we have

$$- \iint\limits_{Q_t} \mu u u_x u_{xx\tau} dx d\tau = \frac{1}{2} \int\limits_0^1 a u_{xx}^2 \big|_{\tau=0}^t dx$$

$$+ \frac{1}{2} \iint\limits_{Q_t} \left(u_{x\tau}^2 + u_{x\tau} a u_{xxx} \right) dx d\tau. \qquad (5.17)$$

Thus, from (5.16), (5.17) one can obtain the inequality

$$\int\limits_0^1 a(u_x) u_{xx}^2 dx + \iint\limits_{Q_t} a^2 u_{xxx}^2 dx d\tau$$

$$\leq \int\limits_0^1 a(u_0')(u_0'')^2 dx + 5 \iint\limits_{Q_t} \mu^2 u_x^4 dx d\tau - \int\limits_0^1 \mu \left(u_x^3 - (u_0')^3 \right) dx. \qquad (5.18)$$

Let us multiply both sides of (5.1) by u_{xx} and integrate over Q_t. We easily get

$$\iint\limits_{Q_t} a(u_x) u_{xx}^2 dx d\tau = \iint\limits_{Q_t} u_\tau u_{xx} dx d\tau - \iint\limits_{Q_t} 2\mu u u_x u_{xx} dx d\tau$$

$$= -\frac{1}{2} \int\limits_0^1 \left(u_x^2 - (u_0')^2 \right) dx + \iint\limits_{Q_t} \mu u_x^3 dx d\tau. \qquad (5.19)$$

Then, from (5.18), (5.19) (see also (1.4)) we derive

$$\int\limits_0^T \iint\limits_{Q_t} a^2 u_{xxx}^2 \, dx d\tau dt \le K \left(\| u_0 \|_{W_2^2(0,1)}, \sup_t \| u_x \|_{L_4(0,1)} \right). \qquad (5.20)$$

Applying (3.1), (5.20) yields the inequality

$$\iint\limits_{Q_{T-\varepsilon}} a^2 u_{xxx}^2 \, dx dt \le K \varepsilon^{-1}, \qquad \forall \varepsilon \in (0,T), \qquad (5.21)$$

where K is a positive constant depending only on the data.

Using (5.9), we obtain the equality (taking into account (5.6))

$$\frac{d}{dt} \int\limits_0^1 a(u_x) u_{xx}^2 \, dx = - \int\limits_0^1 \left(u_{xt}^2 + a^2 u_{xxx}^2 \right) dx - \int\limits_0^1 2\mu u u_x u_{xxt} \, dx$$

$$- \int\limits_0^1 a a' u_{xx}^2 u_{xxx} \, dx - \int\limits_0^1 \left(2\mu u u_{xx} + 2\mu u_x^2 \right) a u_{xxx} \, dx. \qquad (5.22)$$

Notice that (5.17) can be rewritten as follows

$$- \iint\limits_{Q_t} 2\mu u u_x u_{xx\tau} \, dx d\tau = \iint\limits_{Q_t} u_{x\tau}^2 \, dx d\tau$$

$$+ \iint\limits_{Q_t} u_{x\tau} \left(u_{x\tau} - a' u_{xx}^2 - 2\mu u u_{xx} - 2\mu u_x^2 \right) dx d\tau$$

$$+ \int\limits_0^1 a u_{xx}^2 \big|_{\tau=0}^t \, dx = \int\limits_0^1 a u_{xx}^2 \big|_{\tau=0}^t \, dx + 2 \iint\limits_{Q_t} u_{x\tau}^2 \, dx d\tau$$

$$- \iint\limits_{Q_t} u_{x\tau} a' u_{xx}^2 \, dx d\tau + \iint\limits_{Q_t} 2\mu u u_x u_{xx\tau} \, dx d\tau.$$

Thus, it is easy to see that

$$- \iint\limits_{Q_t} 2\mu u u_x u_{xx\tau} \, dx d\tau = \iint\limits_{Q_t} u_{x\tau} d\tau + \frac{1}{2} \int\limits_0^1 a u_{xx}^2 \big|_{\tau=0}^t \, dx$$

$$-\frac{1}{2} \iint\limits_{Q_t} u_{x\tau} a' u_{xx}^2 dx d\tau = \iint\limits_{Q_t} u_{x\tau}^2 dx d\tau + \frac{1}{2} \int\limits_0^1 a u_{xx}^2 |_{\tau=0}^t dx$$

$$-\frac{1}{2} \iint\limits_{Q_t} aa' u_{xx}^2 u_{xxx} dx d\tau - \frac{1}{2} \iint\limits_{Q_t} \left(a' u_{xx}^2 \right)^2 dx d\tau$$

$$-\frac{1}{2} \iint\limits_{Q_t} \left(2\mu u u_{xx} + 2\mu u_x^2 \right) a' u_{xx}^2 dx d\tau,$$

and, by using this chain of identities, we get (taking into account (5.22))

$$\frac{1}{2} \int\limits_0^1 a u_{xx}^2 |_{\tau=0}^t dx = -\frac{1}{2} \iint\limits_{Q_t} \left[2a^2 u_{xxx}^2 + \left(a' u_{xx}^2 \right)^2 \right] dx d\tau$$

$$-\frac{2}{3} \iint\limits_{Q_t} aa' u_{xx}^2 u_{xxx} dx d\tau - \iint\limits_{Q_t} \left(2\mu u u_{xx} + 2\mu u_x^2 \right) a u_{xxx} dx d\tau$$

$$-\frac{1}{2} \iint\limits_{Q_t} \left(2\mu u u_{xx} + 2\mu u_x^2 \right) a' u_{xx}^2 dx d\tau. \qquad (5.23)$$

Let us consider the following equalities

$$-\iint\limits_{Q_t} 2\mu u u_{xx} a u_{xxx} dx d\tau = -\frac{1}{2} \iint\limits_{Q_t} 2\mu u a(u_x) \frac{d}{dx} u_{xx}^2 dx d\tau$$

$$= \frac{1}{2} \iint\limits_{Q_t} 2\mu u u_{xx} a' u_{xx}^2 dx d\tau + \frac{1}{2} \iint\limits_{Q_t} 2\mu u_x u_{xx}^2 a(u_x) dx d\tau$$

$$= \frac{1}{2} \iint\limits_{Q_t} 2\mu u u_{xx} a' u_{xx}^2 dx d\tau + \frac{2}{3} \iint\limits_{Q_t} \mu^2 u_x^4 dx d\tau$$

$$-\frac{1}{6} \int\limits_0^1 \mu \left(u_x^3 - (u_0')^3 \right) dx, \qquad (5.24)$$

where we have used (5.13).

Using the well-known inequality

$$2ab \le \varepsilon a^2 + b^2/\varepsilon \qquad (\forall a, b \in R, \quad \forall \varepsilon > 0),$$

from (5.23) we derive the following (taking into account (5.24))

$$\int\limits_0^1 a u_{xx}^2 dx \leq \int\limits_0^1 a(u_0')(u_0'')^2 dx + \frac{1}{3}\int\limits_0^1 \mu\left((u_0')^3 - u_x^3\right) dx$$

$$-\frac{1}{4}\iint\limits_{Q_t}\left(a'u_{xx}^2\right)^2 dx d\tau + K\iint\limits_{Q_t}\left(a^2 u_{xxx}^2 + \mu^2 u_x^4\right) dx d\tau.$$

By considering (1.4), (5.21), (5.19) it is easy to realize that the following inequality holds

$$\iint\limits_{Q_{T-\varepsilon}}\left(a'(u_x)u_{xx}^2\right)^2 dx d\tau \leq K\varepsilon^{-1}, \qquad \forall \varepsilon \in (0, T). \tag{5.25}$$

Recalling §A.4 and the properties of function $a(\xi)$ (see (5.4)), the inequalities (1.4), (5.25) yield

$$\iint\limits_{Q_{T-\varepsilon}} u_{xx}^2 dx dt \leq K\varepsilon^{-1}, \qquad \forall \varepsilon \in (0, T). \tag{5.26}$$

Moreover, we obtain the following (taking into account (1.4), (5.21), (5.25), (5.26))

$$\iint\limits_{Q_{T-\varepsilon}} u_{xt}^2 dx dt \leq K\varepsilon^{-1} \tag{5.27}$$

Finally, estimates (5.26), (5.27) complete the proof.

§A.6   Setting of the model problem. Some solution estimates.

Now we are going to prove an existence theorem for special non-local initial conditions.

In the domain

$$\Omega_T = \{(y, t) \in [y_1, y_2] \times [0, T]\}$$

consider the problem

$$v_t = \frac{\omega'(y)}{v_y^2} v_{yy} - \frac{\omega''(y)}{v_y}, \tag{6.1}$$

$$v\big|_{y=y_1} = \varphi(t), \quad v\big|_{y=y_2} = \nu, \quad v\big|_{t=0} = v_0(y), \tag{6.2}$$

where

$$v_0 \in C^{2+\alpha}[y_1, y_2], \quad 0 < \alpha < 1, \quad \varphi \in C^2[0, T], \quad \nu = \text{const},$$

$$\omega \in C^4[y_1, y_2].$$

Let the following conditions

$$\omega'(y) > 0, \qquad y_1 < y \le y_2,$$

$$\omega'(y_1) = 0, \qquad \omega''(y_2) > 0, \tag{6.3}$$

$$v_0'(y) \ge \delta > 0, \tag{6.4}$$

$$v_0(y_1) = \varphi(0), \qquad v_0(y_2) = \nu,$$

$$\varphi'(t) < -\frac{\omega''(y_1)}{\delta} \mu \tag{6.5}$$

hold, where $\mu > 1$.

Equation (6.1) is degenerate both on the solution of the problem (for $\ln|v_y| \to \infty$) and in the boundary of $\Omega_T : y = y_1$. Instead of (6.1) we shall state its regularization

$$v_t^\varepsilon = (\omega'(y) + \varepsilon) \frac{v_{yy}^\varepsilon}{f^2(v_y^\varepsilon)} - \frac{\omega''(y)}{f(v_y^\varepsilon)}. \tag{6.6}$$

Here $f \in C^2$ is the monotone function

$$f(\xi) = \begin{cases} M + 1, & \xi \ge M + 1, \\ \xi, & \delta - \delta_1 \le \xi < M, \\ \delta - 2\delta_1, & \xi < \delta - 2\delta_1, \end{cases} \tag{6.7}$$

where $0 < 2\delta_1 < \delta < M$; δ is taken from (6.4).

Theorem 6.1. Let (6.3)-(6.5), (6.7) hold. Then for each $\varepsilon > 0$ problem (6.6), (6.2) has a unique solution

$$v^\varepsilon(y,t) \in H^{1+\alpha}(\Omega_T),$$

where

$$v^\varepsilon(y,t) \in C^{3,2}, \qquad v_t^\varepsilon \in L_2(\Omega_T) \quad \text{for} \quad t > 0.$$

Proof. The solvability of (6.6), (6.2), the smoothness of its solution for $t > 0$, and $v_t^\varepsilon \in L_2(\Omega_T)$ follow from the general theory of quasilinear parabolic equations (Ladyzhenskaya *et al.*, 1968).

Taking into account the estimates for solutions of linear parabolic equations in the spectral weight classes obtained by Belonosov (1981), we may easily obtain that $v_y^\varepsilon \in C^\alpha$ up to $t = 0$.

Further we shall asume that if $\omega'''(y)$ attains positive values anywhere in $[y_1, y_2]$, then

$$T < \frac{\delta_1(\delta - \delta_1)}{\sup\limits_{y_1 \leq y \leq y_2} \omega'''(y)} \,. \tag{6.8}$$

Theorem 6.2. Let $v^\varepsilon(y,t)$ solve (6.6), (6.2) and (6.3)-(6.5), (6.8) hold. Then for

$$\delta_1 < \frac{1}{2}\delta\left[1 - \frac{1}{\mu}\right] \tag{6.9}$$

and for each $\varepsilon > 0, M$ in (6.7), everywhere in Ω_T the estimate

$$v_y^\varepsilon(y,t) \geq \delta - \delta_1 \tag{6.10}$$

holds.

Proof. Suppose that inequality (6.10) does not hold somewhere in Ω_T. Denote by (y^*, t^*) the minimum of v_y^ε in Ω_T. By (6.4), $t^* > 0$. Show that

$$y_1 < y^* < y_2. \tag{6.11}$$

Really, assume, first, that $y^* = y_2$. Then, since $v_y^\varepsilon(y^*, t^*)$ is the minimum, $v_{yy}^\varepsilon(y_2, t^*) \leq 0$.

However, (6.6) and (6.2), (6.3) yield

$$v_{yy}^\varepsilon(y_2, t^*) = \frac{\omega''(y_2)}{\omega'(y_2) + \varepsilon} f\left(v_y^\varepsilon(y_2, t^*)\right) > 0.$$

Hence follows that $y^* < y_2$. If $y^* = y_1$, then, analogously, it should be that $v_{yy}^\varepsilon(y_1, t^*) \geq 0$. But, since (6.6), (6.2) hold, we have

$$\varphi'(t^*) \geq -\frac{\omega''(y_1)}{f\left(v_y^\varepsilon(y_1, t^*)\right)} \ . \tag{6.12}$$

We shall see that by (6.3), (6.9), inequality (6.12) contradicts (6.5). Thus, $y^* > y_1$.

Consider the case when for $y \in [y_1, y_2]$

$$\omega'''(y) \leq 0. \tag{6.13}$$

Fix $\lambda > 0$ and set

$$V(y, t) = \left(v_y^\varepsilon(y, t) - (\delta - \delta_1)\right) e^{-\lambda t} \ . \tag{6.14}$$

This new function $V(y, t)$, evidently, satisfies the equation

$$V_t = \frac{\omega'(y) + \varepsilon}{f^2(\xi)} V_{yy} + \frac{\omega''(y)}{f^2(\xi)} (1 + f'(\xi)) V_y$$
$$- 2\frac{\omega'(y) + \varepsilon}{f^3(\xi)} f'(\xi) V_y^2 e^{\lambda t} - \frac{\omega'''(y)}{f(\xi)} e^{-\lambda t} - \lambda V. \tag{6.15}$$

Here $\xi = V(y, t)e^{\lambda t} + \delta - \delta_1$.

By theorem 6.1, the function v_y, and therefore, $V(y, t)$ are continuous in Ω_T. Further, (6.11) and (6.14) denote that $V(y, t)$ can not attain its minimum for $y = y_1, y_2; \ t > 0$. Thus, if we assume that (6.10) fails, then $V(y, 0) > 0$, $V(y^*, t^*) < 0$ by condition (6.4). Hence it follows that such $(y^{**}, t^{**}) \in \Omega_T$ exists where V attains its negative minimum and $y_1 < y^{**} < y_2, \ t^{**} > 0$. By the well-known relations, in this point (y^{**}, t^{**}) (note, that by theorem 6.1 the function V is sufficiently smooth for $t > 0$) $V_y = 0$, $V_{yy} \geq 0$, $V_t \leq 0$, and taking into account (6.15), we have

$$-\lambda V\left(y^{**}, t^{**}\right) \leq \frac{\omega'''(y^{**})}{f(\xi)} e^{-\lambda t^{**}}\ .$$

This inequality contradicts (6.13) and presupposed negativeness of $V(y^{**}, t^{**})$.

Now, consider the case when (6.13) fails in certain points. Set

$$V(y, t) = v_y^\varepsilon(y, t) + \lambda t - \delta. \tag{6.16}$$

This function V satisfies the equation

$$V_t = \frac{\omega'(y) + \varepsilon}{f^2(\xi)} V_{yy} + \frac{\omega''(y)}{f^2(\xi)} \left(1 + f'(\xi)\right) V_y$$
$$-2 \frac{\omega'(y) + \varepsilon}{f^3(\xi)} f'(\xi) V_y^2 - \frac{\omega'''(y)}{f(\xi)} + \lambda,$$

where $\xi = V(y, t) - \lambda t + \delta$. By the analogous considerations, we see that for

$$\lambda > \frac{1}{\delta - \delta_1} \sup_{y_1 \leq y \leq y_2} \omega'''(y)\ . \tag{6.17}$$

we have $V \geq 0$, or, by (6.16)

$$v_y^\varepsilon(y, t) \geq \delta - \lambda t.$$

Choosing in the latter inequality λ sufficiently close to its lower boundary (6.17) and setting $t = T$, we see that (6.8) provides estimate (6.10). The theorem is proved.

Corollary 6.1. For each ε, M, the solution of (6.6), (6.2) satisfies the estimate

$$\varphi(t) \leq v^\varepsilon(y, t) \leq \nu. \tag{6.18}$$

This statement immediately follows from theorem 6.2 and from the monotonicity of $v^\varepsilon(y, t)$ by y for each t.

§A.7 The general theorem on the estimate for solution derivative for the mixed problem.

Before we to obtain the estimate for $v_y^\varepsilon(y, t)$ from above (problem (6.6), (6.2)), we establish a result which will be applied in some later theorems.

Consider in Ω_T the family of problems depending on the parameter $\varepsilon > 0$

$$w_t = a\Big(\varepsilon, t, y, w, w_y\Big) w_{yy} + b\Big(\varepsilon, t, y, w, w_y\Big),$$

$$w\Big|_{y=y_i} = \beta_i(t), \quad i = 1, 2,$$

$$w\Big|_{t=0} = w_0(y). \tag{7.1}$$

Here $\beta_i(t) \in C^2[0, T]$, functions a, b are twice continuously differentiable by their arguments; all their derivatives are uniformly bounded for $\varepsilon > 0$ for bounded values of their arguments. Let, besides,

$$a \geq \delta(\varepsilon) > 0 \tag{7.2}$$

and $\delta(\varepsilon) \to 0$ for $\varepsilon \to 0$.

We shall assume that the solution of the Cauchy problem corresponding to the "stationary" equation (7.1)

$$a\Big(\varepsilon, t, y, z, z_y\Big) z_{yy} + b\Big(\varepsilon, t, y, z, z_y\Big) = 0,$$

$$z(y_0) = z_0, \qquad z_y(y_0) = z_1 \tag{7.3}$$

may be extended up to a certain point $y = y^*$ for all admissible values of the parameters $\varepsilon > 0$, $0 < t \leq T$ and for all data $y_1 \leq y_0 \leq y_2$, $|z_0| < M_1$, $M_2 < z_1 < M_3$. The numbers M_i will be determined further.

The considerations below will be based on the special properties of Cauchy problem (7.3) solutions. In §8 these properties will be verified for problem (6.6), (6.2). As in §6 we have proved (6.10), (6.18) and, also, that (see theorem 6.1) $v^\varepsilon \in C^{1,2}$ for each $\varepsilon > 0$, we shall be interested in the values

$$|w(y,t)| < M_1, \quad M_2 < w_y(y,t) < M_3(\varepsilon),$$

$$|z_0| < M_1, \quad M_2 < z_1 < M_3(\varepsilon), \tag{7.4}$$

where M_1, M_2 do not depend on ε, and M_3 does not depend on ε.

Theorem 7.1. Let the solution $w(y,t)$ of (7.1) be such that $w \in C^2$ for $t > 0$, $w \in C^{1,2}(\Omega_T)$, $w_t \in L_2(\Omega_T)$, and (7.4) hold. Suppose, besides, that there exists such function $\Psi_1(\varepsilon, t, A, B)$ (here A, B are constructed by the Cauchy problem (7.3) solution with restrictions (7.4) corresponding to (1.32)-(1.33) chapter 2) that
1) for all $\varepsilon > 0$, $0 \leq t \leq T$, $|\xi| < M_1$, $N_1 < \eta < M_3(\varepsilon)$

$$\left| \frac{d\Psi_1}{dt} \right| \leq K_1(N_1)\Psi_1; \tag{7.5}$$

2) for all $\varepsilon > 0$, $0 \leq t \leq T$, $|\xi| < M_1$, $M_2 < \eta < N_2$

$$0 < \Psi_1 < K_2(N_2), \tag{7.6}$$

$$\frac{d\Psi_1}{d\eta} \neq 0; \tag{7.7}$$

3) there exists such $\lambda \in [y_1, y_2]$ that for each $\eta(\varepsilon) : \eta(\varepsilon) \underset{\varepsilon \to 0}{\longrightarrow} \infty$, $\eta(\varepsilon) < M_3(\varepsilon)$,

$$\Psi_1\big(\varepsilon, t, A, B(\varepsilon, t, y, \xi, \eta(\varepsilon))\big)\big|_{y=\lambda} \to \infty, \quad \text{for} \quad \varepsilon \to 0 \tag{7.8}$$

uniformly by $|\xi| < M_1$, $0 \le t \le T$. Assume, also, that for $i = 1$ and $i = 2$ either

$$(-1)^i \beta_i'(t) \le 0, \tag{7.9}$$

or

$$\left| w_y(y, t) \right|_{y=y_i} \le K_3; \tag{7.10}$$

where the numbers K_i do not depend on ε.

Then for $y = \lambda$ the estimate

$$\left| w_y(y, t) \right|_{y=\lambda} \le K_4 \tag{7.11}$$

holds, where K_4 depends on the problem (7.1) data, on constants (7.5)-(7.10), $\| w_0 \|_{C^1}$ and does not depend on ε.

Proof. Denote by ρ_n, Φ_n the functions constructed according as in (1.28)-(1.33) chapter 2, where

$$\Psi = \Psi_1^{2n}. \tag{7.12}$$

It follows from (1.29) chapter 2 that $\rho_n \ge 0$, i.e., (1.27) chapter 2 provides

$$\int_{y_1}^{y_2} \Phi_n \, dy \le \int_{y_1}^{y_2} \Phi_n \bigg|_{t=0} dy + \int_0^t \int_{y_1}^{y_2} \frac{\partial \Phi_n}{\partial \tau} dy \, d\tau + \int_0^t \left(\frac{\partial \Phi_n}{\partial w_y} w_\tau \right) \bigg|_{y=y_1}^{|y_2} d\tau. \tag{7.13}$$

Obtain the estimate from above for the integrals from the right side of (7.13). It is easy to show that (1.28)-(1.33) chapter 2, (7.12), by the property (7.6), by the property (7.6), guarantee that

$$\int_{y_1}^{y_2} \Phi_n \bigg|_{t=0} dy \le K_5(\varepsilon) K_6^{2n}, \tag{7.14}$$

where K_j do not depend on n; K_6 depends on $\| w_0 \|_{C^1}$ and does not depend on ε. The multiplier arises because of F (1.31) chapter 2, which, by assumptions (7.2), may increase by the modulus for $\varepsilon \to 0$.

As from (1.28)-(1.29) chapter 2, (7.12) it follows that for $w_y > M_2$

$$\frac{\partial \Phi_n}{\partial w_y} \geq 0,$$

for each $i = 1, 2$, we shall have

a) in the case of (7.9)

$$(-1)^i \left(\frac{\partial \Phi_n}{\partial w_y} w_\tau \right) \Bigg|_{y=y_i} \leq 0; \tag{7.15}$$

b) in the case of (7.10), analogous to (7.14),

$$\left| \left(\frac{\partial \Phi_n}{\partial w_y} w_\tau \right) \Bigg|_{y=y_i} \right| \leq K_7(\varepsilon) K_8^{2n}, \tag{7.16}$$

where, as before, K_j do not depend on n; K_8 does not depend on ε. The remaining term of the right side (7.13) will be considered more thoroughly. From (1.28)-(1.33) chapter 2, (7.12) we obtain

$$\frac{\partial \Phi_n}{\partial t} = \int_{M_2}^{\eta} (\eta - \zeta) \frac{\partial \rho_n}{\partial t} a \, d\zeta + \int_{M_2}^{\eta} (\eta - \zeta) \rho_n \frac{\partial a}{\partial t} d\zeta + \frac{\partial \Phi_{1n}}{\partial t}. \tag{7.17}$$

Property (7.6), formula (1.30) chapter 2, and estimate (7.4) yield

$$\left| \frac{\partial \Phi_{1n}}{\partial t} \right| \leq 2n K_9(\varepsilon) K_{10}^{2n}, \tag{7.18}$$

where K_j do not depend on n; K_{10} does not depend on ε. For each $\varepsilon > 0$, by (7.2), (7.4), we have

$$\left| \frac{\partial a}{\partial t} \right| \leq K_{11}(\varepsilon) a. \tag{7.19}$$

Introduce in (1.29), (1.31) chapter 2 the notation

$$F_1(\varepsilon, t, y, \xi, \eta) = \exp\left\{ -\int_{y_0}^{y} F \, dy \right\}\bigg|_{\substack{y_0 = y^* \\ z_0 = A(\varepsilon, t, y, \xi, \eta) \\ z_1 = B(\varepsilon, t, y, \xi, \eta).}} \tag{7.20}$$

Differentiating (1.29) chapter 2 by the parameter t, we see that

$$\frac{\partial \rho_n}{\partial t} = 2n\rho_n \frac{\partial \Psi_2}{\partial t}\bigg|\Psi_2 + \rho_n \frac{\partial F_1}{\partial t}\bigg| F_1,$$

where

$$\Psi_2\big(\varepsilon, t, y, \xi, \eta\big) = \Psi_1\big(\varepsilon, t, A, B(\varepsilon, t, y, \xi, \eta)\big).$$

Taking into account (7.20), estimates (7.4)-(7.6), and (1.31) chapter 2, we can see that

$$\left|\frac{\partial \rho_n}{\partial t}\right| \leq 2nK_1\rho_n + K_{12}(\varepsilon)\rho_n + K_{13}(\varepsilon)K_{14}^{2n}; \tag{7.21}$$

where K_j do not depend on n; K_1 is from (7.5); K_{14} does not depend on ε.

Combining (7.4), (7.18), (7.19), (7.21), from (7.16), (1.28) chapter 2 we may easily obtain that

$$\left|\frac{\partial \Phi_n}{\partial t}\right| \leq \big(2nK_1 + K_{15}(\varepsilon)\big)\Phi_n + 2nK_{16}(\varepsilon)K_{17}^{2n}, \tag{7.22}$$

where K_j do not depend on n; K_{17} does not depend on ε. Note that according to (7.4) the integral term in the right side of (1.28) chapter 2, by (7.2), (7.12), is not negative. Applying (7.14)-(7.16), (7.22) to estimate (7.13), we see that

$$\int_{y_1}^{y_2} \Phi_n \, dy \leq 2nK_{18}(\varepsilon)K_{19}^{2n} + \big(2nK_1 + K_{15}(\varepsilon)\big) \int_{0}^{t}\int_{y_1}^{y_2} \Phi_n \, dy \, d\tau.$$

By the well-known integral Gronuall lemma, the estimate then follows

$$\int_{y_1}^{y_2} \Phi_n \, dy \leq 2nK_{18}(\varepsilon)K_{19}^{2n}\left(1 + e^{(2nK_1 + K_{15}(\varepsilon))T}\right). \tag{7.23}$$

We need the estimate from below for the left side of (7.23). Relations (1.28), (1.29) chapter 2 and (7.20), (7.12) yield

$$\Phi_n\left(\varepsilon, t, y, \xi, \eta\right) = \int\limits_{M_2}^{\eta} (\eta - \zeta)\Psi_1^{2n} F_1 a d\zeta + \Phi_{1n}. \qquad (7.24)$$

Assumptions (7.2), (7.4), formulas (1.31) chapter 2, (7.18), (7.20), the smoothness of all the functions in question and conditions (7.6), (7.7) allow us to obtain from (7.24) the inequality

$$\Phi_n \geq K_{20}(\varepsilon, M_2) \int\limits_{M_2M_2}^{\eta}\int\limits^{\zeta} \Psi_1^{2n-1}\frac{d\Psi_1}{dp} dp d\zeta - K_{21}(\varepsilon, M_2)K_{22}^{2n},$$

whence follows

$$\Phi_n \geq \frac{K_{23}(\varepsilon, M_2)}{(4n)^2}\Psi_1^{2n} - K_{24}(\varepsilon, M_2)K_{25}^{2n},$$

where all K_j do not depend on n; K_{25} does not depend on ε.

Combining the latter estimate with (7.23), we see that

$$\int\limits_{y_1}^{y_2} \Psi_2^{2n}\left(\varepsilon, t, y, w, w_y\right) dy \leq (2n)^3 K_{26}(\varepsilon)K_{27}^{2n}\left(1 + e^{(2nK_1+K_{15}(\varepsilon))T}\right),$$

$$(7.25)$$

where, as before,

$$\Psi_2\left(\varepsilon, t, y, \xi, \eta\right) = \Psi_1\left(\varepsilon, t, A, B(\varepsilon, t, y, \xi, \eta)\right).$$

Raise inequality (7.25) to the degree $1/2n$ and pass to the limit for $n \to \infty$. As is known for a continuous function f

$$\lim_{p\to\infty} \| f(y) \|_{L_p(y_1,y_2)} = \sup_{y_1 \leq y \leq y_2} |f(y)|;$$

therefore, we obtain

$$\sup_{y_1 \leq y \leq y_2} \Psi_2\big(\varepsilon, t, y, w, w_y\big) \leq K_{28}\left(1 + e^{K_1 T}\right),$$

where K_1, K_{28} do not depend on ε.

As in the latter inequality the left side is estimated from below by the value of Ψ_2 for $y = \lambda$, then condition (7.8) and (7.4) give the uniform for $\varepsilon > 0$ estimate (7.11). The theorem is proved.

§A.8 The uniform by the regularization parameters derivative estimate for the model problem and its corollaries.

We shall try to obtain from (6.7) the estimate from above for $v_y^\varepsilon(y, t)$ (problem (6.6), (6.2)) which is not dependent on ε, M. Consider the set of functions

$$w_\gamma(y, t) = v^\varepsilon(y, t) + \gamma'(y)t, \qquad (8.1)$$

where $\gamma(y) \in C^{3+\alpha}, v^\varepsilon$ solves (6.6), (6.2).

By theorems 6.1, 6.2, we have

$$\delta - \delta_1 \leq v_y^\varepsilon \leq K(\varepsilon). \qquad (8.2)$$

Setting formally in (6.7), $M = K(\varepsilon)$, we see that in (6.6) $f(v_y^\varepsilon) = v_y^\varepsilon$, i.e., each function w_γ solves the problem

$$w_{\gamma t} = -\frac{\partial}{\partial y}\left(\frac{\omega'(y) + \varepsilon}{w_{\gamma y} - \gamma''(y)t}\right) + \gamma'(y),$$

$$w_\gamma\Big|_{y=y_1} = \varphi(t) + \gamma'(y_1)t, \qquad w_\gamma\Big|_{y=y_2} = \nu + \gamma'(y_2)t,$$

$$w_\gamma\Big|_{t=0} = v_0(y). \qquad (8.3)$$

Remark 8.1.   Strictly speaking, in (8.2) the number $K(\varepsilon)$ depends also on M from (6.7); therefore, we may pass to problem (8.3) only if we obtain the estimate from above for v_y^ε not dependent on M.

Remark 8.2.   For the above assumptions, the functions w_y have sufficient smoothness to apply theorem 7.1. Further, choosing the special form of $\gamma(y)$, we shall not specially note this fact.

Consider for $y, y_0 \in [y_1, y_2]$ the corresponding Cauchy problem for "stationary" equation (8.3)

$$\frac{d}{dy}\left(\frac{\omega'(y) + \varepsilon}{z_y - \gamma''(y)t}\right) = \gamma'(y), \quad z\Big|_{y=y_0} = z_0, \quad z'\Big|_{y=y_0} = z_1. \qquad (8.4)$$

Lemma 8.1.   Let for $y_0 \leq y \leq y_2$

$$0 < \delta - \delta_1 \leq z_1 - \gamma''(y_0)t \leq K(\varepsilon), \qquad (8.5)$$

$$\gamma(y_0) - \gamma(y) \leq \frac{\varepsilon}{2K(\varepsilon)}. \qquad (8.6)$$

Then the Cauchy problem (8.4) solution may be extended up to the point $y = y_2$.

Proof.   After integrating equation (8.4) we obtain

$$z_y - \gamma''(y)t = \frac{\omega'(y) + \varepsilon}{\gamma(y) - \gamma(y_0) + \frac{\omega'(y_0) + \varepsilon}{z_1 - \gamma''(y_0)t}}. \qquad (8.7)$$

For the above mentioned y the function $\omega'(y)$ is nonnegative (6.3); therefore, by (8.5), (8.6), from (8.7) the estimate

$$z_y - \gamma''(y)t \leq 2\frac{\omega'(y) + \varepsilon}{\omega'(y_0) + \varepsilon}\left(z_1 - \gamma''(y_0)t\right)$$

follows. This inequality denotes that for assumptions (8.5), (8.6) equation (8.4) does not degenerate on the solution.

By integrating (8.7), we see that the solution z of the Cauchy problem (8.4) is bounded. From here also follows that $|z_y|$ is bounded from below. It is known that *a priori* estimates gurantee the existence of the general solution of the degenerated equation. This means that the solution may be extended up to the point $y = y_2$. The lemma is proved.

Theorem 8.1. Let

$$\gamma(y) = -K_1(y - y_2)^2, \tag{8.8}$$

where

$$K_1 = \sup \frac{|\varphi'(t)|}{2(y_2 - y_1)} . \tag{8.9}$$

Then the function w_γ constructed by (8.1) satisfies the estimate

$$\left| w_{\gamma y}(y, t) \right|_{y = y_2} \le K_2, \tag{8.10}$$

where K_2 does not depend on ε.

Proof. By theorems 6.1, 6.2 and corollary 6.1, the function w_γ satisfies restrictions (7.4). Setting in (1.33) chapter 2 $y^* = y_2$, from (8.7) we obtain

$$B\left(\varepsilon, t, y, w_\gamma, w_{\gamma y}\right) = -2K_1 t + \frac{\omega'(y_2) + \varepsilon}{K_1(y - y_2)^2 + \frac{\omega'(y) + \varepsilon}{w_{\gamma y} + 2K_1 t}} \ ; \tag{8.11}$$

moreover, by lemma 8.1, this function is determined for all $y_1 \le y \le y_2$. Denote

$$\Psi_2\left(\varepsilon, t, y, w_{\gamma y}\right) = \Psi_1\left(\varepsilon, t, A, B\right) = 2K_1 t + B \tag{8.12}$$

and show that Ψ_2 satisfies the properties (7.5)-(7.8). We have

$$\frac{\partial \Psi_2}{\partial t} = \frac{2K_1(\omega'(y) + \varepsilon)(w_{\gamma y} + 2K_1 t)^{-2}}{K_1(y - y_2)^2 + \frac{\omega'(y) + \varepsilon}{w_{\gamma y} + 2K_1 t}} \Psi_2. \tag{8.13}$$

Denote

$$\sigma = w_{\gamma y} + 2K_1 t, \qquad \mu = (\omega'(y) + \varepsilon)^{-1}.$$

Then the multiplier before Ψ_2 in (8.13) is as follows

$$\frac{2K_1}{\sigma\Big(K_1(y-y_2)^2\sigma\mu+1\Big)} . \qquad (8.14)$$

In this case, by (8.1), (8.2), (8.8) we have $\sigma \geq \delta - \delta_1$, and, as it is easy to show, for $\varepsilon \leq 1$

$$\mu \geq \left(\sup_{y_1 \leq y \leq y_2} \omega'(y)+1\right)^{-1}.$$

Therefore, the value (8.14) uniformly by $0 < \varepsilon \leq 1$ is bounded from above, i.e., (8.13) yields (7.5).

Now check condition (7.6). For the values $w_{\gamma y}$ entering in this inequality we have

$$0 < \Psi_2 < \frac{\omega'(y_2)+\varepsilon}{K_1(y-y_2)^2 + \frac{\omega'(y)+\varepsilon}{N_2+2K_1T}} .$$

Hence follows that for $y_1 \leq y \leq (y_2+y_1)/2$

$$\Psi_2 < \frac{4}{K_1} \cdot \frac{\omega'(y_2)+1}{(y_2-y_1)^2} ,$$

and for $(y_2+y_1)/2 \leq y \leq y_1$

$$\Psi_2 < (N_2 + 2K_1T)(\omega'(y_2)+1)/\ae$$

where

$$\ae = \inf_{(y_2+y_1)/2 \leq y \leq y_2} \omega'(y).$$

Thus, the function Ψ_2 satisfies property (7.6).

Taking the derivative

$$\frac{\partial \Psi_2}{\partial w_{\gamma y}} = \frac{\omega'(y_2)+\varepsilon}{\Big(K_1(y-y_2)^2 + \frac{\omega'(y)+\varepsilon}{w_{\gamma y}+2K_1t}\Big)^2} \cdot \frac{\omega'(y)+\varepsilon}{(w_{\gamma y}+2K_1t)^2},$$

we see that Ψ_2 satisfies inequality (7.7). From (8.11), (8.12) immediately follows that for $\lambda = y_2$ we have (7.8). In order to use theorem 7.1, we need to verify conditions (7.9), (7.10). By (8.8), (8.9), $\gamma'(y_2) = 0$, $\gamma'(y_1) = \sup_t |\varphi'(t)|$; therefore, in (8.8)

$$(-1)^i \frac{\partial}{\partial t} \left(w_\gamma(y,t)\big|_{y=y_i} \right) \le 0,$$

i.e., boundary conditions (8.3) satisfy (7.9). If we apply to solution of problem (8.3) theorem 7.1 for $\gamma = -K_1(y - y_2)^2$, we see that (8.10) holds. The theorem is proved.

Theorem 8.2. For each $y : y_1 < y \le y_2$ the estimate

$$\left| v_y^\varepsilon(y,t) \right| \le K_3 \tag{8.15}$$

holds, where K_3 depends on the difference $y - y_1$ and does not depend on ε.

Proof. Let $\lambda \in (y_1, y_2)$ be arbitrary. Construct the function $\gamma(y)$ as follows

$$\gamma(\lambda) = \gamma(y_2), \tag{8.16}$$

$$\gamma(y) - \gamma(y_2) \le \varepsilon / \left(2K(\varepsilon) \right), \tag{8.17}$$

$$\gamma'(y_1) = 2(y_2 - y_1)K_1, \tag{8.18}$$

$$\| \gamma \|_{C^2(y_1,y_2)} \le K_4, \tag{8.19}$$

where $K(\varepsilon)$ satisfies (8.2); K_1 is taken from (8.9); K_4 depends on $\lambda - y_1$ and does not depend on ε. Such function γ, evidently, exists for each λ.

Note that taking into account (8.19), it suffices to prove for each $\lambda \in (y_1, y_2)$ the inequality

$$\left| w_{\gamma y}(y, t) \right|_{y=\lambda} \leq K_5, \qquad (8.20)$$

where w_γ is connected with solution of problem (7.6), (7.2) by formula (8.1).

The further considerations are almost the same as in the proof of theorem 8.1, and we shall omit them. Note only that for this case from (1.33) chapter 2 we obtain the following form for B

$$B\left(\varepsilon, t, y, w_\gamma, w_{\gamma y}\right) = \gamma''(y_2)t + \frac{\omega'(y_2) + \varepsilon}{\gamma(y_2) - \gamma(y) + \frac{\omega'(y)+\varepsilon}{w_{\gamma y}-\gamma''(y)t}}.$$

The function B is determined for all $y_1 \leq y \leq y_2$. This is provided by lemma 8.1, and estimate (8.17). Properties (8.17), (8.19) show that the function $\Psi_1 = B - \gamma''(y_2)t$ satisfies restrictions (7.5)-(7.7). By (8.16), for $y = \lambda$ we have (7.8). Finally, (8.18) together with theorem 8.1 show that for $y = y_i$ either (7.9) or (7.10) holds.

Applying theorem 7.1, we obtain inequality (8.20), which was to be proved. The theorem is proved.

Corollary 8.1. For each $0 < \Delta < \min\{y_2 - y_1, T\}$ solution of (6.6), (6.2) in

$$\Omega_\Delta = \left\{(y, t) \in [y_1 + \Delta, y_2] \times [\Delta, T]\right\}$$

satisfies the estimate

$$\| v^\varepsilon(y, t) \|_{H^{2+\alpha(\Omega_\Delta)}} \leq K_6, \qquad (8.21)$$

where K_6 does not depend on ε.

This statement follows from Ladyzhenskaya *et al.*(1968) and theorems 6.2, 8.1, 8.2.

Corollary 8.2. For each $\varepsilon > 0$, everywhere in $[y_1, y_2] \times [0, T]$ we have

$$\left|v_y^\varepsilon(y,t)\right| \le K(\varepsilon). \qquad (8.22)$$

The proof of this statement is almost the same as that of theorem 8.2. The only difference is that K_4 from (8.19) depends on ε, and the parameter λ may be equal to y_1. Estimate (8.20) and, therefore, (8.19) hold for each $\varepsilon > 0$ but in the closed interval $y_1 \le y \le y_2$.

Thus, the investigation of (8.3) instead of (6.6) is completely justified.

§A.9 The existence theorems for the model and basic problems. The uniqueness condition.

In §A.6-A.9 we have proved that if Ω_T satisfies (6.8), then, under the conditions of theorem 6.1 the problem (6.6), (6.2) solution has the properties

$$\varphi(t) \le v^\varepsilon(y,t) \le \nu,$$

$$\delta - \delta_1 \le v_y^\varepsilon(y,t) \le M(\lambda), \quad y_1 < \lambda \le y \le y_2,$$

$$\| v^\varepsilon(y,t) \|_{H^{2+\alpha(\Omega_\Delta)}} \le K, \qquad (9.1)$$

where K, M do not depend on ε.

According to (9.1), by tending ε to zero we may obtain the existence theorem for the limit problem (6.1), (6.2), but we do not know anything about the limit function $v(y,t)$ for $y = y_1$.

Make the change of variables and the function

$$x = v^\varepsilon(y,t), \quad t' = t, \quad u_x^\varepsilon(x,t') = y. \qquad (9.2)$$

By (9.1), the domain $\Omega_T^0 = (y_1, y_2] \times (0, T]$ transforms one-to-one into the domain

$$Q_t^\varphi = \left\{(x,t') : 0 < t' \leq T, \quad \varphi(t') < x \leq \nu\right\},\tag{9.3}$$

the closure of which will be the domain of definition of $u^\varepsilon(x,t')$. By differentiating (9.2), we obtain

$$v_y^\varepsilon(y,t) = \left(u_{xx}^\varepsilon(x,t')\right)^{-1},$$

$$v_{yy}^\varepsilon(y,t) = -\left(u_{xx}^\varepsilon(x,t')\right)^{-3} u_{xxx}^\varepsilon(x,t'),\tag{9.4}$$

$$v_t^\varepsilon(y,t) = -\left(u_{xx}^\varepsilon(x,t')\right)^{-1} u_{xt'}^\varepsilon(x,t').$$

Further we shall omit the prime superscript of the variable t. Rewrite equation (6.6) in terms of the function u^ε by means of (9.4)

$$u_{xt}^\varepsilon = \frac{\partial}{\partial x}\left((\omega'(u_x^\varepsilon) + \varepsilon)u_{xx}^\varepsilon\right).\tag{9.5}$$

Besides, taking into account (9.2), (9.3), the initial boundary-value conditions (6.2) for the function u^ε have the form

$$u_x^\varepsilon\big|_{x=\varphi(t)} = y_1, \qquad u_x^\varepsilon\big|_{x=\nu} = y_2,\tag{9.6}$$

$$u_x^\varepsilon\big|_{t=0} = u_0'(x),\tag{9.7}$$

where $u_0'(x) = v_1(x)$, $v_1(v_0(y)) = y$. Note, that conditions (6.3) for the function $v_0(y)$ mean some properties for the function $u_0(x)$. We shall not write them for now. Introduce the notation

$$Q_\Delta^\varphi = \left\{(x,t) : \Delta \leq t \leq T, \varphi(t) + \Delta \leq x \leq \nu\right\},\tag{9.8}$$

where $\Delta \leq \min\{T, \nu - \varphi(0)\}$.

Reformulate the result of §6-8 in new notations.

There exists the function $u^\varepsilon(x,t)$ with the following properties.

1) All derivatives entering in (9.5) are continuous by Hölder in Q_+^φ; for $t > 0$ the function $u_t^\varepsilon(x,t)$ is continuous by Hölder up to the boundary $x = \varphi(t)$, and the uniform for $\varepsilon > 0$ estimates

$$\| u_x^\varepsilon \|_{H^{2+\alpha}(Q_\Delta^\varphi)} \le K_1,$$

$$y_1 \le u_x^\varepsilon(x,t) \le y_2, \qquad (x,t) \in \bar{Q}_+^\varphi, \qquad (9.9)$$

$$0 < u_{xx}^\varepsilon(x,t) \le 1/(\delta - \delta_1), \qquad (x,t) \in Q_+^\varphi,$$

hold.

2) Everywhere inside Q_+^φ, the function $u_\varepsilon(x,t)$ satisfies (9.5) and attains the values (9.6), (9.7) in the boundary.

Before we pass to the limit for $\varepsilon \to 0$, integrate equation (9.5) by x

$$u_t^\varepsilon - (\omega'(u_x^\varepsilon) + \varepsilon)u_{xx}^\varepsilon = \psi_\varepsilon(t), \qquad (9.10)$$

where we, without loss of generality, set

$$\psi_\varepsilon(t) = \left(u_t^\varepsilon - (\omega'(u_x^\varepsilon) + \varepsilon)u_{xx}^\varepsilon \right)\Big|_{x=\varphi(t)}. \qquad (9.11)$$

As we determine by $v^\varepsilon(y,t)$ only the derivative u_x^ε, in order to calculate $u^\varepsilon(x,t)$ we have to make an additional condition. Set

$$u^\varepsilon\Big|_{x=\varphi(t)} = y_1\varphi(t). \qquad (9.12)$$

Differentiating (9.12) by t, taking into account the smoothness of u^ε and conditions (9.6) for $t > 0$, we obtain

$$u_t^\varepsilon(x,t)\Big|_{x=\varphi(t)} + y_1\varphi'(t) = y_1\varphi'(t),$$

i.e., for $t > 0$

$$u_t^\varepsilon(x,t)\Big|_{x=\varphi(t)} = 0. \qquad (9.13)$$

Remark 9.1. It is easy to see that setting for the function u^ε instead of (9.12) additional condition (9.13) and taking $u^\varepsilon(\varphi(0),0) = y_1\varphi(0)$, we obtain (9.12) by integrating $\frac{d}{dt}u^\varepsilon(\varphi(t),t)$ along the curve $x = \varphi(t)$.

Substituting (9.13) into (9.11), by estimates (9.9), we obtain that

$$\lim_{\varepsilon \to 0} \psi_\varepsilon(t) = 0 \qquad (9.14)$$

uniformly for $0 < t \le T$.

Note, that taking into account (9.9), (9.14), after multiplying (9.10) by u_t^ε, integrating and doing rather straightforward transformations, we may obtain the estimate

$$\| u_t^\varepsilon \|_{L_2(Q_+^\varphi)} \le K_2,$$

where K_2 does not depend on ε. Therefore, the described smoothness of the set of functions $\{u^\varepsilon\}$, estimates (9.9) yield the weak compactness of this set in the space

$$W_2^{2,1}(\bar{Q}_+^\varphi) \bigcap H^{1+\alpha}(\bar{Q}_+^\varphi) \bigcap H^{2+\alpha}(Q_\Delta^\varphi)$$

(see (9.3), (9,8)). Then, by equations (9.10), (9.12), (9.14), passing to the limit for $\varepsilon \to 0$, we obtain the following theorem.

Theorem 9.1. Under the conditions (6.3)-(6.5), (6.8), there exists the function

$$u(x,t) \in H^{1+\alpha}(\bar{Q}_+^\varphi) \bigcap H^{2+\alpha}(Q_\Delta^\varphi)$$

which has properties (9.6), (9.7), (9.12), (9.13) and everywhere in Q_+^φ satisfies the equation

$$u_t = \omega'(u_x)u_{xx}. \qquad (9.15)$$

Remark 9.2. We have established the solvability of (9.15), (9.6), (9.7). Moreover, since the second derivative u_{xx} (9.9) is bounded, this solution satisfies (9.13), whence, as follows from remark 9.1, (9.12) holds.

Replace the data for $t = 0$ (9.7) as follows

$$u\big|_{t=0} = u_0(x). \qquad (9.16)$$

In this case, by the continuity of $u(x,t)$ in $\bar{Q}^\varphi_+$, additional condition (9.12), and therefore, condition (9.13) with the restriction $u(\varphi(0),0) = y_1\varphi(0)$ allow us to recover the function $u_0(x)$ by $u'_0(x)$ identically, i.e., by means of $v_0(y)$.

Theorem 9.2. Solution of the problem (9.15), (9.16), (9.6) from the space pointed out in theorem 9.1 is unique.

The proof may be done according to the following scheme. We take the equation for the difference of two solutions. Then we multiply it by this difference. Finally, we get the standard estimate.

Consider in the domain

$$Q^T = [a, b] \times [0, T] \qquad (9.17)$$

the problem

$$u_t = \omega'(u_x)u_{xx},$$

$$u_x\big|_{x=a} = y_0, \qquad u_x\big|_{x=b} = y_2, \qquad (9.18)$$

$$u\big|_{t=0} = u_0(x), \qquad x_0 \leq x \leq b,$$

$$u\big|_{t=T} = u_T(x), \qquad a \leq x \leq x_T. \qquad (9.19)$$

For clearness we shall cite here all sufficient conditions for the problem data. Some of this conditions follow from (6.3)-(6.5), (6.8); other conditions are there restated with new independent variables and for a new unknown function (9.2).

1) The function $\omega(\xi)$ belongs to C^4, where such $y_1 \in (y_0, y_2)$ exists that

$$\omega'(y) > 0 \qquad y \in (y_1, y_2],$$

$$\omega'(y) < 0 \qquad y \in [y_0, y_1), \tag{9.20}$$

$$\omega''(y_i) > 0 \qquad i = 0, 2.$$

2) The functions $u_0(x)$, $u_\tau(x)$ belong to $C^{3+\alpha}$, where

$$u_0''(x) \geq \delta > 0,$$

$$u_\tau''(x) \geq \delta > 0,$$

$$u_0'(x_0) = u_\tau'(x_\tau) = y_1, \tag{9.21}$$

$$u_0(x_0) = y_1 x_0, \qquad u_0'(b) = y_2,$$

$$u_\tau(x_\tau) = y_1 x_\tau, \qquad u_\tau'(a) = y_0.$$

3) The numbers $a < x_\tau < x_0 < b$ are such that

$$x_\tau \leq x_0 - \Delta \omega''(y_1) T \lambda, \tag{9.22}$$

where

$$\lambda > 1, \quad \Delta = \max \left\{ \sup_{x_0 \leq x \leq b} u_0''(x), \sup_{a \leq x \leq x_\tau} u_\tau''(x) \right\}.$$

4) If $\omega'''(y) > 0$ for certain $y \in [y_1, y_2]$, or $\omega'''(y) < 0$ for certain $y \in [y_0, y_1]$, then T satisfies the inequality

$$T < \text{æ} \, (1/\Delta - 1/\Delta_1) / \Delta_1, \tag{9.23}$$

where Δ is taking according as in (9.22), $\Delta_1 > 2\Delta$,

$$\text{æ} = \max \left\{ \sup_{y_1 \leq y \leq y_2} \omega'''(y), \sup_{y_0 \leq y \leq y_1} (-\omega'''(y)) \right\}.$$

Remark 9.3. Note that (9.20) yields

$$\omega''(y_1) \geq 0. \qquad (9.24)$$

Thus, equality (9.22) is the additional restriction on T in the case $\omega''(y_1) > 0$.

For each function $\varphi(t) \in C^2(0,T)$ such that

$$\varphi(0) = x_0, \quad \varphi(T) = x_T, \quad a < \varphi(t) < b, \qquad (9.25)$$

denote

$$Q_+^\varphi = \left\{ (x,t) : 0 < t \leq T, \quad \varphi(t) < x \leq b \right\},$$

$$Q_-^\varphi = \left\{ (x,t) : 0 \leq t < T, \quad a \leq x < \varphi(t) \right\}, \qquad (9.26)$$

$$Q_T^\varphi = Q^T \backslash \{ (x_0, 0) \bigcup (x_T, T) \}.$$

Definition 9.1. We shall call the function

$$u(x,t) \in C^{1,1}(Q_T^\varphi) \bigcap C^{1,0}(Q^T)$$

the solution to problem (9.18), (9.19), if there exists such function $\varphi(t)$ satisfying (9.25) that

$$u(x,t) \in H^{2+\alpha}(Q_-^\varphi \bigcup Q_+^\varphi);$$

everywhere inside this domain condition (9.18) holds; and (9.19) holds in the mentioned components of the boundary Q^T (9.17).

Theorem 9.3. Let conditions (9.20)-(9.23) hold. Then there exists a solution of (9.18), (9.19).

Proof.   Inequality (9.22) yields that there exists the function $\varphi(t)$ satisfying (9.25) and (6.5) for $\delta = \Delta^{-1}$. Fix one of such functions $\varphi(t)$ and divide the domain Q^T into two domains $Q^\varphi_\pm$.

By theorems 9.1, 9.2 there exists a unique function $u^+(x,t)$ determined in $\bar{Q}^\varphi_+$, satisfying inside $\bar{Q}^\varphi_+$ equation (9.18), attaining the values (9.6), (9.7), (9.12) for $\nu = b$ in the mentioned components of the boundary. By the theorems 9.1, 9.2, there exists a unique function $u^-(x,t)$ which solves the problem (9.15), (9.6), (9.7), (9.12) in the domain

$$Q = \left\{ (x,t) : (-x, T-t) \in Q^\varphi_- \right\}.$$

Set

$$u(x,t) = \begin{cases} u^+(x,t), & (x,t) \in \bar{Q}^\varphi_+, \\ u^-(x,t), & (x,t) \in \bar{Q}^\varphi_-. \end{cases}$$

According to (9.13), $u(x,t) \in C^{1,1}(Q^T)$. Besides, by the function $u'_0(x)$, since $u_0\varphi(0) = y_1\varphi(0)$, the function $u_0(x)$ may be uniquely recovered. Therefore, u solves (9.18), (9.19). The theorem is proved.

As follows from this proof for each function satisfying (9.25), (6.5) there exists its own solution of problem (9.18), (9.19), i.e., in order to select a unique solution we have to complete the setting. The role of such additional condition may be fulfilled by one of the following

1) there exists a curve $x = \varphi(t)$ satisfying (9.25), (6.5), where $\omega'(u_x)|_{x=\varphi(t)} = 0$;

2) there exists a degeneration curve, where $\omega'(u_x) = 0$; the solution attains in this curve the given values (9.12).

Bibliography

Agmon, S., Douglis, A., and Nirenberg, L. (1959). Estimates near the boundary for solutions of elliptic partial differential equations satisfying general boundary conditions. I. *Comm. Pure Appl. Math.* **12**, 623-727; II, (1964) **17**, 35-92.

Akhmerov, R. R. (1984). On the structure of a set of solutions of one boundary-value problem for a one-dimensional steady-state equation of variable type. *Chislennye Metody Mekhaniki Sploshnoi Sredy* **15** (5), 20-30 (in Russian).

Akhmerov, R. R. (1985). On the Newmann problem for one second-order differential equation with an alternating coefficient multiplying the leading derivative. *Chislennye Metody Mekhaniki Sploshnoi Sredy.* **16** (1), 3-11 (in Russian).

Akramov, T. A. (1976). On stabilizing solutions of the system of partial differential equations describing the chemical kinetics of reversible reactions. *Dinamica Sploshn. Sredy.* **26** , 3-15 (in Russian).

Akramov, T. A. (1979). One mixed problem for the quasilinear parabolic system. *Dokl. Akad. Nauk SSSR* **244** (3), 554-558 (in Russian).

Akramov, T. A. (1984). The qualitative analysis of differential equations describing chemical reactions taking diffusion into account. *Mathematical Modelling of Chemical Reactors.* 102-115, Nauka, Moscow (in Russian).

Akramov, T. A. and Vishnevskii, M. P. (1992). Global solvability of the system reaction - diffusion.*Mat. Model.* **4** (11), 110-120 (in Russian).

Akramov, T. A. and Vishnevskii, M. P. (1995). Some qualitative properties of the reaction - diffusion system. *Sib. Math. J.* **36**, 1 (in Russian).

Akramov, T. A. and Zelenyak, T. I. (1975). On the number of stationary solutions and attraction domains of stable quasilinear parabolic equations. *Mat. Probl. of Chemistry* P.1, 144-150 (in Russian).

Alikakos, N. D. and Hess, P. (1987). On stabilization of discrete monotone dynamical systems. *Israel J. Math.* **59**, 185-194.

Alikakos, N. D. and Hess, P. (1991). Liapunov operators and stabilization in strongly order-preserving dynamical systems. *Diff. Integ. Equat.* **4**, 15-24.

Alikakos, N. D., Hess, P., and Matano, H. (1989). Discrete order-preserving semigroups and stability for periodic parabolic differential equations. *J. Diff. Equat.* , **82**, 322-341.

Alikakos, N. D., Bates, P. W., and Fusco, G. (1991). Slow motion for the Cahn-Hilliard equation in one space dimension. *Ibid,* **90** (1), 81-135.

Amann, H. (1976). Fixed point equations and nonlinear eigenvalue problems in ordered Banach spaces. *SIAM Rev.* **18**, 620-709.

Amann, H. (1985). Global existence for semilinear parabolic systems. *J. Reine Angew. Math.* **360**, 47-83.

Amann, H. (1988). Dynamic theory of quasilinear parabolic equations. 1. Abstract evolution equations. *Nonlinear Analysis.* **12**, 895-919.

Amann, H. (1989). Dynamic theory of quasilinear parabolic equations. 3. Global existence. *Math. J.* **202**, 219-250.

Amann, H. (1990). Dynamic theory of quasilinear parabolic equations. 2. Reaction - diffusion systems. *Diff. Integ. Equat.* **3**, 13-75.

Amann, H. and Crandall, M. (1978). On some existence for semilinear elliptic equations. *Indiana Univ. Math. J.* **27**, 779-790.

Angenent, S. B. (1988). The zero set of a solution of parabolic equation. *J. Reine Angew. Math.* **390**, 79-96.

Angenent, S. B. (1990). Parabolic equation for curves on surface I. *Ann. Math.* 132, 451-483.

Angenent, S. B. (1991). Parabolic equation for curves on surface II. *Ann. Math.* 133, 171-215.

Angenent, S. B. and Fiedler, B. (1996). The dynamics of rotation waves in scalar reaction - diffusion equation (to be published).

Babin, A. V. and Vishik, M. I. (1982). Attractors for quasilinear parabolic equations. *Dokl. Akad. Nauk SSSR* **254** (4), 780-784 (in Russian).

Babin, A. V. and Vishik, M. I. (1983). Attractors for evolutionary equations with partial derivatives and its dimension estimate. *Uspekhi Mat. Nauk* **38** (3), 133-187 (in Russian).

Babin, A. V. and Vishik, M. I. (1985). Maximal attractors of semigroups corresponding to evolutionary differential equations. *Math. Sbornik* **126** (3), 397-419 (in Russian).

Babin, A. V. and Vishik, M. I. (1986). Unstable invariant sets of semigroups of nonlinear operators and their disturbances. *Uspekhi Mat. Nauk* **41** (4), 3-34 (in Russian).

Babin, A. V. and Vishik, M. I. (1987). Unstable invariant sets of semigroups of nonlinear operators and their disturbances.*Izv. Akad. Nauk SSSR*, **51** , 44-78 (in Russian).

Ball, J. M. and Mizel, V. J. (1984). Singular minimizers in the calculus of variations.*Bull Amer. Math. Soc.* (N.S) **11**, 143-146.

Bates, P. W. and Zheng, S. (1992). Inertial manifolds and inertial sets for the phase - field equations. *J. Dynamics Differential Equations* **4** (2), 375-398.

Belonosov, V. S. (1978). Estimates of solutions of nonlinear parabolic systems in the Hölder classes with weight. *Dokl. Akad. Nauk SSSR* **241** (2), 265-268 (in Russian).

Belonosov, V. S. (1981). Estimates of solutions of nonlinear parabolic systems in the Hölder classes with weight and some applications. *Math. USSR Sbornik,* **38** (2), 151-173.

Belonosov, V. S. (1983). On the instability indices for unbounded differential operators. *Dokl. Akad. Nauk SSSR* **273** (1), 11-14 (in Russian).

Belonosov, V. S. (1984). On the instability indices for unbounded differential operators. I. *Some Applications of Mathematical Analysis to the Problems of Mathematical Physics* , Novosibirsk, 25-51 (in Russian).

Belonosov, V. S. (1985). On the instability indices for unbounded differential operators. II. *Functional Analysis and Mathematical Physics,* Novosibirsk, 5-33 (in Russian).

Belonosov, V. S. and Vishnevskii, M. P. (1977). On stability for stationary solutions of nonlinear parabolic systems. *Math. Sbornik* **33** (4), 465-484.

Belonosov, V. S. and Zelenyak, T. I. (1975).*Nonlocal Problems in the Theory of Quasilinear Parabolic Equations* . Izd. Novos. Gos. Univ., Novosibirsk (in Russian).

Belonosov, V. S., Vishnevskii, M. P., Zelenyak, T. I., and Lavrent'ev, M. M. -Jr. (1982). *On the Qualitative Properties of Solutions for Parabolic Equations* . Preprint N466. Izd. Vych. Tsentr Sib. Otdel. Akad. Nauk SSSR, Novosibirsk (in Russian).

Beltramo, A. and Hess, P. (1984). On the principal eigenvalue of periodic - parabolic operators. *Comm. Part. Diff. Equat.* **9**, 919-941.

Bernstein, S. N. (1960). Sur les equations du calcul des variations. *Ann. Ec. Norm.* **29**, 431-485. Russian transl. (1960).

Bers, L., John, F., and Schechter, M. (1964). *Partial Differential Equations* . New York - London - Sydney.

Browder, F. E. (1964). Strongly nonlinear parabolic boundary - value problems. *Amer. J. Math.* **86** (2), 339-357.

Brunovskii, P. and Fiedler, B. (1988). Connecting orbits in scalar reaction - diffusion equations I. *Dynamics Reported* **1**, 57-89.

Brunovskii, P. and Fiedler, B. (1989). Connecting orbits in scalar reaction - diffusion equations II. *J. Diff. Equat.* **81** (1), 106-135.

Brunovskii, P. and Polačik, P. (1987). Generic hyperbolicity for reaction - diffusion equation in symmetric domains. *J. Appl. Math. Phys.* **38**, 172-183.

Brunovskii, P., Polačik, P., and Sandstede, B. (1992). Convergence in general periodic parabolic equations in one space dimension. *Nonlinear Anal* . **18** (3), 209-215.

Cahn, J. W. and Hilliard, J. E. (1958). Free energy of a nonuniform system. I. Interfacial free energy. *J. Chem. Phys.* **28** (2), 258-267.

Cantrell, R. and Schmitt, K. (1986). On the eigenvalue problem for coupled elliptic systems. *SIAM J. Math. Anal.* **17** (4), 850-862.

Chaffee, N. (1974). A stability analysis for a semilinear parabolic partial equation. *J. Diff. Equat.* **15,** 540-552.

Chaffee, N. and Infante, E. (1974). A bifurcation problem for a nonlinear parabolic equation. *J. Appl. Anal.* **4,** 17-37.

Chen, M., Chen, X. - Y., and Hale, J. K. (1991). Structural stability for time - periodic one - dimensional parabolic equations. *J. Diff. Equat.* **94** , 266-291.

Chen, X. - Y. and Matano, H. (1989). Convergence, asymptotic periodicity and finite - point blow - up in one - dimensional semilinear heat equations. *J. Diff. Equat.* **78** (1), 160-190.

Costen, R. and Holland, C. (1978). Instability results for reaction - diffusion equations with Neumann boundary conditions. *J. Diff. Equat.* **27,** 266-273.

Courant, R., Friedrichs, K., and Zewy, H. (1928). Uber der partiellen Differentialgleichungen der mathematischen Physik. *Math. Ann.* **100,** 32-74.

Crandall, M. G., Ishii, H., and Lions, P. L. (1992). User's guide to viscosity solutions of second order partial differential equations. *Bull Amer. Math. Soc.* **27,** 1-67.

Daletskii, Yu. L. and Krein, M. G. (1970). *Stability of Solutions of Differential Equations in Banach Space.* Nauka, Moscow (in Russian).

Danser, E. (1991). On the existence of two-dimensional invariant tori for scalar parabolic equations with time periodic coefficients. *Ann. Scuola Norm. Super Pisa Sci* (4), **28** (3), 445-472.

Danser, E. and Hess, P. (1991). Stability of fixed point for order - preserving discrete-time dynamical systems. *J. Reine Angew. Math.* **419,** 125-139.

Demidovitch, V. M. (1977). *Lectures on the Stability Theory.* Nauka, Moscow (in Russian).

Eidel'man, S. D. (1964). *Parabolic Systems.* Nauka, Moscow, (in Russian).

Elliott, C. M. and French, D. A. (1989). A nonconforming finite-element method for the two - dimensional Cahn - Hilliard equation. *SIAM J. Numer. Anal.* **26** (4), 884-903.

Eltysheva, N. A. (1988). On qualitative properties of solutions of certain hyperbolic systems in the plane. *Math. Sbornik.* **135** (2), 186-209 (in Russian).

Fiedler, B. (1989). Discrete Liapunov functionals and ω-limit sets. *Math. Mod. Num. Anal.* **23**, 415-431.

Fiedler, B. (1994). Global attractors of one-dimensional parabolic equations. *Tatra Mountains Math. Pull.* **4**, 67-92.

Fiedler, B. (1996). Do global attractors depend on their boundary conditions. Preprint 13/96. *Inst. fur Math.* Berlin.

Fiedler, B. and Mallet-Paret, J. (1989). A Poincare - Bendixson theorem for scalar reaction - diffusion equations. *Arch. Rational Mech. Anal.* **107** (4), 325-345.

Fiedler, B. and Rocha, C. (1994). *Heteroclinic Orbits of Semilinear Parabolic Equations.* Preprint 14/94, Freie Universität, Berlin.

Fiedler, B. and Rocha, C. (1996). *Orbit Equivalence of Global Attractors of Semilinear Parabolic Differential Equations.* Preprint 14/96. Inst. für Math. Berlin.

Fiedler, B. and Rocha, C. (to be published). Heteroclinic orbits of semilinear parabolic equations. *J. Diff. Equat.*

Fila, M. and Lieberman, G. (1994). Derivative blow-up and beyond for quasilinear parabolic equations. *Diff. and Integr. Equat.* **7** (314), 811-822.

Filippov, A. F. (1961). On conditions of solution existence for quasilinear parabolic equation. *Dokl. Akad. Nauk SSSR* **141**, 568-570 (in Russian).

Fokin, M. V. (1981). On the limit sets of trajectories for dynamic systems of gradient type. *Math. Sbornik* **115** (4), 502-514 (in Russian).

Friedman, A. (1964). *Partial Differential Equations of Parabolic Type.* Prentice - Hall, Inc.

Friedman, A. (1968). *Equations with Partial Derivatives of Parabolic Type.* Mir, Moscow (in Russian).

Furihata, D., Onda, T., and Mori, M. (1993). A finite difference scheme for the Cahn - Hilliard equation based on a Liapunov functional. *GAKUTO Intern. Series Math. Sci. Appl.* **2**, 347-358.

Fusco, G. and Hale, J. K. (1989). Slow - motion manifolds, dormant instability and singular perturbation. *J. Dyn. Diff. Equat.* **1**, 111-137.

Fusco, G. and Rocha, C. (1991). A permutation related to the dynamics of a scalar parabolic PDE. *J. Diff. Equat.* **91**, 75-94.

Galactionov, V. A. (1980). Two methods of comparing solutions of parabolic equations. *Dokl. Akad. Nauk SSSR* **251** (4), 832-835 (in Russian).

Galactionov, V. A. (1982). On the existence and nonexistence of global solutions of boundary - value problems for quasilinear parabolic equations. *J. Calc. Math. and Math. Phys.* **22** (6), 1369-1385 (in Russian).

Galactionov, V. A. (1983). The critical condition and comparison theorems for difference solutions of nonllinear parabolic equations. *J. Calc. Math. and Math. Phys.* **23** (1), 109-118 (in Russian).

Galactionov, V. A. and Posashkov, S. A. (1986). Application of new comparison theorems for investigating unbounded solutions of nonlinear parabolic equations. *Diff. Equat.* **22** (7), 1165-1173 (in Russian).

Galactionov, V. A., Kurdyumov, S. P., and Samarskii, A. A. (1983). On one parabolic system of quasilinear equations. *Diff. Equat.* **19** (12), 2123-2140 (in Russian).

Galactionov, V. A., Kurdyumov, S. P., and Samarskii, A. A. (1984). On asymptotic stability of invariant solutions to nonlinear heat conductivity equations with a source. *Diff. Equat.* **20** (4), 614-632 (in Russian).

Galactionov, V. A., Kurdyumov, S. P., and Samarskii, A. A. (1985). On one parabolic system of quasilinear equations. *Diff. Equat.* **21** (9), 1544-1559 (in Russian).

Galactionov, V. A., Kurdyumov, S. P., and Samarskii, A. A. (1987). A method of stationary states for nonlinear evolutionary parabolic problems. *Dokl. Akad. Nauk SSSR* **278** (6), 1296-1300 (in Russian).

Gevrey, M. (1913). Sur les équations aux derivées partielles du type parabolique. *J. Math. Pures Appl.* **6** (9), 305-471.

Gokhman, E. N., Tovbin, Yu. K., and Fedyanin, V. K. (1981). Concentration dependence of surface and volume diffusion in regular structures. *Physics of Interphase Phenomena.* Nal'chik, 145-153 (in Russian).

Golovkin, K. K. (1967). New model equations for the motion of a viscous liquid and their unique solvability. *Trudy Inst. Matem. Akad. Nauk SSSR.* **102**, 29-50 (in Russian).

Grebnev, V. N. and Novikov, V. A. (1980). On the properties of one steady state equation of variable type. *Chislennye Metody Mekhaniki Sploshnoi Sredy* **11** (5), 45-53 (in Russian).

Grebnev, V. N. and Novikov, V. A. (1986). On a structure of the set of alternating for a solution of the equation $u_t = [\varphi(u)]_{xx}$. *Dinamica Sploshnoy Sredy.* **76**, 90-100 (in Russian).

Gushchin, A. K. (1968). On stabilizing solutions of boundary - value problems for parabolic equations *Sib. Math. J.* **10** (1), 43-57 (in Russian).

Gushchin, A. K. (1982). On the uniform stabilizing solutions of the second mixed problem for a parabolic equation. *Math. Sbornik* **119** (4), 451-508 (in Russian).

Gushchin, A. K. and Mikhailov, V. P. (1971a). On stabilizing solution of the Cauchy problem for a parabolic equation. *Diff. Equat.* **7** (2), 297-311 (in Russian).

Gushchin, A. K. and Mikhailov, V. P. (1971b). On stabilizing solution of the Cauchy problem with one spatial variable. *Trudy Inst. im. V. A. Steklova,* **112**, 181-202 (in Russian).

Hail, D. (1962). *Ordinary Differential Equations.* Nauka, Moscow (in Russian).

Hale, J. K. (1988). Asymptotic behavior of dissipative systems. *Math. Surveys Monogr.* **25**, Amer. Math. Soc., Providence, R. I.

Hale, J. K. (1994). Numerical Dynamics. *Contemporary Math.* 1972, 1-30.

Hale, J. K. and Rocha, C. (1987). Interaction of diffusion and boundary conditions. *Nonlinear Analysis* **11**, 633-649.

Hale, J. K., Magalhaes, L. L., and Oliva, W. M. (1984). *An Introduction to Infinite Dimensional Dynamical Systems. Geometric Theory.* Springer - Verlag, New York.

Hartman, F. (1970). *Ordinary Differential Equations.* Mir, Moscow (in Russian).

Henry, D. B. (1985a). *Geometric Theory of Parabolic Equations*. Mir, Moscow (in Russian).

Henry, D. B. (1985b,c). Some infinite dimensional Morse - Small systems defined by parabolic partial differential equations. *J. Diff. Equat.* **53**, 401-458; **59**, 165-205.

Herrero, M. and Vazquez, J. (1987). The one-dimensional non-linear heat equation with absorption: regularity of solutions and interfaces. *SIAM J. Math. Anal.* **18** (1), 149-167.

Hess, P. (1983). On the eigenvalue problem for weakly - coupled elliptic systems. *Arch. Ration. Mech. Anal.* **81**, 151-159.

Hess, P. (1987). Spatial homogeneity of stable solutions of some periodic - parabolic problems with Neumann boundary conditions. *J. Diff. Equat.* **68**, 320-331.

Hess, P. and Kato, T. (1980). On some linear and nonlinear eigenvalue problems with an indefinite weight function. *Comm. Per. Diff. Equat.* **5**, 999-1030.

Hille, E. and Fillips, R. (1962). *Functional Analysis and Semigroups*. Mosk. Izdat. Lit., Moscow (in Russian).

Hirsch, M. W. (1988). Stability and convergence in strongly monotone dynamical systems. *J. Reine Angew. Math.* **383**, 1-53.

Hollig, K. (1983). Existence of infinitely many solutions for a forward backward heat equation. *Transl. Amer. Math. Soc.* **278** (1), 299-316.

Il'in, A. M., Kalashnikov, A. S., and Oleinik, O. A. (1962). Second order linear equations of parabolic type. *Uspekhi Mat. Nauk* **17** (3), 3-146 (in Russian).

Ivasishen, S. D. and Eidel'man, S. D. (1970). Investigation of Green matrix for homogeneous parabolic boundary problem. *Trudy Mosk. Math. Obshch.* **23**, 179-234 (in Russian).

Jimbo, S. (1989). The singularly perturbed domain and the characterization for the eigenfunctions with Neumann boundary conditions. *J. Diff. Equat.* **77**, 2, 322-350.

Kalashnikov, A. S. (1982). On the Cauchy problem for second-order parabolic equations with degeneracies which have power-type nonlinearities. *Proceedings of Petrovsky's Seminar.* Moscow State University. **6**, 83-96 (in Russian).

Kato, T. (1965). *Nonlinear Evolutionary Equations in Banach Spaces.* (Preprint).

Kato, T. (1972). *Perturbations Theory for Linear Operators.* Mir, Moscow (in Russian).

Kim, V. F. and Zelinskaya, G. I. (1981). Numerical solutions of some equations with alternating viscosity. *Chislennye Metody Mekhaniki Sploshnoi Sredy* **12** (1), 54-68 (in Russian).

Kishimoto, Yu. S. and Weinberger, H. (1985). The spatial homogeneity of stable equilibria of some reaction - diffusion systems in convex domains. *J. Diff. Equat.* **58**, 15-21.

Kolesov, Yu. S. (1966). Stability of solutions of parabolic equations. *Izv. Akad. Nauk SSSR* **33**, 1356-1372 (in Russian).

Kolesov, Yu. S. (1970). Periodic solutions of quasilinear parabolic equations of the second order. *Trudy Mosk. Mat. Obshch.* **21**, 103-134 (in Russian).

Kolmogorov, A. N., Petrovskii, I. G., and Piskunov, N. S. (1937). Investigation of the diffusion equation connected with the growth of the substance amount and its application to one biological problem. *Bull. Mosk. Gos. Univ. Section A1,* **6**, 1-26 (in Russian).

Krasnoselskii, M. A. (1962). *Positive Solutions of Operator Equations.* Nauka, Moscow (in Russian).

Krein, M. G. and Rutman, M. A. (1948). Linear operators leaving invariant the cone in the Banach space. *Uspekhi Mat. Nauk* **3** (1), 3-95 (in Russian).

Kruzhkov, S. N. (1964). *A priori* estimates and some properties of solutions of elliptic and parabolic equations. *Math. Sbornik* **65** (4), 522-570 (in Russian).

Kruzhkov, S. N. (1967). Nonlinear parabolic equations with two independent variables. *Trudy Mosk. Math. Obshch.* **16**, 329-346 (in Russian).

Kruzhkov, S. N. (1969a). *Nonlinear Equations with Partial Derivatives*. Part I, II. Izd. Mosk. Gos. Univ. Moscow (in Russian).

Kruzhkov, S. N. (1969b). Results on the character of continuity of solutions to parabolic equations and some their applications. *Mat. Zametki* **6** (1), 97-108 (in Russian).

Kruzhkov, S. N. (1969c). On the Cauchy problem for certain classes of quasilinear parabolic equations. *Mat. Zametki* **6** (3), 295-300 (in Russian).

Kruzhkov, S. N. (1972). On basic *a priori* estimate for solutions of quasilinear parabolic equations. *Izv. Akad Nauk Uzb SSR, Ser. Fiz-Mat. Nauk.* **3**, 16-20 (in Russian).

Kruzhkov, S. N. (1979). Quasilinear parabolic equations and systems with two independent variables. *Trudy sem. im. Petrovskogo* **5**, 217-272 (in Russian).

Kruzhkov, S. N. and Oleinik, O. A. (1961). Second order quasilinear equations with many independent variables. *Uspekhi Mat. Nauk* **16** (5), 115-155 (in Russian).

Kruzhkov, S. N. and Peregudov, A. N. (1991). The Cauchy problem for the system of quasilinear parabolic equations of the chemical kinetics type. *Trudy sem. im. Petrovskogo* **1**, 242-261 (in Russian).

Krylov, S. N. (1984). On derivative estimates for solutions to nonlinear parabolic equations. *Dokl. Akad. Nauk SSSR* **274** (1), 23-26 (in Russian).

Krylov, S. N. (1985). *Nonlinear Elliptic and Parabolic Second Order Equations.* Nauka, Moscow (in Russian).

Kvasova, G. E. (1973). On behavior for large time of solutions of boundary-value problems for nonlinear parabolic equations. *Izv. Akad Nauk Kaz. SSR, Ser. Fiz.-Math. Nauk* **1**, 37-40 (in Russian).

Ladyzhenskaya, O. A. (1967). New equations for the description of flow of viscous liquids and the general solvability of boundary-value problems for them. *Trudy Inst. Matem. Akad. Nauk SSSR.* **102**, 85-104 (in Russian).

Ladyzhenskaya, O. A. (1968). Modifications of the Navier-Stokes equations for the case of large velocity gradients. *Trudy Inst. Matem. Akad. Nauk SSSR.* **7**, 126-154 (in Russian).

Ladyzhenskaya, O. A. (1982). On finite dimensionality of invariant sets for Navier - Stokes system and for other dissipative systems. *Zap. Nauch. Sem. LOMI.* **115**, 137-155 (in Russian).

Ladyzhenskaya, O. A. (1986). On attractors of nonlinear evolutionary problems with dissipation. *Zap. Nauch. Sem. LOMI.* **118**, 72-85 (in Russian).

Ladyzhenskaya, O. A. (1987). On determining global attractors for Navier - Stokes equations and for other equations with partial derivatives. *Uspekhi Mat. Nauk* **42**, 6, 25-60 (in Russian).

Ladyzhenskaya, O. A. and Solonnikov, V. A. (1973). On the linearizations principle and invariant manifolds for the problems of magnetic hydrodynamics. *Zap. Nauch. Sem. LOMI.* **38**, 46-93 (in Russian).

Ladyzhenskaya, O. A. and Ural'tseva, N. N. (1973). *Linear and Quasilinear Equations of Parabolic Type.* Nauka, Moscow (in Russian).

Ladyzhenskaya, O. A. and Ural'tseva, N. N. (1986). The survey of results on boundary - value problem solvability for uniformly elliptic and parabolic quasilinear equations which have unbounded singularities. *Uspekhi Mat. Nauk* **41** (5), 59-83 (in Russian).

Ladyzhenskaya, O. A., Solonnikov, V. A., and Ural'tseva, N. N. (1968). *Linear and Quasilinear Equations of Parabolic Type.* Amer. Math. Soc., Providence, RI.

Landis, E. M. (1959). Some problems of qualitative theory of parabolic and elliptic equations. *Uspekhi Mat. Nauk* **14** (1), 21-85 (in Russian).

Landis, E. M. (1971). *Second Order Equation of Elliptic and Parabolic Types.* Nauka, Moscow (in Russian).

Lar'kin, N. A., Yanenko, N. N., and Novikov, V. A. (1983). *Nonlinear Equations of Mixed Type.* Nauka, Novosibirsk (in Russian).

Lavrentiev, M. M. -Jr. (1980). On the properties of approximate solutions of nonlinear equations of variable type. *Sib. Math. J.*, **21** (6), 165-175 (in Russian).

Lavrentiev, M. M. -Jr. (1982). *A priori Estimates and Existence Theorems for Solutions of Nonlinear Parabolic Equations.* Ph. D. Thesis. Novosibirsk State University. Novosibirsk (in Russian).

Lavrentiev, M. M. -Jr. (1987). On the solvability of boundary-value problems for some parabolic equations with degeneracies. *Sib. Math. J.* **28** (2), 79-95 (in Russian).

Lavrentiev, M. M. -Jr. (1989). Estimates for solutions of a variable-type equation. *Mat. Model.* **1** (11), 132-138 (in Russian).

Lavrentiev, M. M. -Jr. (1990). *A priori* smoothness of solutions to some equations of mixed type. *Mat. Model.* **2** (9), 145-153 (in Russian).

Lavrentiev, M. M. -Jr. (1991). The estimate for the higher derivative for some nonlinear parabolic equations. *Sib. Math. J.* **32** (1), 72-81 (in Russian).

Lavrentiev, M. M. -Jr. (1993a). Estimates of a solution of the variable type equation. *Mat. Model. Comp. Exp.* **1** (3), 273-279 (in Russian).

Lavrentiev, M. M. -Jr. (1993b). Solvability of nonlinear parabolic problems. *Sib. Math. J.* **34** (6), 1110-1116 (in Russian).

Leray, J. and Schauder, J. (1934). Topologie et équations functionneles. *Ann. Ec. Sup.* **51**, 45-78.

Lions, P. L. (1984). Structure of the set of the steady - state solutions and asymptotic behavior of semilinear heat equations. *J. Diff. Equat.* **53**, 362-386.

Lopatinskii, Ya. B. (1953). On one way of transforming boundary - value problems for a system of elliptic type equations to systems of regular integral equations. *Ukr. Math. Jurn.* **5**, 123-151.

Mallet-Paret, J. and Sell, G. (1988). Inertial manifolds for reaction-diffusion equations in higher space dimensions. *J. Amer. Math. Soc.* **1**, 805-866.

Mallet-Paret, J. and Smith, H. (1990). The Poincare - Bendixson theorem for monotone cyclic feed back system. *J. Diff. Equat.* **4**, 367-421.

Marsden, J. B. and McCracken, M. (1976). *The Hopf Bifurcation and its Application.* Springer-Verlag, New York, 1976.

Maslova, N. N. and Novikov, V. A. (1980). *Numerical Investigation of a Nonlinear Equation with an Alternating Coefficient Multiplying the Second Derivative.* Preprint N 17 ITPM Siberian Department USSR Acad. Sci. (in Russian).

Maslova, N. N., Novikov, V. A., and Yanenko, N. N. (1984). On the stability of self-similar solutions in the model of a liquid with alternating viscosity. *Chislennye Metody Mekhaniki Sploshnoi Sredy* **15** (2), 98-120 (in Russian).

Matano, H. (1978). Convergence of solutions of one-dimensional semilinear parabolic equations. *J. Math. Kyoto Univ.* **18** (2), 221-227.

Matano, H. (1979). Asymptotic behavior and stability of solutions of semilinear diffusion equations. *Publ. Res. Inst. Math. Sci.* **15**, 401-454.

Matano, H. (1982). Non-increase of the lap-number of a solution for a one-dimensional semilinear parabolic equation. *J. Fac. Sci. Univ. Tokyo. Sect. IA Math.* **29** (2), 401-441.

Matano, H. (1984). Existence of nontrivial unstable sets for equilibriums of strongly order - preserving systems. *J. Fac. Sci. Univ. Tokyo. Sect. IA Math.* **30**, 645-673.

Matano, H. (1986). Strongly order - preserving local semi - dynamical systems - theory and applications. *Semigroups, Theory and Application.*Kappel, F; Brezc, H; Grandall, M. G. (eds). John Wiley. New York, 178-185,

Matano, H. (1987). Strong comparison principle in nonlinear parabolic equations. *Nonlinear Parabolic Equations. Qualitative Properties of Solutions* (Rome 1985). Pitman Res. Notes Math. Ser. 141, Longman Sci. Tech., Harlow, 148-155.

Matano, H. (1988). Asymptotic behavior of solutions of semi-linear heat equation on S'. *Nonlinear Diffusion Equations and their Equilibrium States II.* J. Serrin, W M. Ni, L. A. Peletier (eds). Springer - Verlag, New York, 139-162.

Matano, H. and Mimura, M. (1983). Parrent formation in competition - diffusion systems in nonconvex domains. *Publ. Res. Inst. Math. Sci.* **19**, 1049-1079.

Mierczinskii, J. (1994). The C^1 properties of carrying simplices for a class of competitive systems of ODE. *Diff. Int. Equat.* **7** (5/6), 1473-1494.

Mierczinskii, J. (1991). On monotone trajectories. *Proc. Amer. Math. Soc.* **113**, 537-544.

Mikhailov, V. P. (1960). Solving the mixed problem for the parabolic system by the potential method. *Dokl. Akad. Nauk SSSR* **132**, 2, 291-294 (in Russian).

Mikhailov, V. P. (1961). On the Dirichlet problem and on the first mixed problem for a parabolic equation. *Dokl. Akad. Nauk SSSR* **140** (2), 306-309 (in Russian).

Mikhailov, V. P. (1963). On the Dirichet problem for a parabolic equation. Part 1. *Mat. Sbornik* **61** (1), 40-64. Part 2. *Mat. Sbornik* **62** (2), 140-159 (in Russian).

Mikhailov, V. P. (1976). *Differential Equations with Partial Derivatives.* Nauka, Moscow (in Russian).

Mikhailov, V. P. (1983). *Equations of Mathematical Physics.* Nauka, Moscow (in Russian).

Miranda, C. (1970). *Equazioni alle Derivate Parziali di Tipo Ellittico.* Springer-Verlag, Berlin (1955). English translated (1970).

Nagumo, M. (1929). Uber die gleichmassige Summierbarkeit and ihre Anwendung auf ein Variationsproblema. *Japan J. Math.* **6**, 173-182.

Nash, J. (1958). Continuity of solutions of parabolic and elliptic equations. *Amer. J. Math.* **80,** 931-954.

Nemytskii, V. V. and Stepanov, V. V. (1949). *Qualitative Theory of Differential Equations.* Gos. Izd. Tehn. -Teor. Lit., Moscow (in Russian).

Novikov, V. A. and Shvab, I. V. (1985). On nonuniqueness of stability of some solutions of parabolic equations with changing parabolic direction. *Chislennye Metody Mekhaniki Sploshnoi Sredy* **16** (4), 53-76 (in Russian).

Oleinik, O. A. (1961). Boundary - value problems for linear equations of elliptic and parabolic types with discontinuous couficients. *Izv. Akad. Nauk SSSR,* **25** , 9-20 (in Russian).

Payne, L. (1976). Some remarks on maximum principles. *J. Anal. Math.* **30,** 421-433.

Peregudov, A. N. (1981). The second boundary - value problem for quasilinear parabolic system of the chemical kinetics type. *J. Calc. Math. and Math. Phys.* **21** (1), 18-28 (in Russian).

Peterson, L. D. and Maple, C. G. (1966). Stability of solutions of nonlinear diffusion problem. *J. Math. Anal. and Appl.* **14** (2), 221-241.

Petrovskii, I. G. (1938). On the Cauchy problem for linear equations system with partial derivatives in the domain of nonanalytic functions. *Bull. Mosk. Gos. Univ.* (A) **1** (7), 1-72 (in Russian).

Petrovskii, I. G. (1986). *Selected Works. Systems of Partial Differential Equations.* Vol. I, Nauka, Moscow (in Russian).

Petrovskii, I. G. (1987). *Selected Works. Differential Equations.* Vol. II. Nauka, Moscow (in Russian).

Pokhozhaev, S. I. (1965). On the eigenvalues of the equation $\Delta u + \lambda f(u) = 0$. *Dokl. Akad. Nauk SSSR* **165** (1), 36-39 (in Russian).

410   *T.I.Zelenyak, M.M.Lavrentiev-Jr. and M.P.Vishnevskii*

Pokhozhaev, S. I. (1970). On the eigenvalues of quasilinear elliptic problems. *Mat. Sbornik* **82** (2), 192-212 (in Russian).

Pokhozhaev, S. I. (1978). The problems on nonexistence of solutions to nonlinear boundary - value problems. *Trudy. Vses. Konf. po Dif. Uravn.* Izd. Mosk. Gos. Univ., Moscow (in Russian).

Pokhozhaev, S. I. (1980). On the equations $\Delta u = f(x, u, \nabla u)$. *Math. Sbornik* **133** (2), 324-338 (in Russian).

Polačik, P. (1988). Generic properties of strongly monotone semiflows defined by ordinary and partial differential equations. *Qualitative Theory of Diff. Equat.* (Szeged) Collog. Math. Soc. Janos. Bolycei, 53, North - Holland, Amsterdam, 519-530.

Polačik, P. (1989a). Convergence in smooth strongly monotone flows defined by semilinear parabolic equations. *J. Diff. Equat.* **79** (1), 89-110.

Polačik, P. (1989b). Domain of attraction of equilibria and monotonicity properties of convergent trajectories in parabolic systems. *J. Reine Angew. Math.* **400,** 32-56.

Polačik, P. (1992). Imbedding of any vector field in scalar semilinear parabolic equation. *Proc. Amer. Math. Soc.* **115,** 1001-1008.

Polačik, P. (1995). Hight - dimensional ω - limit sets and chaos in scalar parabolic equations. *J. Diff. Equat.* **119,** 24-53.

Polačik, P. and Rubakovskii, K. (1994). *Nonconvergent Bounded Trajectories in Semilinear Heat Equations.* (Preprint Universiteta degli Studi di Trieste).

Polačik, P. and Tereščak, I. (1991). Convergence to the cycles as typical asymptotic behavior in smooth strongly monotone discrete - time dynamical system. *Arch. Rat. Mech. Anal.* **116,** 339-360.

Polačik, P. and Tereščak, I. (1993). Exponential separation and invariant boundless for maps in ordered Banach spaces with application to parabolic equations. *J. Dynamics Diff. Equat.* **5**, 279-303.

Pontryagin, L. S. (1965). *Ordinary Differential Equations.* Nauka, Moscow (in Russian).

Podgaev, A. G. (1982). On some-well-posed problems for equations of variable type. *Dinamica Sploshnoy Sredy.* **55**, 143-153 (in Russian).

Protter, M. H. and Weinberger, H. F. (1966). On the spectrum of general second-order operators. *Bull. Amer. Math. Soc.* **72** (2), 251-255.

Protter , M. H. and Weinberger, H. F. (1984). *Maximum Principle in Differential Equations.* Berlin, Springer Verlag.

Provorova, O. G. (1969). On the behavior for large time solutions to parabolic equations. *Diff. Equat.* **5** (1), 108-114 (in Russian).

Provorova, O. G. (1973). On qualitative properties of solutions of autonomous and close to them parabolic equations. *Ph. D. Thesis,* Novosibirsk (in Russian).

Pyatkov, S. G. (1987). *Solvability of a Boundary-Value Problems for a Certain Nonlinear Parabolic Equation with Changing Direction of Time.* Preprint N 16. Inst. Math. Siberian Division USSR Acad. Sci, Novosibirsk (in Russian).

Pyatkov, S. G. and Podgaev, A. G. (1987). Solvability of a boundary-value problem for a nonlinear parabolic equation with changing direction of time. *Sib. Math. J.* **28** (3), 498-505 (in Russian).

Quittner, P. (1994). Boundedness of trajectories of parabolic equations and stationary solutions via dynamical methods. *J. Diff. and Int. Equat.* **7** (6), 1547-1556.

Riesz, F. and Szökefalvi-Nagy, B. (1972). *Lesous D'analyse Fonctionnelle*. Akademiai Kiado. Budapest.

Rocha, C. (1991). Properties of the attractor of a scalar parabolic ordinary differential equation. *J. Dynamics Diff. Equat.* **3** , 575-591.

Samarskii, A. A., Galactionov, V. A., and Mikhailov, A. P. (1987). Abruptly Changing Regimes in Problems for Quasilinear Parabolic Equations, Nauka, Moscow (in Russian).

Samarskii, A. A., Galaktionov, V. A., Kurdyumov, S. P., and Mikhailov, A. P. (1987). *Regimes with Peaking in Problems for Quasilinear Parabolic Equations*.Nauka, Moscow (in Russian).

Samarskii, A. A., Galactionov, V. A., Kurdyumov, S. P., and Mikhailov, A. P. (1995). Blow-up in quasilinear parabolic equations. *De Gruyter Expositions in Mathematics*. **19** Berlin.

Slemrod, M. (1991). Dynamics of measured valued solutions to a backward-forward heat equation. *J. Dynamical Diff. Equations.* **3** (1), 1-28.

Smith, H. L. (1991). Periodic solutions of periodic competitive and cooperative systems. *SIAM J. Math. Anal.* **22**, 1081-1100.

Smith, H. L. and Thieme, H. R. (1990). Quasi-convergence and stability for strongly order - preserving semiflows. *SIAM J. Math. Anal.* **21**, 673-692.

Sobolev, S. Z. (1962). *Some Applications of Fuctional Analysis in Mathematical Physics*. Izd. Sib. Otdel. Akad. Nauk SSSR, Novosibirsk (in Russian).

Sobolevskii, P. E. (1961). Equations of parabolic type in the Banach space. *Trudy Mosk. Math. Obshch.* **10**, 297-350 (in Russian).

Solonnikov, V. A. (1965). On the boundary - value problems for linear parabolic systems of differential equations of general form. *Trudy Math. Inst. im. Steklova* **83**, 3-162 (in Russian).

Solonnikov, V. A. and Hachatryan, A. G. (1980). Estimates for solutions of parabolic boundary - value problems in the weight Hölder norms. *Ibid.* **147,** 147-155 (in Russian).

Starr, V. P. (1968). *Physics of Negative Viscosity Phenomena,* Ed. McGraw-Hill.

Sychyev, M. A. (1992). On the regularity of variationaly problems solutions. *Math. Sbornik* **183** (4), 118-142 (in Russian).

Takač, P. (1991). Domains of attraction of generic ω - limit sets for strongly monotone semiflows. *Z. Anal. Anwendungen* **10,** 275-317.

Takač, P. (1992a). Lineary stable subharmonics orbits in monotone time - periodic dynamical systems. *Proc. Amer. Math. Soc.* **115,** 691-698.

Takač, P. (1992b). Domains of attraction of generic ω - limit sets for strongly monotone discrete - time semigroups. *J. Reine Angew. Math.* **423,** 101-173.

Teman, R. (1988). *Infinite Dimensional Dynamical Systems in Mechanics and Physics.* Springer - Verlag, New York.

Tersenov, S. A. (1982). *Introduction to the Theory of Parabolic Equations with a Variable Direction of Time.* Press of Inst. of Math. Sib. Division, USSR, Novosibirsk (in Russian).

Tikhonov, A. N. (1937). On the heat conductivity equation for several variables. *Bull. Mosk. Gos. Univ.* (A) I, **9,** 1-49 (in Russian).

Tikhonov, A. N. and Samarskii, A. A. (1972). *Equations of Mathematical Physics.* Nauka, Moscow (in Russian).

Ventsel, T. D. (1957). The first boundary - value problem and the Cauchy problem for quasilinear equations of parabolic type. *Math. Sbornik* **41** (83), 105-128 (in Russian).

Tonelli, L. (1923). *Fondamenti di Calcolo della Variazioni*, **2**, Zanichelli, Bologna.

Vishik, M. I. and Agranovich, M. S. (1962). On solvability of boundary - value problems for quasilinear equations of high orders. *Math. Sbornik* **59**, 289-325 (in Russian).

Vishik, M. I., Myshkis, A. D., and Oleinik, O. A. (1959). Differential equations with partial derivatives. *Mathematics in USSR during 40 years* (1917-1957). Vol. I Moscow, 563-636 (in Russian).

Vishnevskii, M. P. (1981a). Periodic and other solutions close to them for nonlinear parabolic systems. *Sib. Math. J.* **22** (6), 208-209 (in Russian).

Vishnevskii, M. P. (1981b). Periodic solutions of nonlinear parabolic systems and solutions close to them. Dep. VINITI 1219-1281. from *Sibir. Math. J.* **22** (6) (in Russian).

Vishnevskii, M. P. (1982). Integral sets of nonlinear parabolic systems. *Dinamica Sploshn. Sredy.* **54**, 74-84 (in Russian).

Vishnevskii, M. P. (1984a). Stability criterion for solutions to the mixed problems for parabolic equations. *Boundary - Value Problems for Partial Differential Equations.* Trudy sem. S. L. Soboleva **1**, 5-22 (in Russian).

Vishnevskii, M. P. (1984b). On certain qualitative properties of periodic solutions of quasilinear parabolic equations. *Dinamica Sploshn. Sredy.* **64,** 11-23 (in Russian).

Vishnevskii, M. P. (1984c). The attraction domains of stable stationary solution of parabolic equation. *Dinamica Sploshn. Sredy.* **67,** 3-20 (in Russian).

Vishnevskii, M. P. (1984d). The stability criterion for solutions of mixed problems for parabolic equations. *Trudy sem. acad. Soboleva S.L.* **1**. Inst. Math., Novosibirsk, 5-22 (in Russian).

Vishnevskii, M. P. (1986). Invariant sets of nonlinear parabolic systems. *Some Applications of Functional Analysis to the Mathematical Physics Problems.* Novosibirsk, Inst. of Math. Sib. Otdel. Akad. Nauk SSSR. 32-56 (in Russian)

Vishnevskii, M. P. (1987). Bounded solutions of nonlinear parabolic equations. *Differential Equations with Partial Derivatives.* Novosibirsk, Inst. of Math. Sib. Otdel. Akad. Nauk SSSR, 35-59 (in Russian).

Vishnevskii, M. P. (1988). Asymptotic behavior of solutions to the mixed problems for quasilinear parabolic equations. *Boundary - Value Problems for Partial Differential Equations.* Novosibirsk, Inst. of Math. Sib. Otdel. Akad. Nauk SSSR, 65-80 (in Russian).

Vishnevskii, M. P. (1989). On behavior for large time of solutions to parabolic equations with many spatial variables. *Dinamica Sploshn. Sredy.* **85,** 34-41 (in Russian).

Vishnevskii, M. P. (1990a). On nonlocal behavior of solutions to quasilinear mixed problems for parabolic equations. *Mat. Model.* **2** (4), 67-77 (in Russian).

Vishnevskii, M. P. (1990b).Periodic solutions of autonomous parabolic equations. *Dinamica Sploshn. Sredy.* **98,** 98-106 (in Russian).

Vishnevskii, M. P. (1990c). Attraction domains for periodic solutions of quasilinear parabolic equations. *Some Applications of Functional Analysis in the Problems of Mathematical Physics.* Novosibirsk Novosibirsk, *Izd. Inst. Math. Sib. Otdel. Akad. Nauk SSSR* , (in Russian)

Vishnevskii, M. P. (1992). On stabilizing solutions of weakly connected cooperative parabolic systems. *Math. Sbornik* **183** (10), 45-62 (in Russian)

Vishnevskii, M. P. (1993a). On stable stationary solutions of parabolic equations. *Sib. Math. J.* **34** (2), 42-51 (in Russian).

Vishnevskii, M. P. (1993b). The criterion of exponential stability of the trivial solution of weakly connected parabolic systems. *Sib. Math. J.* **34** (3), 27-42 (in Russian).

Vishnevskii, M. P. (1993c). Monotone solutions of quasilinear parabolic equations. *Sib. Math. J.* **34** (4), 50-60 (in Russian).

Vishnevskii, M. P. (1993d). On stabilizing solutions of boundary - value problems for quasilinear parabolic equations periodically depending on time. *Sib. Math. J.* **34** (5), (in Russian).

Vishnevskii, M. P. (1993e). Solvability as a whole for reaction - diffusion systems. *Abstracts Conf. EQUADIFF,* **8,** Bratislava.

Vishnevskii, M. P., Zelenyak, T. I., and Lavrentiev, M. M. -Jr. (1995). Behavior of solutions to nonlinear parabolic equations at large time. *Sib. Math. J.* **36** (3), 435-453.

Vladimirov, V. S. (1971). *Equations of Mathematical Physics.* Nauka, Moscow (in Russian).

Voronko, V. P. (1981). Difference schemes for nonlinear equations of variable type. *Chislennye Metody Mekhaniki Sploshnoi Sredy* **8** (5), 48-52 (in Russian).

Yudovich, V. I. (1965). On stability of stationary flows of viscous incompressible fluid. *Dokl. Akad. Nauk SSSR* **161** (5), 1037-1040 (in Russian).

Yanenko, N. N. (1980). On nonlinear equations of variable type. *Proceed. Seminars, Leningrad. Division of Steklov Inst. Mathem. (LOMI).* **96,** 294-301 (in Russian).

Yanenko, N. N. and Novikov, V. A. (1973). On the model of a liquid with variable viscosity. *Chislennye Metody Mekhaniki Sploshnoi Sredy.* **4** (2), 142-147 (in Russian).

Yudovich, V. I. (1970). On stability of auto - oscillation of fluid. *Dokl. Akad. Nauk SSSR* **195,** 574-576 (in Russian).

Zagorskii, T. Ya. (1961). Mixed Problems for Systems of Partial Differential Equations. Izd. L'vov. Univ. L'vov (in Russian).

Zelenyak, T. I. (1966). On stability of stationary solutions of one mixed problem. *Dokl. Akad. Nauk SSSR* **171** (2), 266-268 (in Russian).

Zelenyak, T. I. (1967a,b). On the stability of stationary solutions of mixed problems for one quasilinear parabolic equation. *Diff. Equat.* **3** (1), 19-29; **171** (2), 266-268 (in Russian).

Zelenyak, T. I. (1968). On stabilizing solutions of boundary - value problems for a parabolic second order equation with one spatial variable. *Diff. Equat.* **4,** 34-45 (in Russian).

Zelenyak, T. I. (1972). *Qualitative Theory of Boundary - Value Problems for Quasilinear Second Order Equations of Parabolic Type.* Novosibirsk, Izd. Novos. Gos. Univ. (in Russian).

Zelenyak, T. I. (1977). On qualitative properties of solutions to quasilinear mixed problems for parabolic type equations. *Math. Sbornik* **104,** 486-510 (in Russian).

Zelenyak, T. I. and Mikhailov, V. P. (1970). Asymptotic behavior of solutions to certain boundary - value problems of mathematical physics for $t \to +\infty$. *Trudy Konf. Posv. S. L. Sobolev.* 96-118 (in Russian).

Zelenyak, T. I. and Slin'ko, M. G. (1977a, b). Dynamics of catalytic systems I. *Kinetics and Catalysis.* **17** (5), 1235-1248; II *Kinetics and Catalysis.* **18** (6), 1548-1560 (in Russian).

Zelenyak, T. I., Yanenko, N. N., and Novikov, V. A. (1974). On properties of solutions to nonlinear equations of mixed type. *Calc. Methods of Solid Medium Mechanics.* **5** (4), 35-47 (in Russian).